The Millimeter Wave Synthetic Aperture Radar Technology

Hui Wang

The Millimeter Wave Synthetic Aperture Radar Technology

Science Press
Beijing

Springer

Hui Wang
Shanghai Academy of Spaceflight
Technology
Shanghai, China

ISBN 978-981-97-1046-1 ISBN 978-981-97-1044-7 (eBook)
https://doi.org/10.1007/978-981-97-1044-7

Jointly published with Science Press
The print edition is not for sale in China (Mainland). Customers from China (Mainland) please order the print book from: Science Press

Translation from the Chinese Simplified language edition: "Millimeter Wave Synthetic Aperture Radar Technology" by Hui Wang et al., © Science Press 2022. Published by Science Press. All Rights Reserved.

This Springer imprint is published by the registered company Springer Nature Singapore Pte Ltd.
The registered company address is: 152 Beach Road, #21-01/04 Gateway East, Singapore 189721, Singapore

Preface

Synthetic-aperture radar (SAR) technology has been developed rapidly at globally since it was proposed in the 1970s. It has been widely used in important fields such as national economy, and has become an important means of remote sensing observation of the earth.

With the development of SAR remote sensing observation towards higher resolution, higher accuracy, and finer characterization, research on new systems, models, and new frequency bands of SAR technology has been carried out. Among them, millimeter wave has become an important development direction for future SAR remote sensing observation technology due to its larger relative bandwidth, smaller antenna size, shorter interference baseline, and stronger detail description ability, and has received widespread attention from China's and international research institutions. At the same time, the improvement in the level of core power devices in the millimeter wave frequency band has also promoted the acceleration of theoretical and methodological research from the perspective of engineering implementation.

The international research on airborne millimeter wave SAR technology began in the 1990s. After years of theoretical research and flight verification, further research on spaceborne millimeter wave SAR was carried out in the early 21st century, and a large number of research results were achieved.

Based on the development of core power devices in China's millimeter wave frequency band, the author took the lead in proposing the research direction of spaceborne millimeter wave SAR technology in China in 2012. After analyzing the characteristics and application requirements of the millimeter wave frequency band and conducting research and demonstration analysis, two major technical systems were proposed—frequency modulated continuous wave system and pulse system based on digital beam synthesis (DBF) scanning and receiving technology.

The content of this book is based on the author's previous research achievements and strives to incorporate the latest research methods. Based on the characteristics of different platforms and systems such as airborne, spaceborne, and ground-based, as well as pulse and continuous wave systems, typical millimeter wave SAR signal processing and image application processing methods for different application scenarios, such as imaging processing, moving object detection, and interferometry, are discussed separately.

With the continuous deepening of research on millimeter wave remote sensing mechanisms and the continuous promotion of corresponding projects in various countries, spaceborne millimeter wave SAR systems are about to be added to the family of in orbit SAR satellites. The acquisition of a large amount of millimeter wave SAR remote sensing data will undoubtedly promote the rapid progress of millimeter wave SAR processing theory and methods. I hope that the explanation and summary of this book can inspire researchers interested in millimeter wave SAR technology and provide assistance for subsequent work.

This book emphasizes the connection between theoretical and practical processing methods. Most of the processing methods have been verified by actual data, and the content arrangement strives for logical rigor and coherent knowledge. Zheng Shichao, Wu Sili, and others have helped complete the revisions of each chapter, and we would like to express our gratitude. Finally, I would like to thank the millimeter wave imaging technology research team of Shanghai Academy of Spaceflight Technology, whose work and achievements are the basis of the content of this book; Thank you to experts such as Li Chunsheng and Xing Mengdao for their help and support in the research process of millimeter wave SAR technology. Thank you to the Shanghai Key Laboratory of Millimeter Wave Aerospace Information Acquisition and Application Technology for their help and support.

Although we have made efforts in writing this book, the author's level is limited, and any shortcomings in the book are inevitable. We kindly request readers to criticize and correct us.

Shanghai, China Hui Wang

Contents

Chapter 1
Overview

1.1 Principles and Development of SAR

1.1.1 Principles of SAR

Synthetic Aperture Radar (SAR) is an active microwave remote sensing system that utilizes the relative motion between the radar platform and the target to obtain coherent processing of wideband echo signals received at different spatial positions over a certain integration time, resulting in a two-dimensional image of the target.

In the range direction, SAR uses linear frequency modulated signals with large time-bandwidth products for pulse compression to improve resolution, which is consistent with conventional radar. In the azimuth direction, SAR utilizes the relative motion between the target and the radar to form a virtual synthetic aperture, achieving high resolution. During operation, SAR emits and receives pulse signals with a certain pulse repetition frequency on a uniformly moving platform, and after coherent processing of the echo signals, it achieves high azimuth resolution.

The resolution in the range direction of the radar (on the ground) is defined as the minimum distance between two distinguishable points that the system can resolve. If the leading edge of the echo pulse from a distant point arrives later than the trailing edge of the echo pulse from a closer point, each point can be distinguished based on the time of the radar echo. If the pulse duration of the radar is τ_p, the minimum separation between two distinguishable points is given by:

$$\Delta R_g = \frac{\Delta R_s}{\sin\theta} = \frac{c\tau_p}{2\sin\theta} \tag{1.1}$$

where ΔR_s is the range resolution, c is the speed of light, θ is the incidence angle.

The relationship between the imaging swath and the radar beam width on the ground is shown in Fig. 1.1.

H. Wang, *The Millimeter Wave Synthetic Aperture Radar Technology*,
https://doi.org/10.1007/978-981-97-1044-7_1

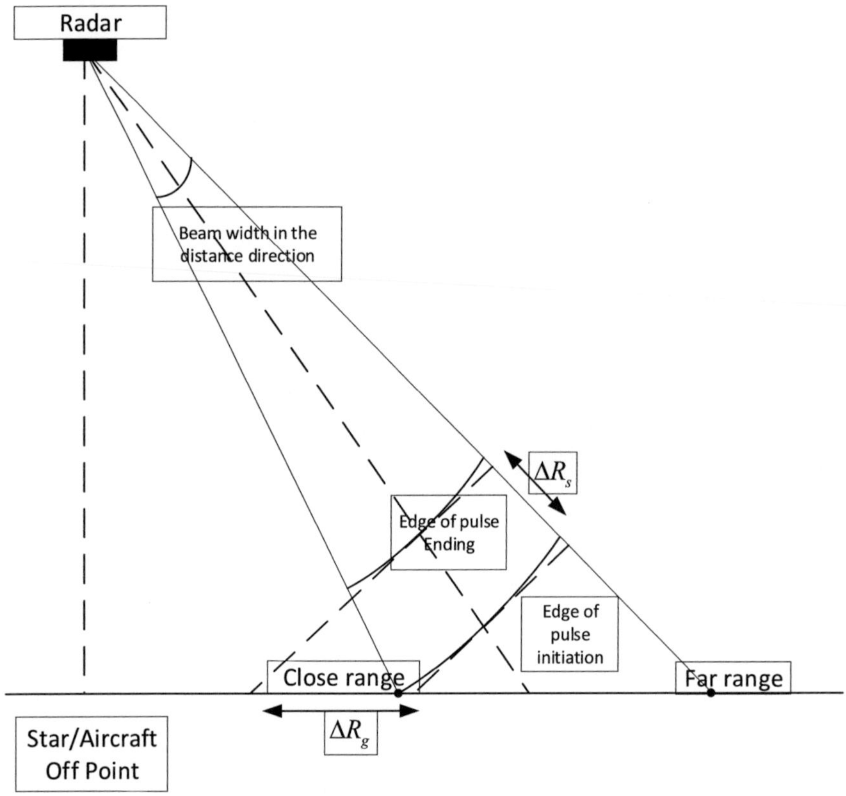

Fig. 1.1 Relationship between the breadth of ground imaging and the beamwidth of radar

To achieve the desired range resolution, it is necessary to have a very short pulse duration τp, which leads to a very low duty cycle and average transmit power, making it difficult to satisfy the target observation requirements in terms of system signal-to-noise ratio. Therefore, SAR systems employ pulse compression techniques to achieve both high resolution (with longer pulses) and high signal-to-noise ratio. By compressing the received pulse signal (matched filtering), the range resolution becomes:

$$R_g = \frac{c}{2B_r \sin\theta} \tag{1.2}$$

where Br is the bandwidth of the transmitted signal.

In SAR systems, high resolution in the azimuth direction (parallel to the platform flight direction) is achieved through synthetic aperture techniques, which is the fundamental difference between SAR systems and other radar systems. SAR systems utilize the motion of the radar antenna to form a virtual antenna array, sequentially acquiring and storing the echo signals during platform motion. During

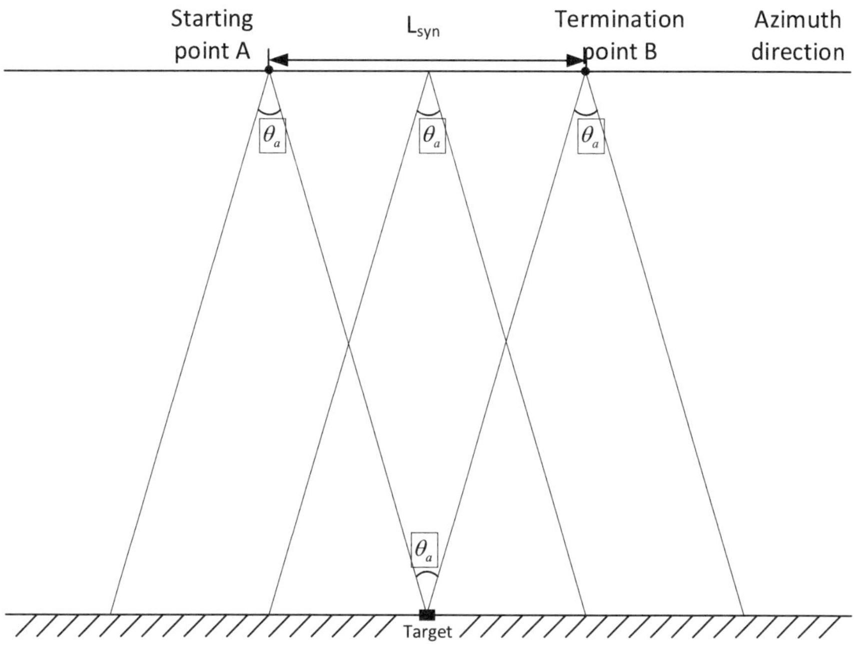

Fig. 1.2 Schematic diagram of azimuth synthetic aperture

signal processing, the phase introduced by the changing distance between the radar and the target is compensated, and the multiple echo signals from the same target are coherently added to achieve high azimuth resolution. The schematic diagram of azimuth synthetic aperture is shown in Fig. 1.2.

During the illumination of the target by the radar antenna beam, the platform moves a distance L_{syn}, and the azimuth size of the antenna is Da. The synthetic aperture length is determined by the azimuth beam width of the actual antenna and the distance R between the radar and the target, given by:

$$L_{syn} = \theta_a \cdot R = 0.886 \cdot \frac{\lambda}{D_{az}} \cdot R \tag{1.3}$$

Considering the round trip propagation of the signal, the phase difference between the antenna and the target during any two acquisitions is twice the phase difference introduced by one-way propagation. Therefore, the effective beam width of the synthetic aperture virtual antenna array becomes:

$$\theta_{az} = 0.886 \cdot \frac{\lambda}{2L_{syn}} \tag{1.4}$$

Thus, after synthetic aperture signal processing, the azimuth resolution of SAR is improved to:

$$\Delta R_{az} = \theta_{az} R = 0.886 \cdot \frac{\lambda R}{2L_{syn}} = 0.886 \cdot \frac{\lambda R}{2} \cdot \frac{D_{az}}{0.886 \cdot \lambda R} = \frac{D_{az}}{2} \quad (1.5)$$

From the above analysis, it can be observed that the azimuth resolution obtained through synthetic aperture processing is independent of the distance between the platform and the target, and depends only on the azimuth size of the antenna.

1.1.2 Development of Synthetic Aperture Radar

In June 1951, Carl Wiley of Goodyear Aerospace in the United States first proposed the frequency analysis method to improve radar angular resolution. At the same time, the Control Systems Laboratory at the University of Illinois independently conducted experiments using non-coherent radar to verify the effectiveness of the frequency analysis method in improving radar angular resolution. Since then, with the strong attention and support from governments around the world, SAR technology has experienced rapid development. On June 27, 1978, the National Aeronautics and Space Administration (NASA) of the United States launched the Seasat-A ocean satellite equipped with SAR payload, marking a crucial transition of SAR from laboratory research to applied research and pioneering the history of spaceborne synthetic aperture radar. The system acquired high-quality image data covering one billion square kilometers of the Earth's surface throughout its flight, demonstrating the capability of synthetic aperture radar to obtain high-resolution images from space and ushering in a new era in SAR technology research and application. Over the past 60 years, SAR technology has become an important means for military reconnaissance, natural resource surveying, and monitoring of natural disasters.

1. **Spaceborne SAR**

Since the successful launch of the first SAR satellite, Seasat-1, by the United States in 1978, spaceborne SAR has received significant attention from many countries worldwide. The former Soviet Union launched the Almaz-1 SAR satellite in 1991, the European Space Agency launched the ERS-1 SAR satellite in 1991, Japan launched its first SAR satellite, JERS-1, in 1992, and Canada launched the Radarsat-1 SAR satellite in 1995. These SAR satellite systems have played important roles in both military and civilian applications. Since the beginning of the twenty-first century, the development of spaceborne SAR technology has accelerated. Italy, Israel, China, India, Russia, South Korea, and Argentina have successively launched their own SAR satellites, expanding the applications of SAR technology.

The Seasat satellite and the SAR image obtained are shown in Fig. 1.3.

Currently, advanced spaceborne SAR systems around the world adopt multi-polarization, multi-mode, and phased array technologies. Typical systems include the TerraSAR-X series satellites from Germany, the Radarsat series satellites from Canada, the Sentinel series satellites from the European Space Agency, the TecSAR

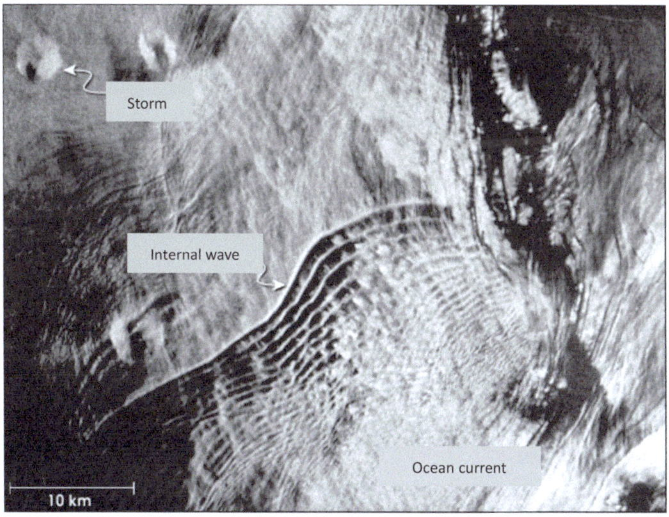

Fig. 1.3 SAR images obtained from the Seasat satellite

satellite from Israel, and the GF-3 (Gaofen-3) satellite from China. Figure 1.4 shows an SAR image (Xiamen) obtained by the GF-3 satellite.

Currently, spaceborne SAR radars mainly operate in the L, C, and X bands. Low-frequency bands such as L and C are commonly used for spaceborne radars with lower resolution requirements, while the X band is often used for high-resolution spaceborne SAR. With the increasing demand for high-precision information acquisition, spaceborne SAR technology is beginning to develop towards higher millimeter-wave frequency bands. Research institutions such as the National Aeronautics and Space

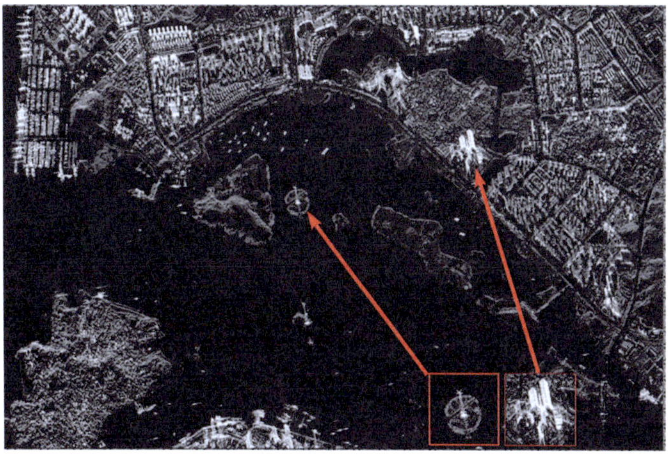

Fig. 1.4 SAR image obtained by the GF-3 satellite (Xiamen)

Administration (NASA) of the United States and the German Aerospace Center (DLR) have successively conducted studies on spaceborne millimeter-wave SAR systems. It can be foreseen that with the continuous development of relevant technologies, spaceborne millimeter-wave SAR will also enter the practical application stage in the near future.

Table 1.1 provides a summary of some typical spaceborne SAR systems launched to date.

2. **Airborne Synthetic Aperture Radar**

In the field of airborne SAR technology, in 1972, the Jet Propulsion Laboratory (JPL) in the United States conducted airborne calibration flights for L-band spaceborne SAR. Today, advanced airborne SAR systems are developed in two directions. On one hand, they are evolving towards high-performance complex systems with multiple frequency bands, multiple channels, multiple polarizations, multiple modes, ultra-high resolution, and three-dimensional observation capabilities. Typical systems include the RAMAES system developed by the French Aerospace Research Center (ONERA), the Lynx SAR system by General Atomics Aeronautical Systems in the United States, and the PAMIR system by FGAN-FHR in Germany. On the other hand, they are developing miniaturized unmanned aerial vehicle (UAV) SAR systems that balance high resolution, moving target detection, and real-time imaging. A typical system is the NanoSAR system developed by ImSAR Corporation in the United States.

(a) The RAMES system developed by ONERA is a multi-frequency, multi-channel system. It achieves a resolution of 0.1 m in the X-band and Ku-band. It utilizes the synthesis of wide bandwidths by transmitting five 300 MHz linearly frequency-modulated sub-pulses. The accurate trajectory information is obtained to achieve a resolution of 0.1 m in the azimuth direction.

(b) Lynx SAR System

The Lynx SAR system operates in the Ku-band and can work within the 3 GHz frequency band from 15.2 to 18.3 GHz. The operating mode can be selected through a ground control station link. The resolution in the spotlight mode is 0.1 m, while in the strip mode, it is 0.3 m, with an operating range of 30 km. The system utilizes real-time motion compensation technology, which allows high-quality imaging even when the UAV is turning or performing other maneuvering flights.

(c) PAMIR System in Germany

The PAMIR system is a wideband phased array SAR/GMTI (Ground Moving Target Indicator) system. It operates at a center frequency of 9.45 GHz with a bandwidth of 1820 MHz. PAMIR has 5 receive channels and supports advanced array signal processing techniques in SAR mode. For example, it can suppress GMTI clutter using STAP (Space–Time Adaptive Processing) and support ECCM (Electronic Counter-Countermeasures) and high-resolution 3-D imaging using InSAR (Interferometric Synthetic Aperture Radar) technology.

Table 1.1 Typical spaceborne SAR systems

The name of the SAR system	Country	Transmission time	Band	Maximum resolution (m)	Imaging width (km)	Incident angle	Working mode
Seasat-1	America	1978	L	25	100	Antenna pointing fixed	Stripmap SAR
SIR-A	America	1981	L	40	50	47°	Stripmap SAR
SIR-B	America	1984	L	25	20–50	15~60°	Stripmap SAR
SIR-C/X	America	1994	C/X	15	15–90	15~60°	Stripmap SAR
ERS-1/2	Europe	1991, 1994	C	30	56–100	23°	Stripmap SAR
JERS-1	Japan	1992	L	20	75	35°	Stripmap SAR
ALMAZ-1	Russia	1991	C	10	40	30~60°	Stripmap SAR
Radarsat-1	Canada	1995	C	9	50	20~45°	Stripmap SAR, scan SAR
ENVISAT	Europe	2002	C	30	50–100	15~45°	Stripmap SAR, scan SAR
ALOS	Japan	2006	L	7	40–350	20~45°	Stripmap SAR
SAR-Lupe	Germany	2006	X	≤1 m	5.5–60	Antenna pointing fixed	Stripmap SAR, spotlight SAR
Radarsat-2	Canada	2007	C	3	20–500	20~60°	Stripmap SAR, scan SAR
TerraSAR	Germany	2007, 2010	X	1	10–100	20~55°	Stripmap SAR, scan SAR, spotlight SAR
TecSAR	Israel	2008	X	1	10–100	Antenna pointing fixed	Stripmap SAR, spotlight SAR, Mosaic SAR

(continued)

Table 1.1 (continued)

The name of the SAR system	Country	Transmission time	Band	Maximum resolution (m)	Imaging width (km)	Incident angle	Working mode
Sentinel	Europe	2014	C	5	80–400	20~44°	Stripmap SAR, TOPS
GF-3	China	2017	C	1	10–650	17~60°	Stripmap SAR, scan SAR, spotlight SAR

The system adopts advanced phased array antenna technology and a multi-carrier transmit waveform and receive scheme. SAR images obtained by the PAMIR system are shown in Fig. 1.5.

(d) NanoSAR System in the United States
The NanoSAR system is developed by ImSAR Corporation, a developer of miniature synthetic aperture radar. The first-generation system is NanoSAR A, and the second-generation system is NanoSAR B. The NanoSAR system can be mounted on the "ScanEagle" and E-BUSTER (BLACKLIGHT)™ small unmanned aerial vehicles.

NanoSAR A is the world's smallest synthetic aperture radar system with a payload weight of 0.9 kg. It includes a navigation system, antenna, cables, real-time imaging processor, and receiver. The system performs real-time imaging processing onboard and provides NTSC video for real-time image analysis. The NanoSAR-A system and the SAR image obtained are shown in Fig. 1.6.

Fig. 1.5 SAR images obtained by the PAMIR system

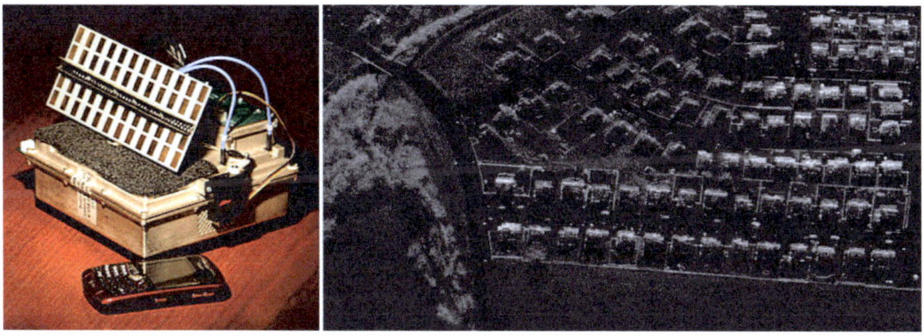

Fig. 1.6 NanoSAR-A system and SAR image obtained

Table 1.2 provides a summary of some typical airborne SAR systems.

3. **Future Development Trends**

With the development of SAR technology and the continuous improvement of SAR system performance worldwide, SAR will find even wider applications and evolve towards providing more extensive, rich, and detailed target information. Based on the current trends, the future development of SAR technology will focus on high-frequency bands, high-resolution wide-swath imaging, polarization diversity, multi-mode operation, multi-platform integration, and 3D information acquisition.

(1) High-resolution wide-swath SAR technology:
High-resolution wide-swath imaging is an important direction for the development of spaceborne SAR. On one hand, remote sensing applications such as disaster assessment, target detection and recognition, and military strike assessment require SAR images with high spatial resolution. On the other hand, applications such as maritime target surveillance and emergency exploration in disaster areas require radar systems with wide-swath imaging capabilities. However, traditional spaceborne SAR systems face a trade-off between spatial resolution and swath width, making it difficult to achieve both high spatial resolution and wide coverage in ground observations. To address this challenge, extensive research and practical experiments have been conducted on various technical approaches, including multi-channel transceivers, agile parameterization, and compressed sensing. Innovative working modes are also being explored.
(2) 3D terrain mapping SAR technology:
As an important branch of SAR technology, Interferometric Synthetic Aperture Radar (InSAR) utilizes the differences in antenna perspectives to extract terrain elevation information, enabling the acquisition of 3D terrain models for target areas. InSAR is widely used in surface deformation monitoring, mapping, glacier research, and other fields. However, traditional InSAR systems face challenges in achieving high-precision and high-stability 3D mapping in cases of abrupt terrain changes, low coherence regions, and areas with overlapping signals. To overcome these limitations, SAR imaging systems and modes

Table 1.2 Typical airborne synthetic aperture radar systems

The name of the SAR system	Country	Band	Maximum Resolution (m)	Working mode	Flying platform	Main features
NanoSAR	America	X	1 m	Stripmap SAR	Scan eagle and E-BUSTER (BLACKLIGHT) ™ small drone	The world's smallest synthetic aperture radar
Lynx	America	Ku	0.1 m	Stripmap SAR, spotlight SAR, GMTI, coherent change detection	Predator drone, I-GNAT drone, MQ-9 reaper drone	Developed by general atoms
HISAR	America	X	1.8 m	Stripmap SAR, spotlight SAR, MTI, wide area search, sea surface surveillance	Global Hawk unmanned reconnaissance aircraft, RC-7B low altitude multifunctional reconnaissance aircraft	Air ground surveillance, military reconnaissance, maritime patrols, ground surveying and imaging, border surveillance, environmental resource management, transportation, agriculture, and forest monitoring
ARL-M	America	X	1.8 m	Wide area moving target display, spotlight sar	ARL low altitude reconnaissance aircraft	Low altitude airborne reconnaissance
TESAR	America	Ku	0.3 m	Stripmap SAR, spotlight SAR, GMTI	Predator drone	Developed by Northrop Grumman
STacSAR	America	Ku	0.5 m	Stripmap SAR, spotlight SAR, GMTI	–	Real time imaging processing can be carried out on board, and the obtained images can be transmitted to the ground station in real time
TUSAR	America	Ku	0.3 m	Stripmap SAR, spotlight SAR, GMTI	–	Northrop Grumman developed a project for the American Army

(continued)

Table 1.2 (continued)

The name of the SAR system	Country	Band	Maximum Resolution (m)	Working mode	Flying platform	Main features
IFSARE	America	X	3 m	–	ERIM's Learjet 36 aircraft	Capable of providing high-precision 3D terrain images
AER-I/II	Germany	X	1 m	SAR imaging, SAR/GMTI, 4-channel polarization, 4-channel interference, multi view, spotlight SAR	–	Verify the application prospects of active phased array antennas in airborne environments
PAMIR	Germany	X	0.1 m	Stripmap SAR, spotlight SAR, Sliding spotlight SAR, scan-MTI, InSAR, ISAR	–	Next generation products of AER and AER-II
MISAR	Germany	Ka	0.5 m	SAR IMAGING, MTI	LUNA drone	Weight less than 4 kg, with real-time processing capability
MEMPHIS	Germany	Ka, W	0.2 m	SAR IMAGING, Insar	–	Multi baseline interference
MIRANDA	Germany	W	0.15 m	SAR IMAGING	NEO S-300 drone	Equipped with real-time processing capabilities
MiSAR	The Netherlands	Ka	0.3 m	–	Stemme S10 light electric glider	Developed by delft University of Technology in the Netherlands
RAMSES	France	P, L, S, C, X, Ku, Ka, W	0.11 m	Quick look, high resolutionstripmap SAR, fast vision imaging, InSAR, PolInSAR	Transall C160 (platform retired)	The original intention of the design is to provide a testing platform for military radars, gradually applied to new technology testing in the scientific research field

(continued)

Table 1.2 (continued)

The name of the SAR system	Country	Band	Maximum Resolution (m)	Working mode	Flying platform	Main features
SETHI	France	P, L, X, UHF/VHF	0.1 m	High resolution imaging, InSAR, PolInSAR	Falcon 20	Civil airborne SAR system for scientific research
CV-580	Canada	C, X	6 m	Wide viewing angle, stripmap SAR, stripmap SAR	Convair-580 aircraft	The first international remote sensing aircraft to conduct imaging flights within China
EL/M-2055	Israel	–	–	Stripmap SAR, spotlight SAR, GMTI	UAV	Mainly used for military reconnaissance
Sea Dragon	Russia	X	–	SAR, ISAR	Il-38 and Tu-204P maritime patrol aircraft, Tu-142 M-Z anti-submarine warfare aircraft	Equipped with the Russian and Indian navies
ASTOR	Britain	X	0.5 m	Stripmap SAR, spotlight SAR, wide area search GMTI	BD-700, Sentinel R1	Capable of coordinating operations with the US military's JSTARS system
QuaSAR	Britain	L, S, C, X, Ku	0.3 m	Stripmap SAR, spotlight SAR, GMTI	UAV	Adopting universal radar components that can be combined in different ways
PI-SAR	Japan	X, L	1.5 m	Interference, polarization interference	–	Used for global environmental and disaster monitoring

with 3D resolution capabilities are being developed, significantly expanding the application range of SAR mapping. Currently, different techniques such as multi-baseline tomography, array spotlight imaging, and circular trajectory imaging have been validated in airborne and spaceborne experiments.

(3) Multi-baseline SAR technology:
Multi-baseline SAR enhances the capabilities of traditional single-baseline SAR systems by flexibly configuring the relative positions of the transmitting and receiving ends. Multi-baseline spaceborne SAR systems form multiple baselines with different lengths by adjusting the vertical separations between satellites, combining the advantages of long and short baselines to achieve higher accuracy in height measurements. The most successful multi-satellite SAR system to date is the German TerraSAR-X/TanDEM-X constellation, which consists of two X-band spaceborne SAR systems. During a single flight, multiple interferometric height maps can be obtained. The working principle of TanDEM-X is shown in Fig. 1.7.

(4) Polarimetric SAR technology:
Polarimetric SAR technology utilizes the full-vector properties of electromagnetic waves to extract target polarization scattering characteristics. The polarization scattering matrix contains rich information about the target, as it is sensitive to the target's dielectric constant, physical features, and geometric shape. Polarimetric measurements are widely applied in agricultural land classification, forest species identification, geological structure description, soil moisture estimation, sea ice monitoring, oceanography, and military target detection

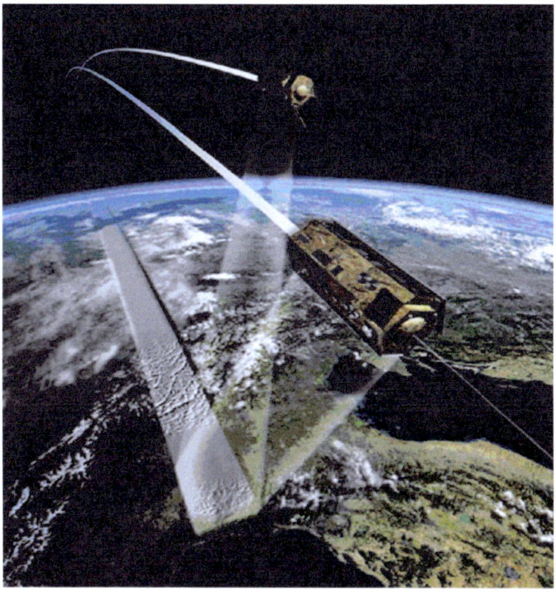

Fig. 1.7 Working principle of TanDEM-X

and recognition. Considerable research has been conducted on target-oriented polarimetry, polarimetric decomposition, polarimetric image classification, and polarimetric SAR target detection and recognition.

As SAR technology rapidly advances, numerous new ideas, modes, and methods continue to emerge. The development trend is shifting from breakthroughs in individual technologies to conceptual framework updates, from single-source information applications to multi-source information applications, and from supporting strategic applications to supporting tactical applications. As the engineering and application challenges associated with these new ideas, modes, and methods are overcome, SAR technology will enter a new era of significant development.

1.2 Millimeter Wave Characteristics

Millimeter waves are electromagnetic waves that fall between microwaves and optical waves, with a frequency range of 30 to 300 GHz and wavelengths ranging from 1 to 10 mm. The development of millimeter wave radar technology began during World War II to improve radar accuracy and resolution. However, the progress of millimeter wave radar technology was hindered by the adverse effects of atmospheric propagation, such as rain and dust weather, as well as the rapid development of laser and infrared detection technologies. It was not until the late 1980s that breakthroughs in key millimeter wave radar technologies revived its development, with developed countries designating millimeter wave technology as a key research project in the military domain. In recent years, the increasing demand for high precision in military applications has further promoted the development and wider application of millimeter wave radar technology.

1.2.1 Millimeter Wave Atmospheric Propagation Characteristics

The main atmospheric effects on millimeter waves are caused by gases (water vapor and oxygen), clouds, and rain. Oxygen and water vapor are the major gas components in the atmosphere that can significantly influence electromagnetic wave propagation within the frequency range of 20 to 300 GHz. Oxygen absorption of electromagnetic waves is due to the resonance of oxygen molecules, particularly in the frequency range of 50–70, 118.7, and above 300 GHz. Oxygen attenuation depends on air pressure and temperature. Water vapor absorption occurs due to the interaction of electromagnetic waves with water molecules, resulting in absorption lines at typical frequencies of 22.253, 183.3, and 324.4 GHz. Water vapor absorption is influenced by meteorological parameters such as surface pressure, temperature, and total water vapor content (WVTC). In cases of rainfall, severe attenuation-induced interruption losses can reach several percent of the total time per year.

In 2008, Salvatore D'Addio et al. analyzed the performance of airborne Ka-band SAR based on scanning receiving antenna beams. The simulations demonstrated that the scanning receiving technique could effectively reduce the attenuation and clutter effects caused by rain in the Ka-band, even under 10 times higher rainfall rates. Thus, the scanning receiving technique can mitigate the impact of rain on Ka-band SAR performance.

In 2009, A. Danklmayer and M. Chandra quantitatively assessed the impact of precipitation on SAR imaging in the X-band and Ka-band using actual data from the German satellite TerraSAR-X and synchronous weather radar measurements. They constructed rainfall area models to simulate the relationship between path attenuation in the Ka-band (35 GHz) and rainfall rates. A comparison between TerraSAR-X images and simulated Ka-band SAR images generated using weather radar measurements revealed a significant impact of rainfall rates of 10 mm/hour on Ka-band SAR images.

In 2009, A. Danklmayer and M. Chandra quantitatively estimated the impact of precipitation on X-band and Ka band SAR imaging based on actual data collection from the German satellite TerraSAR-X and synchronous weather radar measurements [1]. The rainfall area model is shown in Fig. 1.8.

Figures 1.8 show the constructed rainfall area model, assuming a height of 4 km and uniform rainfall in the area. By simulating the relationship between bidirectional path attenuation and rainfall in the Ka band (35 GHz) using the model shown in the figure, and comparing the TerraSAR-X image with the simulated Ka band SAR image generated through weather radar measurements, it was found that a rainfall rate of 10 mm per hour can have a significant impact on the Ka band SAR image.

In 2009, Salvatore D'Addio et al. analyzed the influence of precipitation on the performance of Ka-band InSAR. They proposed an extended form of interferogram coherence to evaluate the effect of rain on the random height error of InSAR, demonstrating that Ka-band InSAR can meet the random height error requirements compatible with the HRTI-3 specification under moderate rain conditions.

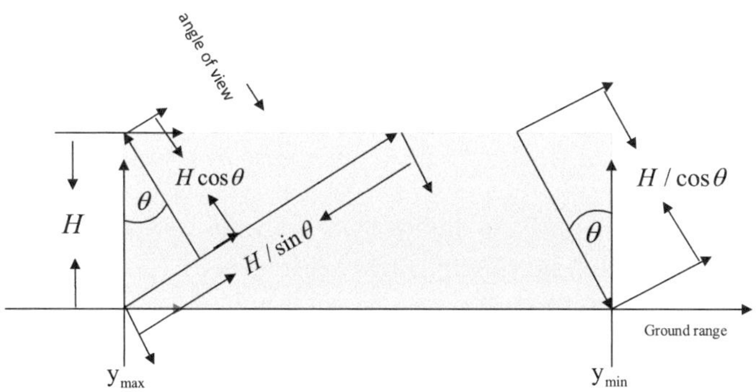

Fig. 1.8 Rainfall area model

In 2012, Carlo Capsoni et al. designed a compact high-resolution Ka-band SAR system and studied the effect of weather on SAR imaging quality under clear sky conditions and uniform cloud coverage up to rainfall. Numerical estimations of attenuation and cloud backscattering were performed using a layered model.

According to statistical analyses based on ITU regulations, the two-way attenuation of signals for Ka-band observations caused by atmospheric effects is around 2 dB in about 90% of the time and can go up to 5 dB in as much as 95% of the time.

1.2.2 Millimeter Wave Backscattering Characteristics of Objects

The electromagnetic characteristics of targets, namely their scattering properties, include the amplitude, phase, frequency, polarization, and angle scintillation of the target's scattering field. The most important characteristic is the radar cross-section (RCS), which represents the effective area of the target for reflection. A significant feature of the reflected signal occurs when the incident wave is perpendicular to a close-to-circular surface, resulting in a strong reflection. Depending on the incident angle, the reflected signal from the target may vary rapidly, exhibiting variations in radar cross-section as a function of the incident angle.

In 2005, researchers from the Massachusetts Institute of Technology (MIT) conducted research and comparative analysis on the clutter backscattering characteristics of X-band, Ka-band, and W-band, based on experimental verifications using 160, 520 GHz, and 1.56 THz 1/16-scale field ISAR imaging.

The experiments revealed that under the HH polarization condition, the Ka-band exhibited a stronger backscattering coefficient compared to the X-band by approximately 2.5 to 5.5 dB (with a significant difference at large incident angles). Under the VV polarization condition, the Ka-band exhibited a stronger backscattering coefficient compared to the X-band by approximately 2 to 4 dB (with a significant difference at large incident angles). The normalized backscattering coefficients at different frequency bands (HH polarization) are shown in Fig. 1.9, while the normalized backscattering coefficients at different frequency bands (VV polarization) are shown in Fig. 1.10.

Table 1.3 compares the backscattering coefficients of the Ka-band and X-band.

In 2012, the European Space Agency (ESA) conducted the Ka SAR-2012 project research to provide data support for the development of subsequent satellite-borne Ka-band SAR engineering. The project involved the analysis of backscattering characteristics of different objects in the Ka-band at various incident angles based on extensive data obtained from airborne flight experiments.

The statistical comparison of measured average backscattering coefficients in different frequency bands for agricultural areas is shown in Fig. 1.11. The project's statistical analysis of average backscattering coefficients in agricultural areas indicated that within the incident angle range of 10° to 70°, the average backscattering

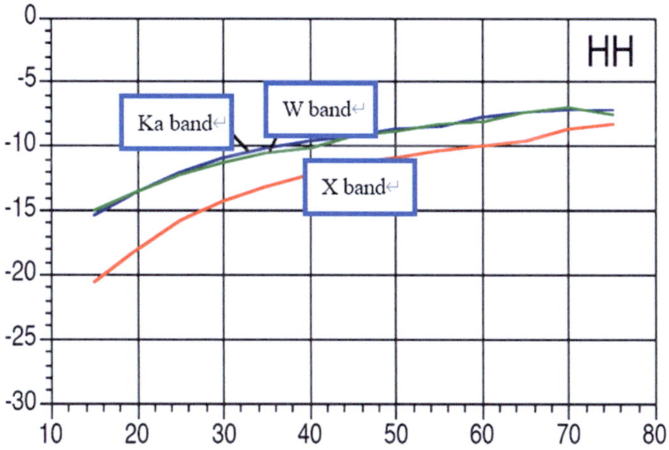

Fig. 1.9 Normalized backscattering coefficients at different frequency bands (HH polarization)

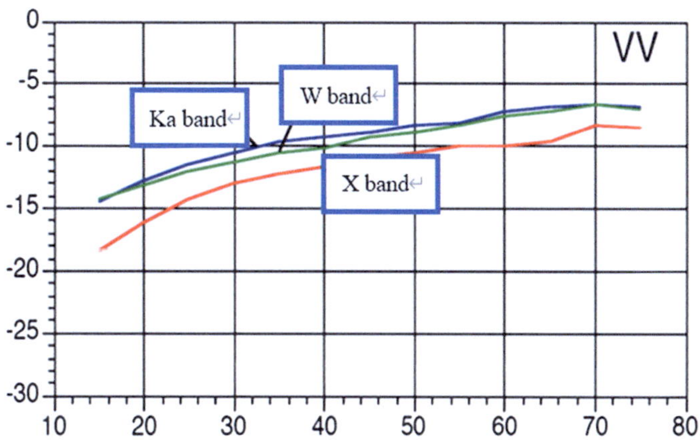

Fig. 1.10 Normalized backscattering coefficients at different frequency bands (VV polarization)

Table 1.3 Comparison of backscatter coefficients between Ka band and X band

Band	Polarization mode	Incident angle 70°		Incident angle 30°	
		σ° (theoretical value) (dB)	σ° (measured value) (dB)	σ° (theoretical value) (dB)	σ° (Measured value) (dB)
X band	HH	−18	−17	−10	−10
	VV	−17	−16	−11	−10
Ka band	HH	−15	−14	−6	−8
	VV	−14	−13	−7	−7

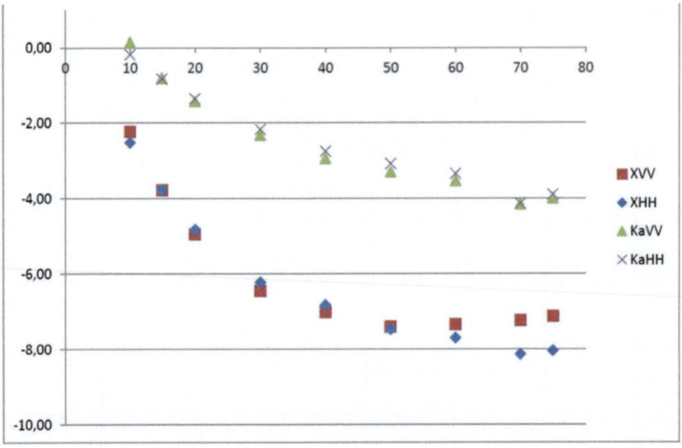

Fig. 1.11 Comparison of measured average backscattering coefficients in different frequency bands for agricultural areas

coefficient in the Ka-band is approximately 0 dB to −4 dB, significantly higher than the average backscattering coefficient in the X-band (−2 dB to −8 dB) within the same incident angle range.

1.3 Development of Millimeter Wave SAR Technology

Compared with low-frequency SAR, millimeter wave SAR has the following advantages.

1. Higher resolution can be achieved. The available bandwidth in the millimeter wave band is large, and the millimeter wave antenna size is smaller under the same beam width, so millimeter wave SAR can easily obtain higher range and azimuth resolution than low-frequency SAR.
2. Smaller volume and lighter weight. The size of the millimeter wave band antenna and RF module is much smaller than that of the low-frequency band, so the millimeter wave SAR system has the characteristics of small size and light weight.
3. The interference baseline is shorter and the millimeter wave length is longer, therefore obtaining the same interference measurement accuracy. The baseline of millimeter wave InSAR is much smaller than that of low-frequency SAR.
4. The target contour effect is more pronounced. The higher the radar operating frequency band, the more obvious the target contour effect. In Ka band, the contour of the target's geometric shape can produce stronger echoes, which is conducive to obtaining clear geometric characteristics of the target.

5. Weak penetration. The penetration of millimeter waves through vegetation, snow, soil, etc. is weak, and signal reflections mostly occur at the top of snow and tree crowns. Therefore, millimeter wave SAR can achieve accurate measurement of ice and snow.
6. Stronger anti-interference ability. Due to the fact that the operating frequencies of a large number of electronic warfare equipment have not fully covered the millimeter wave frequency band, the electronic countermeasures performance of millimeter wave SAR is relatively stronger.

With the continuous improvement of core power device level, in recent years, the United States and European countries have begun to study the application of Ka and W bands in the SAR field. In terms of airborne technology, the development and application of Ka band and W band systems have been completed; In the aspect of spaceborne, the interest of technicians mainly focuses on the research of Ka band SAR technology which is more engineering realizability.

1.3.1 Development of Airborne Millimeter Wave SAR Technology

The development of airborne millimeter wave SAR technology began in the 1990s. The Lincoln Laboratory of Massachusetts Institute of Technology in the United States first published their achievements in airborne Ka band SAR technology in 1991. They completed the development and flight tests of a fully polarized high-resolution Ka band SAR system and obtained 0.3 m resolution Ka band SAR images for conducting research on millimeter wave imaging mechanisms; In the same period, Martin Marietta (the predecessor of Lockheed Martin) developed the 35 GHz airborne dual polarization SAR system in 1992 and carried out flight tests, obtaining Ka band SAR images with 1 m resolution. The development and successful testing of these two sets of airborne millimeter wave SAR systems have laid the foundation for the subsequent development of airborne millimeter wave SAR technology.

In the twenty-first century, airborne millimeter-wave SAR technology has undergone rapid development. In 2003, EADS Germany introduced the first-generation Miniature Synthetic Aperture Radar (MISAR) sensor system. MISAR is a highly miniaturized Ka-band frequency-modulated continuous-wave (FMCW) SAR installed on unmanned aerial vehicles (UAVs) weighing less than 5 kg. It demonstrated the capability to obtain high-quality imaging on extremely small platforms. Based on multiple flight tests and development experience, EADS Germany developed the second-generation MISAR system. The MISAR system prototype and SAR images obtained are shown in Fig. 1.12.

In 2004, Sandia National Laboratories in the United States conducted flight tests of a Ka-band SAR system and publicly released a large number of high-quality Ka-band SAR images with a resolution of 0.1 m. These images demonstrated the rich optical-like and detailed features of Ka-band SAR compared to lower-frequency

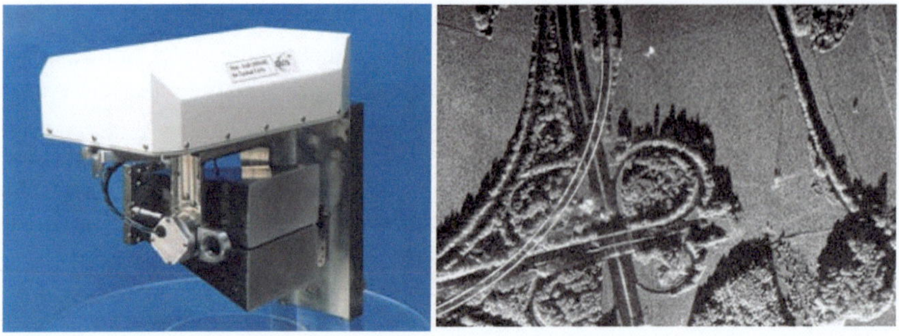

Fig. 1.12 MISAR system prototype and obtained SAR images

SAR images, highlighting the advantages of millimeter-wave SAR in high-quality fine observation.

In 2004, the Southwest Institute of Electronic Technology in China published their research results in airborne millimeter-wave SAR technology. Their system operated in the Ka-band and utilized a focused feed (traveling-wave tube) and parabolic antenna. The system had real-time imaging capabilities and obtained SAR images with a resolution of 1 m through flight tests.

Since 2005, the German FGAN-FHR has conducted a large amount of research on airborne millimeter wave SAR technology, and has developed multiple sets of Ka and W-band SAR systems and completed multiple flight tests: (1) The Ka/W dual frequency multi baseline interferometric SAR system MEMPHIS, which works simultaneously at 35 and 94 GHz, can achieve high-resolution images with range resolution better than 0.2 m and azimuth resolution better than 0.05 m, and achieve height measurement accuracy of about 0.2 m; (2) The MIRANDA system in the W-band (94 GHz) is used on unmanned aerial vehicle platforms and can achieve high-resolution imaging of 0.15 m; (3) The MIRANDA-35 system in the Ka band (35 GHz) is mainly used for coherent change detection research, and can also obtain 0.15 m SAR images. The flight test of the NEO S-300 unmanned aerial vehicle MIRANDA system is shown in Fig. 1.13, and the MIRANDA imaging results are shown in Fig. 1.14.

In 2011, the Institute of Electronics, Chinese Academy of Sciences, developed China's first airborne millimeter-wave three-baseline InSAR system. The system operated in the Ka-band, used horn antennas, and employed millimeter-wave switches to switch different channel RF signals for consistency between channels. The system obtained SAR images with a resolution of 0.5 m and elevation measurement accuracy better than 1 m through flight tests. The digital surface model (DSM) of terrain with elevation variations and the annotation of actual scenes are shown in Fig. 1.15.

After nearly 30 years of development, airborne millimeter-wave SAR technology has evolved from a simple system with centralized power amplifiers and horn antennas to a complex system with solid-state distributed power amplifiers and

Fig. 1.13 Flight tests of the NEO S-300 UAV with the MIRANDA system

Fig. 1.14 Imaging results of MIRANDA

phased array antennas. The systems have become more functional and their performance has continuously improved. The development and flight tests of different systems have overcome key technical challenges in millimeter-wave SAR system development and data processing and deepened the understanding of millimeter-wave remote sensing mechanisms, greatly promoting the progress of millimeter-wave SAR technology.

1.3.2 Development of Spaceborne Millimeter-Wave SAR Technology

Due to the high atmospheric attenuation in the millimeter-wave frequency band compared to traditional low-frequency bands, the early development of spaceborne millimeter-wave SAR was limited. In recent years, with the advancement of millimeter-wave SAR technology research and corresponding technological levels, spaceborne millimeter-wave SAR has entered a rapid development stage.

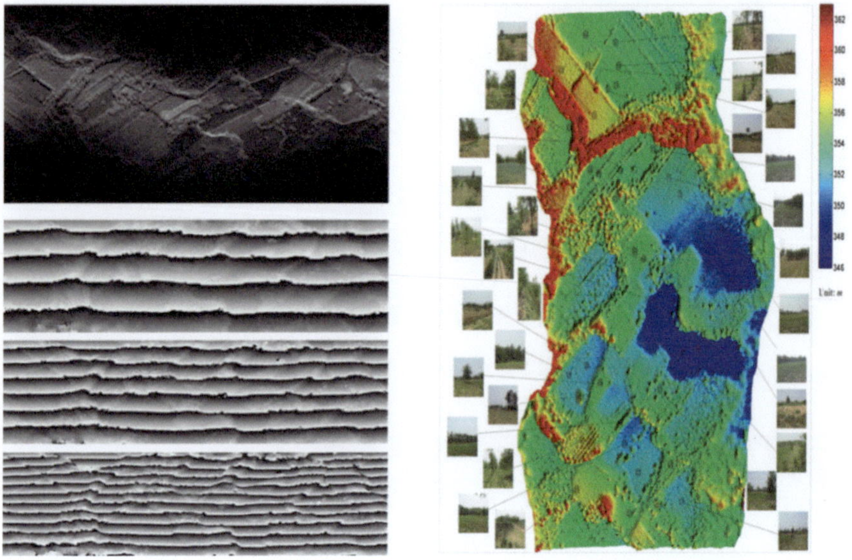

Fig. 1.15 Digital surface model (DSM) of terrain with elevation variations and annotation of actual scenes

In 2003, Lockheed Martin Corporation proposed the design concept of a reconfigurable aperture synthesis SAR in the Ka-band for spaceborne applications. The satellite was set at a height of 700 km, and the system employed high-power solid-state transceiver (TR) array devices. By switching a portion of the TR modules, the SAR aperture could be reconstructed as the incidence angle varied from 15 to 70°. The system design achieved a distance and azimuth resolution of 1 m.

In 2005, EADS Astrium proposed the SIGNAL system (SAR for Ice Glacier and global Dynamics) as part of the Earth Explorer 8 program. SIGNAL is an innovative Ka-band SAR system designed to accurately estimate glaciers, ice caps, polar changes, altimetry, and flow velocity. It aims to provide scientific research support for better understanding the hydrology of glacier basins and the water cycle in the polar regions and fill in the gaps in the global database for glacier balance and dynamics. The system uses a pair of formation-flying satellites to obtain the required long baselines for high sensitivity and measurement stability.

In 2006, the Jet Propulsion Laboratory at the California Institute of Technology proposed the Glacier and Land Ice Surface Topography Interferometer (GLISTIN) system for mapping glaciers and ice surfaces in the Ka-band. The system features high spatial resolution, high vertical measurement accuracy, and is not affected by cloud cover. GLISTIN is a single-platform interferometric synthetic aperture radar (InSAR) system. The 8 mm wavelength in the Ka-band minimizes snow penetration and takes into account atmospheric losses. Compared to lidar, the SAR in this frequency band is not sensitive to cloud cover, providing sufficient mapping bandwidth, subseasonal revisit frequency, and inherent variable spatial resolution. In coastal and glacier

Fig. 1.16 Schematic diagram of the GLISTIN system

areas, GLISTIN can achieve meter-level vertical accuracy, while in inland areas, it can achieve decimeter-level vertical accuracy.

To address the limitations of wide-area mapping and high transmitting power, GLISTIN adopts Digital Beamforming (DBF) technology, making it the first proposed system with such a design concept in a spaceborne system. The system achieves high-resolution elevation measurement accuracy of less than 10 cm over inland surfaces and several tens of centimeters in coastal areas. The system operates at an orbital height of approximately 600 km with a 92-degree inclination non-sun-synchronous orbit. The ground mapping swath width is 70 km. The system provides complete coverage of Antarctica and other mid-latitude regions, offering sufficient mapping bandwidth, a subseasonal revisit frequency, and inherent variable spatial resolution. The schematic diagram of the GLISTIN system is shown in Fig. 1.16.

In 2007, the National Research Council of the United States released the first ten-year space observation plan for Earth science and applications, with SWOT (the surface water and ocean topography) being a scientific research project related to surface water and ocean topography [2–9], originally scheduled for launch in 2016 (currently postponed). One of the main payloads of SWOT is the Ka band InSAR, which operates at a frequency of 35 GHz and has a signal bandwidth of 200 MHz. It can achieve a slant range resolution of 0.75 m and a ground distance resolution of 5 m. The antenna size of the SWOT system is 4×0.2 m, with an interference baseline length of 10 m. By utilizing this system, high-precision and wide swath measurements of the sea surface can be achieved, thereby studying changes in small and medium-sized ocean ranges; At the same time, it can also measure the height of land water bodies and study the spatial and temporal distribution caused by the storage and loss of land water bodies. The SWOT system model diagram is shown in Fig. 1.17.

The SWOT system adopts a dual mapping strip mode for the sub satellite point, and simultaneously observes both sides of the sub satellite point. The width of each mapping strip is 60 km, with a total of 120 km on both sides. The beams of the two mapping strips are H-polarization and V-polarization, respectively. Using a biased

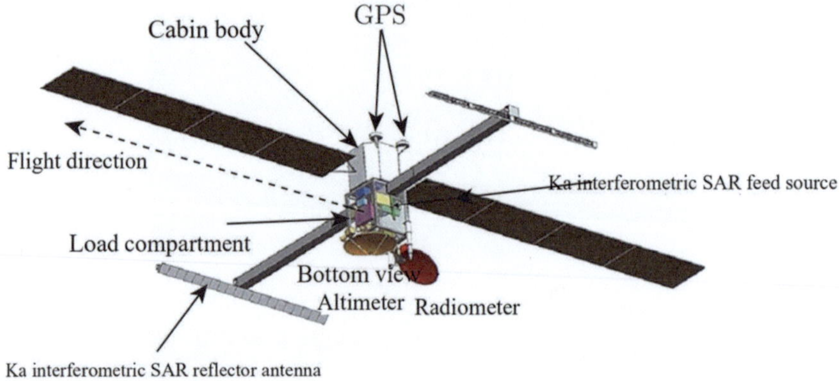

Fig. 1.17 SWOT system model diagram

transmission array antenna (phased array), the antenna is directed by adjusting the RF network phase shifter, while transmitting H-polarized and V-polarized beams to point to different mapping bands. At the same time, the substellar point altimeter has a very narrow path at the substellar point for altitude measurement. The schematic diagram of the SWOT task is shown in Fig. 1.18.

In 2008, ESA proposed a design scheme for Ka band spaceborne SAR based on Scan on Receive (SCORE) technology [10, 11]. Considering that the system components are prone to significant losses in the high-frequency range, a separate antenna architecture for transmitting and receiving is adopted in the design: a high gain reflector antenna is selected for transmitting, an 8-channel phased array antenna is selected for receiving, and a scanning and receiving technology based on DBF is adopted. Adopting this system architecture can effectively reduce the demand for transmission power, and using scanning and receiving technology can also reduce the interference of rainwater backscattering on imaging performance.

In 2010, the European Space Agency conducted verification work on Ka band SAR to achieve single platform high-resolution interference. The system also used scanning and receiving technology based on high gain antennas [12]. The system operates at a frequency of 35.75 GHz, has a transmission bandwidth of 300 MHz, and utilizes 2 m × A 0.35 m transmitting antenna and a 2.25 m diameter receiving antenna can obtain 1 m × The resolution is 1 m, and the radar has an interference baseline length of 12 m. The experimental results indicate that a 12 m baseline in the Ka band can achieve very high elevation measurement accuracy. The hardware configuration diagram of ESA Ka InSAR is shown in Fig. 1.19.

In 2012, Alenia Aerospace Company of Italy proposed a single satellite SAR interferometer operating in the Ka frequency band [13, 14], which was designed using an improved bistatic scheme to meet the needs of single satellite interferometry measurement. Three pairs of antennas are used, one of which is only used for transmission and installed on the satellite body. Two pairs of receiving antennas are

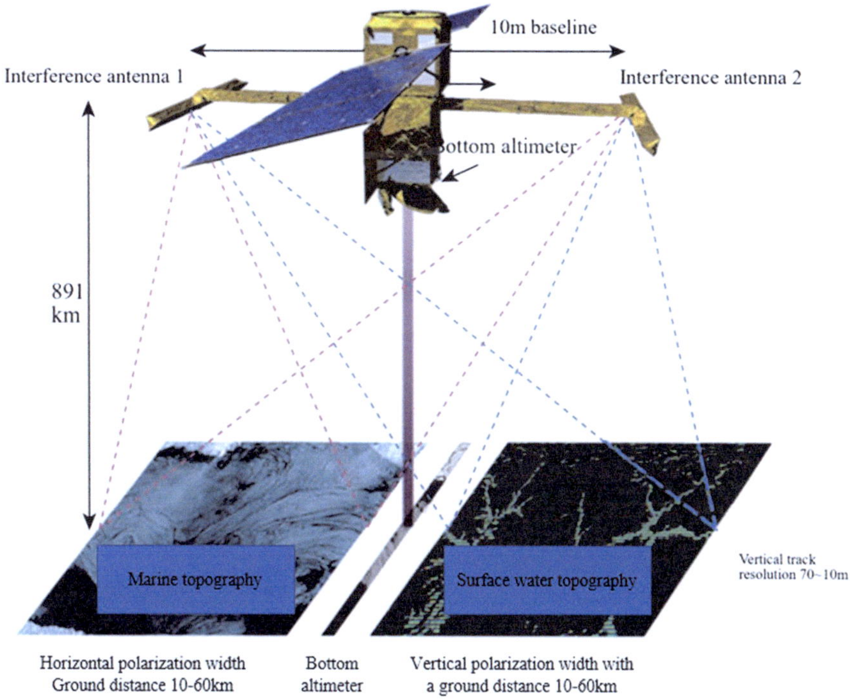

Fig. 1.18 Schematic diagram of SWOT tasks

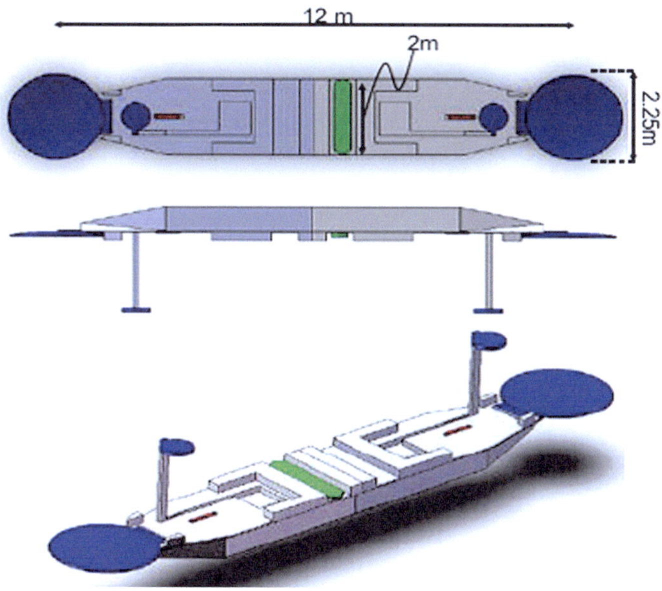

Fig. 1.19 Hardware configuration diagram of ESA Ka InSAR

installed on the top of the boom, and the antenna design has the ability to provide cross orbit interference and along orbit interference.

In 2013, DLR, with funding support from the European Space Agency, also conducted research on the Ka band interferometric SAR system and proposed a design proposal. In this scheme, the satellite extends 10 m of antenna support arms on both sides, with two transmitting antennas installed at the end of the antenna support arm and two receiving antennas installed on the satellite body. The interference baseline length is about 10 m, and both transmitting and receiving antennas are parabolic antennas. The system uses DBF technology in azimuth direction with two transmissions and two receivers, and adopts multi aperture processing method (MAPS) in data processing to reduce PRF and obtain a wide surveying band; SCORE technology is used to achieve high gain in the distance direction. The system structure can be adjusted to a structure with four phase centers, which can simultaneously achieve interference and GMTI. The schematic diagram of the Ka InSAR system is shown in Fig. 1.20.

In 2014, the Dutch Space Agency and German Aerospace proposed a 5-channel Ka band High Resolution Wideband (HRWS) SAR system with GMTI function,

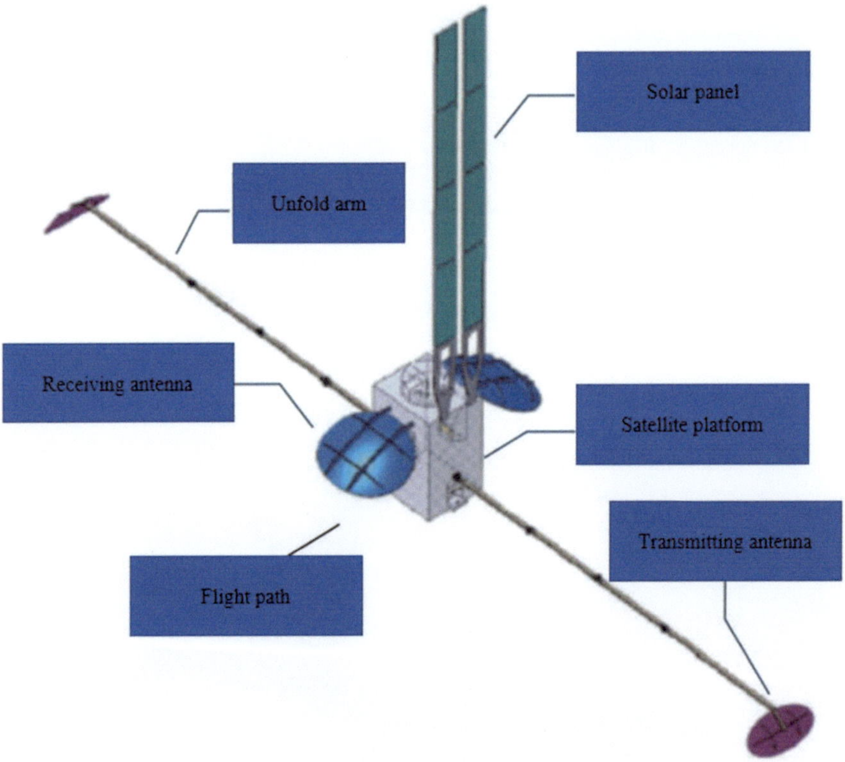

Fig. 1.20 Schematic diagram of Ka InSAR system

Fig. 1.21 Schematic diagram of Ka band single channel cross orbit spaceborne interferometric SAR

which can operate at 1.5 m × Achieving target scene observation with a width of 50 km at a resolution of 1.5 m, the system sensitivity is better than −18 dB, and both AASR and RASR are better than −20 dB. Based on the HRWS-GMTI algorithm, GMTI with a speed measurement accuracy of 10 km/h can be achieved.

In 2016, Deutschland Aerospace proposed a single channel cross orbit spaceborne interferometric SAR scheme operating in the Ka band [15–18]. The single channel cross orbit interferometer has a height scanning beam with a side view angle of 25°, and a transmitting antenna is a single biased reflector fed by a feeding array located in the focal plane. This antenna can generate 8 scanning beams through a feed array and matrix switch. There are two receiving antennas (MFAs) at the receiving end, each consisting of three offset reflectors fed by a feed array also located in the focal plane of the reflector system to achieve scanning receive operation (SCORE). Each receiving antenna can generate 56 elevation beams, each beam generated by 3 adjacent feed sources. The main reflector of each receiving antenna is located at the tip of the deployable truss mast, while a fixed dual imaging reflector system consisting of two hyperbolic toroidal reflectors is accommodated on the ground platform of the satellite platform. The schematic diagram of Ka band single channel cross orbit spaceborne interferometric SAR is shown in Fig. 1.21.

After more than ten years of development, the satellite-based millimeter wave SAR technology has evolved from the system exploration and demonstration stage of pure technology verification to the system design and development stage with clear application and engineering development plans [19–31]. With the continuous deepening of research on the mechanism of space-based millimeter wave remote sensing and the continuous promotion of corresponding projects in various countries, satellite borne millimeter wave SAR systems will undoubtedly join the family of in orbit satellite borne SAR satellites in the near future.

References

1. Danklmayer A, Chandra M. Comparison of precipitation effects in space-borne X- and Ka-band SAR imaging[C]. Geoscience and Remote Sensing Symposium,2009 IEEE International,igarss. IEEE, 2009:IV-909–IV-912.
2. Cao F, Tupin F, Nicolas J M, et al. Extraction of water surfaces in simulated Ka-band SAR images of KaRIn on swot[C]. Geoscience and Remote Sensing Symposium. IEEE, 2011:3562–3565.
3. Fjortoft R, Gaudin J M, Pourthie N, et al. KaRIn on SWOT: Characteristics of Near-Nadir Ka-Band Interferometric SAR Imagery[J]. IEEE Transactions on Geoscience & Remote Sensing, 2013, 52(4):2172-2185.
4. Moller D, Esteban-Fernandez D. Near-Nadir Ka-band Field Observations of Freshwater Bodies[M]. Remote Sensing of the Terrestrial Water Cycle. John Wiley & Sons, Inc, 2014:143–155.
5. Vaze P, Kaki S, Limonadi D, et al. The Surface Water and Ocean Topography Mission[C]. IEEE Aerospace Conference. IEEE, 2018.
6. Fjørtoft R, Gaudin J M, Pourthie N, et al. KaRIn - the Ka-band radar interferometer on SWOT: Measurement principle, processing and data specificities[C]. Geoscience and Remote Sensing Symposium. IEEE, 2010:4823–4826.
7. Wu X, Hensley S, Rodriguez E, et al. Near nadir Ka-band sar interferometry: SWOT airborne experiment[C]. Geoscience and Remote Sensing Symposium. IEEE, 2011:2681–2684.
8. E. Rodriguez, "Surface Water and Ocean Topography mission (SWOT) science requirements document," Jet Propulsion Laboratory, California Inst. Technol., Pasadena, CA, USA, Tech. Rep., Oct. 2011 [Online]. Available: http://swot.jpl.nasa.gov/science/
9. L. Fu, D. Alsdorf, R. Morrow, E. Rodriguez, and N. Mognard, SWOT: The Surface Water and Ocean Topography Mission: Wide-Swath Altimetric Measurement of Water Elevation on Earth, Pasadena, CA, USA: Jet Propulsion Laboratory, 2012.
10. Ludwig M, D'Addio S, Saameno-Perez P. Ka-Band SAR for Spaceborne Applications based on Scan-on- Receive Techniques[C]. Synthetic Aperture Radar (EUSAR), 2008 7th European Conference on. VDE, 2008:1–4.
11. Ludwig M, D'Addio S, Aguirre M, et al. Imaging Ka-band SAR interferometer[C]. International Asia-Pacific Conference on Synthetic Aperture Radar. IEEE, 2011:1–4.
12. Ludwig M, D'Addio S, Engel K, et al. A Spaceborne Ka-Band SAR Interferometer Concept based on Scan-on-Receive Techniques[C]. European Conference on Synthetic Aperture Radar. VDE, 2010:1–4.
13. Germani C, Venturini R, Ludwig M, et al. Ka-band SAR interferometer[C]. Geoscience and Remote Sensing Symposium. IEEE, 2012:5598–5601.
14. Schaefer C, Lopez-Dekker P. Interferometric Ka-band SAR with DBF capability[C]. European Conference on Synthetic Aperture Radar. VDE, 2012:7–10.
15. Lori M, Zolla P, Pfeiffer E, et al. Design and performance of a spaceborne Ka-band interferometrie SAR antenna system[C]. IEEE International Symposium on Antennas and Propagation. IEEE, 2016.
16. Jacobs S, Bekers D, Monni S, et al. Introducing sparsity in a spaceborne Ka-band SAR antenna[C]. European Conference on Antennas and Propagation. IEEE, 2016:1–2.
17. Mao C X, Gao S, Tienda C, et al. X/Ka-Band Dual-Polarized Digital Beamforming Synthetic Aperture Radar[J]. IEEE Transactions on Microwave Theory & Techniques, 2017, 65(11):4400-4407.
18. Bekers D J, Jacobs S, Bolt R J, et al. Design of a Ka-band sparse array antenna for spaceborne SAR applications[C]. European Microwave Conference. 2017:117–120.
19. Wang H, Jiang M, Zheng S. Research on Bi-satellite Ka-band FMCW SAR Design and Imaging[C]. Eusar 2016:, European Conference on Synthetic Aperture Radar, Proceedings of. VDE, 2016.
20. Wang H, Chen Z, Zheng S. Preliminary Research of Low-RCS Moving Target Detection Based on Ka-Band Video SAR[J]. IEEE Geoscience & Remote Sensing Letters, 2017, 14(6):811-815.

21. D'Addio S, Ludwig M. Rain Impact on Sensitivity of Ka-band Scan-on-Receive Synthetic Aperture Radars[C]. Geoscience and Remote Sensing Symposium, 2008. IGARSS 2008. IEEE International. IEEE, 2009:III-1174–III-1177.
22. B. Waske and M. Ludwig, "Modelling and analysis of rain effect on Ka-band single pass InSAR Performance[C]. Geoscience and Remote Sensing Symposium. IEEE, 2009, pp. IV-913–IV-916.
23. Capsoni C, Guarnieri A M, Riva C, et al. Impact of atmospheric propagation in a Ka-band spaceborne SAR for imaging and interferometry[C]. Geoscience and Remote Sensing Symposium. IEEE, 2012:3815–3818.
24. Ulaby F T, Dobson M C. Handbook of Radar Scattering Statistics for Terrain[J]. 1989.
25. Nashashibi A, Ulaby F T, Sarabandi K. Measurement and modeling of the millimeter-wave backscatter response of soil surfaces[J]. IEEE Transactions on Geoscience & Remote Sensing, 1996, 34(2):561-572.
26. D. Vandermark, B. Chapron, J. Sun, G. H. Crescenti, and H. C. Graber, "Ocean wave slope observations using radar backscatter and laser altimeters," J. Phys. Oceanogr., vol. 34, no. 12, pp. 2825–2842, 2004.
27. S. Tanelli and S. L. Durden, "Simultaneous measurements of Ku- and Ka-band sea surface cross sections by an airborne radar," IEEE Geosci. Remote Sens. Lett., vol. 3, no. 3, pp. 359–363, Jul. 2006.
28. Moller, Heavey D K, BrandonHodges, et al. The Glacier and Land Ice Surface Topography Interferometer (GLISTIN): a novel Ka-band digitally beamformed interferometer[J]. 2006.
29. Moller D, Hensley S, Sadowy G A, et al. The Glacier and Land Ice Surface Topography Interferometer: An Airborne Proof-of-Concept Demonstration of High-Precision Ka-Band Single-Pass Elevation Mapping[J]. IEEE Transactions on Geoscience & Remote Sensing, 2011, 49(2):827-842.
30. Ludwig M, Daganzo-Eusebio E, Davidson M. Ka-Band radar missions for earth observation[C]. Geoscience and Remote Sensing Symposium. IEEE, 2014:2289–2292.
31. Tienda C, Younis M, Lopez-Dekker P, et al. Ka-band reflect array antenna system for SAR applications[C]. European Conference on Antennas and Propagation. IEEE, 2014:1603–1606.
32. Patyuchenko A, Younis M, Krieger G. Compact X/Ka-band dual-polarization spaceborne digital beamforming Synthetic Aperture Radar[C]. Radar Symposium. IEEE, 2015:1–3.
33. Jacobs S, Bekers D, Monni S, et al. Design considerations on a sparse array antenna for Ka-band spaceborne SAR applications[C]. Radar Conference. IEEE, 2015:409–412.

Chapter 2
Continuous Wave Regime Airborne Millimeter Wave Synthetic Aperture Radar Technology

2.1 Overview

The miniaturization of SAR system is an important development direction of SAR technology. The miniaturization of SAR system is to reduce the size, mass and power consumption of the radar system as much as possible without affecting the imaging quality. The miniaturized SAR system can be installed on smaller and more flexible flight platforms, which will help to reduce the load requirement of the platform and expand the use of SAR.

The small size of millimeter-wave Ka-band and W-band antennas and their RF modules can enhance the integration of SAR systems, and significantly reduce the system size and mass. Based on this, the millimeter wave SAR combined with the FMCW has become the first choice for the international airborne miniaturized SAR system.

FMCW SAR is a fusion of FMCW technology and SAR technology. Compared with the traditional pulse-based SAR system, the peak transmit power of FMCW SAR system is greatly reduced while achieving the same average transmit power. Meanwhile, the FMCW SAR system adopts the dechirp reception method to reduce the data size of the echo signal. However, since FMCW SAR is based on continuous transmission and reception, the pulse duration is very long. The radar movement during transmission and reception cannot be ignored. The "go-stop" assumption set in the traditional pulse-impulse SAR imaging processing no longer holds, so the improvement of the imaging method needs to be completed in a targeted manner.

The FMCW SAR's nearly 100% transmit duty cycle can significantly reduce the number of amplifier components, which is particularly advantageous when applied to a spaceborne system to reduce costs. In contrast, a sub-meter high resolution imaging capability cannot be obtained with a spaceborne system using only strip mode. By scanning the radar beam from front to back in azimuth, the sliding spotlight can obtain a longer synthetic aperture, thus breaking the limitation of antenna azimuth size and achieving higher resolution, which has become the most common mode to achieve

H. Wang, *The Millimeter Wave Synthetic Aperture Radar Technology*,
https://doi.org/10.1007/978-981-97-1044-7_2

high-resolution imaging. Therefore, it is important to carry out the verification of airborne FMCW SAR sliding spotligh imaging technology for the future engineering implementation of FMCW SAR.

In addition, with the rapid development of the UAV remote sensing industry, the need for high-performance miniaturized sensors has become increasingly urgent. Small UAV-based millimeter wave SAR can work all day and provide optical-like SAR image, which can well meet the needs of multi-industry applications. In order to improve the observation efficiency and application effectiveness, real-time image generation capability is required, and how to solve the problem of the impact of large motion errors of small UAV platforms on W-band SAR images becomes the key to determine the prospect of small UAV-borne millimeter wave SAR applications.

To address the above situation, this chapter firstly introduces the basic strip imaging method of airborne millimeter wave FMCW SAR, based on which an improved two-step imaging method is proposed to realize the sliding spotlight imaging of airborne millimeter wave FMCW SAR for subsequent on-board applications. Finally, we introduce the real-time imaging method for W-band small UAVborne FMCW SAR, taking into account the unstable attitude and error-sensitive characteristics of W-band.

2.2 Airborne FMCW SAR Strip Mode Imaging

In this section, we first start from the radar equation to explore the mechanism of FM continuous wave synthetic aperture radar in order to reveal the essential reasons why it can simplify the system design and realize the miniaturization, light weight and low cost of the system [1], and establish the theoretical basis of FMCW SAR. On this basis, the principle of range resolution capability of FMCW signal is investigated. After that, the imaging algorithm of airborne FMCW SAR system is discussed, including range Doppler algorithm and wave number domain imaging algorithm. Finally, the airborne millimeter wave FMCW SAR data processing is introduced to lay the research foundation and theoretical support for the subsequent development of new FMCW SAR systems and signal processing techniques.

2.2.1 Airborne FMCW SAR Strip Imaging Model

1. Theoretical basis of FMCW SAR

According to the radar equation, the signal-to-noise ratio (SNR) of the received signal of a synthetic aperture radar system can be expressed as

$$SNR^1 = \frac{P_t G^t G^r \lambda^3 \cdot PRF \cdot T_p \cdot c \cdot \sigma_0}{4 \cdot (4\pi)^3 R^3 \cdot v_s \sin(\eta) FkT B_r} \tag{2.1}$$

where P_t is the peak power of the system transmit signal; T_p is the pulse duration; G^t and G^r are the transmit and receive antenna gains; σ_0 is the backscattering coefficient of the detection target; R is the range of the detection target; F is the noise factor; k is the Boltzmann constant equal to 1.38×10^{-23} J/K; T is the system ambient temperature; B_r is the signal bandwidth; PRF is the pulse repetition frequency; η is the incident angle of the SAR illuminated target.

For a given system operating environment, increasing the antenna size and decreasing the signal bandwidth can improve the SNR, but at the same time, it can also reduce the resolution of SAR images and affect the imaging quality. Increasing the peak power of the transmit signal will increase the system pressure, improve the complexity of the system design, and reduce the performance and reliability of the system. Increasing the pulse duration can effectively improve the signal-to-noise ratio of the system without degrading any other performance metrics of the system. From another perspective, increasing the pulse duration can reduce the peak power of the transmit signal without reducing the signal-to-noise ratio or affecting the resolution capability of the system, thus greatly simplifying the system design, reducing the system stress, improving the reliability and stability of the system, and thus significantly reducing the power consumption, cost, size and quality of the system.

Conventional SAR systems mainly use the pulse system, which emits high-energy narrow pulses in each pulse repetition cycle, and the time width of the narrow pulses is a few microseconds to tens of microseconds, which is much smaller than the pulse repetition cycle in milliseconds. The pulsed SAR system can increase the range of the radar by increasing the transmitting power to achieve long-range detection. FMCW SAR mainly transmits FMCW signals with large time width and duty cycle close to 100%, and the peak transmit power of FMCW system is much lower than that of pulse SAR at the same range, which greatly reduces the device requirements and cost of SAR system. This greatly reduces the device requirements of the SAR system and lowers the cost [2]. FMCW SAR uses a dechirp reception method, and the bandwidth of the mixed signal depends on the modulation frequency of the transmit signal and the mapping bandwidth, which is usually much smaller than the transmit signal bandwidth, which greatly reduces the acquisition pressure of the receiver and has a limited impact on the range-oriented resolution, as discussed in the FMCW SAR range direction characteristics below.

2. FMCW SAR range direction characteristics

The most common FMCW signal is a sawtooth wave, and this section analyzes the range direction characteristics of FMCW SAR with a positive FM sawtooth wave signal. Figure 2.1 shows the time–frequency diagram of the FMCW SAR echo signal. Figure 2.2 shows the time–frequency diagram of the FMCW SAR echo signal received by dechirp. In Fig. 2.1: t is the global time variable; f_r is the range-oriented frequency variable; f_c is the carrier frequency; B_t is the transmit signal bandwidth; T_{AD} is the maximum duration of each receiver A/D acquisition; $\Delta\tau_\omega$ is the time delay difference between the nearest echo signal and the farthest echo signal of the received signal.

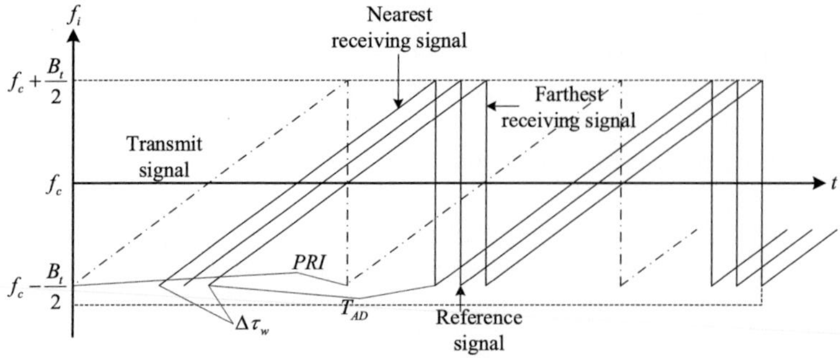

Fig. 2.1 Time–frequency diagram of FMCW SAR echo signal

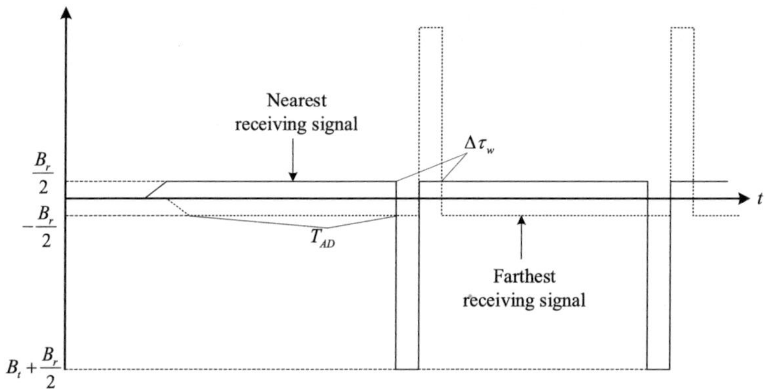

Fig. 2.2 Time–frequency diagram of the FMCW SAR echo signal received by dechirp

As can be seen from Fig. 2.2, the dechirp reception method makes the bandwidth of the received signal much lower, so the pressure on the receiver is greatly reduced, and the received bandwidth of the dechirp received signal is

$$B_{\mathrm{r}} = K_{\mathrm{r}}\Delta\tau_{\omega} = K_{\mathrm{r}}\frac{2\triangle R_{\mathrm{S}}}{c} \ll B_{\mathrm{t}} \tag{2.2}$$

where $\triangle R_{\mathrm{S}}$ is the range difference between the near and far ranges of the imaged scene; K_{r} is the transmit signal tuning frequency; and c is the speed of light. The receive signal bandwidth B_{r} is much smaller than the transmit signal bandwidth B_{t}, which greatly reduces the receive pressure for A/D sampling. The effective bandwidth of the dechirp receive is

$$B_{\mathrm{u}} = K_{\mathrm{r}}T_{\mathrm{AD}} = K_{\mathrm{r}}(T_{\mathrm{p}} - \frac{2\triangle R_{\mathrm{S}}}{c}) = B_{\mathrm{t}}(1 - \frac{2\triangle R_{\mathrm{S}}}{cT_{\mathrm{p}}}) \tag{2.3}$$

where T_p is 1/PRF, the time width of one sweep cycle. $2\triangle R_s/c/T_p$ is typically around 1%, losing 1% of the transmit signal bandwidth, resulting in a loss of 1% of the range-to-resolution. This effect is perfectly acceptable compared to the reduced stress in data acquisition at the receiver caused by dechirp reception.

Decomposing the global time variable t into a sum of the bearing time variable t_a and the range time variable t_r takes the form of

$$t = t_a + t_r = (n + \frac{1}{2})T_p + t_r \tag{2.4}$$

where $n \in Z$; $-0.5T_p \leq t_r < 0.5T_p$.

In a swept period, where n is a fixed value, the expression for the FMCW transmit signal is

$$s_0(t_r) = \exp\{j2\pi f_0 t_r\} \exp\left\{j\pi K_r t_r^2\right\} \tag{2.5}$$

The echo signal delay of an ideal point target with range $R(t_r, t_a)$ in the observation scene is $\tau(t_r, t_a) = 2R(t_r, t_a)/c$, then its echo signal is

$$\begin{aligned} s_r(t_r, t_a) = \mathrm{rect}[\tfrac{t}{T_{AD}}]\exp\{j2\pi f_c[t_r - \tau(t_r, t_a)]\}\times \\ \exp\left\{j\pi K_r[t_r - \tau(t_r, t_a)]^2\right\} \end{aligned} \tag{2.6}$$

Dechirp reception mixes the echo signal with the reference signal and then receives it. The mixed reference signal is used as the time delay signal for the transmit signal, with a time delay of $t_{ref} = 2R_{ref}/c$. The reference signal is

$$s_{ref}(t_r) = s_0(t_r - t_{ref}) \tag{2.7}$$

The received signal at the receiver is the conjugate multiplication of the echo signal and the reference signal:

$$\begin{aligned} s_d(t_r, t_a) &= s_r(t_r, t_a) \times s_{ref}^*(t_r) \\ &= \mathrm{rect}[\tfrac{t}{T_{AD}}]\exp\{-j2\pi f_c[\tau(t_r, t_a) - \tau_{ref}]\}\times \\ &\quad \exp\{-j2\pi K_r t_r[\tau(t_r, t_a) - \tau_{ref}]\}\times \\ &\quad \exp\left\{j\pi K_r[\tau(t_r, t_a) - \tau_{ref}]^2\right\} \end{aligned} \tag{2.8}$$

In the FMCW SAR, due to the large signal time width, the motion of the antenna phase center within a sweep period is not negligible, so $\tau(t_r, t_a)$ is a variable quantity, and the specific effect will be discussed in Sect. 2.3. The discussion makes the approximation of $\tau(t_r, t_a) = \tau(0, t_a)$. For a point target, Eq. (2.8) is a single-frequency signal with frequency $-K_r\tau(0, t_a)$, so that for the dechirp received signal, a single fast Fourier transform (FFT) gives a result similar to the pulse compression in the pulse SAR signal processing. The actual bandwidth of the received signal B_e is

$$B_e = K_r T_{AD} = \frac{B_t}{T_p} T_{AD} \tag{2.9}$$

Correspondingly, the range to the resolution is

$$\rho_r = \frac{c}{2B_e} = \frac{T_p}{T_{AD}} \cdot \frac{c}{2B_t} \tag{2.10}$$

The width of the scene of the FMCW SAR system is mostly in the order of several kilometers, and the corresponding $1 - T_{AD}/T_p < 1\%$, so the bandwidth is currently approximated by the transmit signal bandwidth B_t in the imaging process.

3. FMCW SAR echo signal model

In order to deeply understand the signal characteristics of FMCW SAR and develop efficient and accurate signal processing methods, its point target echo signal model and point target reference spectrum model are essential foundations in FMCW SAR signal processing research.

The imaging geometry of FMCW SAR is shown in Fig. 2.3. In the conventional pulse SAR, the radar is considered to be stationary during the transmitting and receiving of a pulse. The platform travels a negligible range between transmitting and receiving a pulse due to the short pulse duration, which is the widely used "go-stop" assumption. However, FMCW SAR operates based on continuously transmitting and receiving signals with long pulse durations, so in the continuous wave case, the radar travels during transmitting and receiving signals can no longer be ignored. In the FMCW SAR regime, the "go-stop" assumption is no longer valid.

The radar signal return delay refers to the time course of the radar transmitted signal reflected by the ground target and then received, and this parameter determines the phase characteristics of the received signal. Due to the failure of the "go-stop" assumption, the return delay in FMCW SAR cannot be described directly as in pulse radar, and in order to build a time domain model of the return signal, the return delay of FMCW SAR has to be solved accurately.

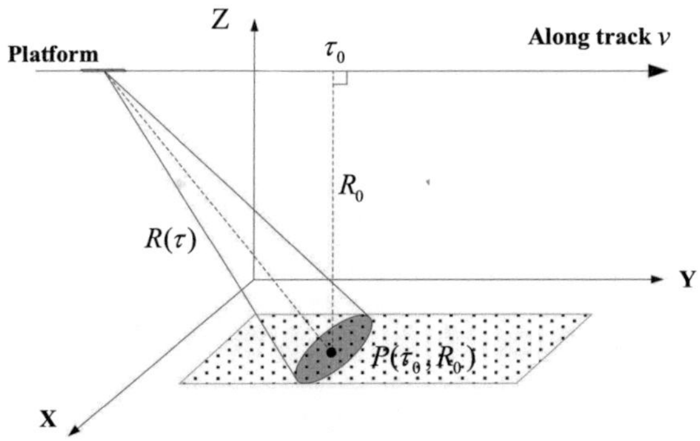

Fig. 2.3 The imaging geometry of FMCW SAR

Taking a point target as an example, the shortest range from the radar track is R_0, the zero Doppler time is τ_0, and the backscatter coefficient is σ_0. Assuming that τ_d is the two-way delay of signal propagation, then if the radar transmits a LFM signal at any time τ, it will receive the reflected echo from the above point target at time $\tau + \tau_d$. Then the range $R(\tau)$ of the signal from transmitting to reaching the target can be expressed as

$$R(\tau) = \sqrt{R_0^2 + v^2(\tau - \tau_0)^2} \tag{2.11}$$

Where, v is the radar platform speed. The slant range course of the signal from being reflected by the target to being received by the antenna can be expressed as

$$R(\tau + \tau_d) = \sqrt{R_0^2 + v^2(\tau_d - \tau_0)^2} \tag{2.12}$$

Meanwhile, the relationship between the signal echo delay time and the slope range course can be expressed as

$$\tau_d = \frac{R(\tau) + R(\tau + \tau_d)}{c} \tag{2.13}$$

By moving the term to the right of the equal sign to the left of the equation, and squaring both sides of the equation, we obtain a quadratic equation for τd. Solving this equation yields

$$\tau_d = 2\alpha\left[\frac{R(\tau)}{c} + \frac{v^2}{c^2}((\tau - \tau_0))\right] \tag{2.14}$$

Where, α is the Doppler parameter, and the expression is

Assuming that the system transmits a linear FM signal expressed in the form $s(t) = \exp\left(j\pi K_r t^2\right)$, the echo signal of the above point target can be expressed as

$$\alpha = \frac{c^2}{c^2 - v^2} \tag{2.15}$$

The reference function for the signal demodulation operation is defined as $s(t) = \exp\left(j\pi K_r t^2\right)$, the echo signal of the point target can be represented as

$$g_l(\tau, t) = \sigma(\tau_0, r_0)s(t - \tau_d)\exp[j2\pi f_0(t - \tau_d)] \tag{2.16}$$

The reference function for signal demodulation operation is defined as

$$g_{ref}(\tau, t, \tau_c) = s^*(t - \tau_c)\exp[-j2\pi f_0(t - \tau_c)] \tag{2.17}$$

where $s^*(t)$ denotes the conjugate function of $s(t)$; τ_c is the reference delay of the signal demodulation operation, defined as $\tau_c = \frac{2ar_c}{c}$, where the reference range r_c is generally chosen as the shortest range from the scene center to the radar track, then the dechirp signal can be expressed as

$$\begin{aligned} g_d(\tau, t) &= g_l(\tau, t) \times g_{ref}(\tau, t, \tau_c) \\ &= \sigma(\tau_0, R_0)\exp[-j2\pi f_0(\tau_d - \tau_c)] \\ &\times \exp\left[-j2\pi K_r(t - \tau_c)(\tau_d - \tau_c)\exp\left[j\pi K_r(\tau_d - \tau_c)^2\right]\right] \end{aligned} \tag{2.18}$$

Equation (2.18) is the exact expression of the FMCW SAR echo signal after the dechirp operation. As can be seen, dechirp operation reduces the echo data while producing the last term of the signal, commonly referred to as the residual audio phase (PVP). Accurate focusing of the signal requires that the RVP term be removed first, which can be achieved by both time-domain and frequency-domain algorithms at [3].

After removing the RVP term, the expression of the signal is

$$g(\tau, t) = \sigma(\tau_0, R_0)\exp[-j2\pi f_0(\tau_d - \tau_c)]\exp[-j2\pi K_r(t - \tau_c)(\tau_d - \tau_c)] \tag{2.19}$$

At this point, the signal has been transformed into the range-frequency domain to the azimuth-time domain, so it is necessary to perform the azimuth Fourier transform to transform the signal into the two-dimensional frequency domain.

Unlike conventional pulse radar, where τ only represents the azimuth time, for FMCW SAR, τ is also influenced by the range time, which can be expressed intuitively as $\tau = \tau_n + t$, where τ_n represents the discrete azimuth time, and the signal is expressed as

$$g(\tau, t) = \sigma(\tau_0, R_0)\exp\left\{-j4\pi\alpha(f_0 + f)\left(\frac{R(\tau_n + t)}{c} + \frac{v^2}{c^2}(\tau_n + t - \tau_0) - \frac{r_c}{c}\right)\right\} \tag{2.20}$$

After the azimuthal Fourier transform of the azimuthal time τ_n, according to the principle of stationary phase (POSP), the first order derivative of the phase is zero.

$$\left.\frac{d\Phi_1(f_\tau, f)}{d\tau_n}\right|_{\tau_n = \tau_s} = 0 \tag{2.21}$$

Solving the above equation yields

$$\tau_s = \tau_0 - \tau - \frac{\frac{r_0}{v}\left[\frac{v}{c} + \frac{cf_\tau}{2av(f_0+f)}\right]}{\sqrt{1 - \left[\frac{v}{c} + \frac{cf_\tau}{2av(f_0+f)}\right]^2}} \tag{2.22}$$

Substituting Eq. (2.22) into Eq. (2.20), the spectrum model of FMCW SAR can be established as

$$G(f_\tau, f) = \sigma(\tau_0, R_0)\exp[-j\Phi(f_\tau, f)] \tag{2.23}$$

Among them

$$\begin{aligned}\Phi(f_\tau, f) = &\frac{4\pi\alpha R_0}{c}\sqrt{(f_0+f)^2 - \left[\frac{v}{c}(f_0+f) + \frac{cf_\tau}{2av}\right]^{-2}} \\ &- \frac{2\pi f_\tau f}{K_r} + 2\pi f_\tau \tau_0 - 4\pi(f_0+f_\tau+f)\tau_c\end{aligned} \tag{2.24}$$

This spectrum is based on the range of the target and zero Doppler time, and no approximation is used in the derivation, so the model is an accurate model of the FMCW SAR echo signal spectrum at any squint angle.

4. FMCW SAR Residual Audio Phase

The azimuthal characteristics of the SAR echo signal depend on the variation of the delay time $\tau(t_r, t_a)$, i.e., the variation of the range $R(t_r, t_a)$. Figure 2.4a illustrates the slope range model for the "go-stop" model in the pulse SAR, and Fig. 2.4b illustrates the slope range model for the "non-go-stop" model in the FMCW SAR.

As shown in Fig. 2.4a, the carrier signal time width of a conventional pulse SAR system ranges from a few microseconds to tens of microseconds. The position of the target and the phase center of the radar antenna are approximately constant during the time of a single pulse action on the target, as if the platform is not moving. For a point target $Q(X_0, R_0)$ in the observation scene, its range to the platform is

$$R(t_a) = \sqrt{R_0^2 + (vt_a - X_0)^2} \tag{2.25}$$

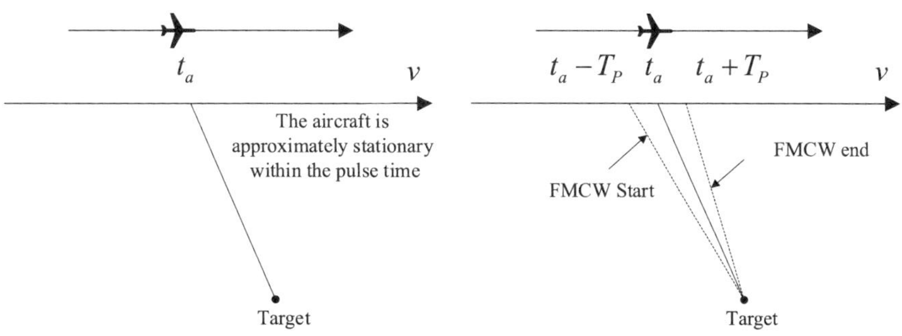

(a) "go-stop" model (b) "non-go-stop" mode l

Fig. 2.4 Comparison of range changes between pulse and FMCW SAR

The "go-stop" model is the basis of the pulse SAR, which greatly simplifies the theoretical analysis of the pulse SAR. While the FMCW SAR transmits signals with a large time width and a duty cycle close to 100%. The major difference of FMCW SAR compared to pulsed SAR is the failure of the "go-stop" model [4]. As shown in Fig. 2.4b, during the time when the transmitted FMCW signal acts on the target, the change of the instantaneous range between the target and the antenna phase center is no longer negligible compared to the pulse regime radar system, i.e. the slope range $R(t_r, t_a)$ is not approximated as equal to $R(0, t_a)$ in one sweep cycle, and the range becomes a variable quantity with respect to t_r. Therefore, it is necessary to reanalyze and model the FMCW SAR echo signal. In the FMCW SAR, Eq. (2.25) becomes

$$R(t_r, t_a) = \sqrt{R_0^2 + (v(t_r + t_a) - X_0)^2} \tag{2.26}$$

where the effect of the fast time variable t_r on the echo signal is called the fast time walk. The fast time walk introduces an additional amount of range variation, which causes a change in the azimuthal Doppler characteristics of the echo signal. In a swept period of an FMCW signal, a Taylor series expansion of Eq. (2.26) at $t_r = 0$, keeping only one term.

$$\begin{aligned} R(t_r, t_a) &\approx \sqrt{R_0^2 + (vt_a - X_0)^2} + \frac{vt_r(vt_a - X_0)}{\sqrt{R_0^2 + (vt_a - X_0)^2}} \\ &= R(0, t_a) - \frac{\lambda}{2} f_{\mathrm{d}} t_r \end{aligned} \tag{2.27}$$

where $f_{\mathrm{d}} = -\frac{2v}{\lambda}\frac{vt_a - X_0}{R(0,t_a)}$ is the instantaneous Doppler term introduced by the fast time walk. Substituting Eq. (2.27) into Eq. (2.8), we get

$$\begin{aligned} s_d(t_r, t_a) = \mathrm{rect}[\tfrac{t_r}{T_{\mathrm{AD}}}] &\exp\left\{-\mathrm{j}2\pi f_{\mathrm{c}}[\tau(0, t_a) - \tfrac{f_{\mathrm{d}}}{f_{\mathrm{c}}} t_r - \tau_{\mathrm{ref}}]\right\} \times \\ &\exp\left\{-\mathrm{j}2\pi K_{\mathrm{r}} t_r [\tau(0, t_a) - \tfrac{f_{\mathrm{d}}}{f_{\mathrm{c}}} t_r - \tau_{\mathrm{ref}}]\right\} \times \\ &\exp\left\{\mathrm{j}\pi K_{\mathrm{r}}[\tau(0, t_a) - \tfrac{f_{\mathrm{d}}}{f_{\mathrm{c}}} t_r - \tau_{\mathrm{ref}}]^2\right\} \end{aligned} \tag{2.28}$$

Compared to the non-go-stop model, the received signal of the pulsed SAR system under the go-stop model is

$$\begin{aligned} s_1(t_r, t_a) = \mathrm{rect}[\tfrac{t_r}{T_{\mathrm{AD}}}] &\exp\{-\mathrm{j}2\pi f_{\mathrm{c}}[\tau(0, t_a) - \tau_{\mathrm{ref}}]\} \times \\ &\exp\{-\mathrm{j}2\pi K_{\mathrm{r}} t_r [\tau(0, t_a) - \tau_{\mathrm{ref}}]\} \times \\ &\exp\{\mathrm{j}\pi K_{\mathrm{r}}[\tau(0, t_a) - \tau_{\mathrm{ref}}]^2\} \end{aligned} \tag{2.29}$$

The signal in the "go-stop" mode does not contain the $f_{\mathrm{d}} t_r / f_{\mathrm{c}}$ term. Equation (2.28) is divided by Eq. (2.29) to compare the phase terms of the two models:

$$\frac{s_{\mathrm{d}}(t_r,t_a)}{s_1(t_r,t_a)}=\exp\left\{\mathrm{j}2\pi\left\{f_{\mathrm{d}}-K_{\mathrm{r}}[\tau(0,t_a)-\tau_{\mathrm{ref}}]\frac{f_{\mathrm{d}}}{f_{\mathrm{c}}}\right\}t_r\right\}\times \exp\left\{\mathrm{j}2\pi K_{\mathrm{r}}[\frac{f_{\mathrm{d}}}{f_{\mathrm{c}}}+(\frac{f_{\mathrm{d}}}{f_{\mathrm{c}}})^2]t_r^2\right\} \tag{2.30}$$

Equation (2.30) contains the primary and secondary terms of t_r. The primary term indicates that a signal frequency shift is generated in each sweep cycle, and the fast time walk introduces additional range migration; the secondary term indicates that the linear modulation frequency of the signal changes, and the additional modulation frequency is related to $f_{\mathrm{d}}/f_{\mathrm{c}}$, which causes distortion of the range-compressed signal obtained by direct Fourier transform of the range.

It is therefore necessary to correct the fast time walk term by transforming $s_{\mathrm{d}}(t_r, t_a)$ into $s_1(t_r, t_a)$ in the form of the echo signal assumed by the "walk-stop" model, such that $t_a = t_a - t_r$ and there exists $\tau(t_r, t_a - t_r) = \tau(0, t_a)$, which yields

$$s_{\mathrm{d}}(t_r, t_a - t_r) = \mathrm{rect}[\frac{t_r}{T_{\mathrm{AD}}}]\exp\{-\mathrm{j}2\pi f_{\mathrm{c}}[\tau(0, t_a) - \tau_{\mathrm{ref}}]\}\times \exp\{-\mathrm{j}2\pi K_{\mathrm{r}}t_r[\tau(0, t_a) - \tau_{\mathrm{ref}}]\}\times \exp\left\{\mathrm{j}\pi K_{\mathrm{r}}[\tau(0, t_a) - \tau_{\mathrm{ref}}]^2\right\} \tag{2.31}$$

where, f_a is the azimuth-frequency variable. Equation (2.31) converts the echo signal of the FMCW SAR into the same form as the echo signal of the pulsed regime SAR, and achieves pulse-like processing of the FMCW SAR signal.

2.2.2 *Airborne FMCW SAR Strip Imaging Algorithm*

1. Range Doppler algorithm

The range Doppler domain is a commonly used domain in SAR signal processing where the trajectories of all targets located on the same range gate coincide. Therefore, in the range Doppler domain, the correction of one migration curve is equivalent to the correction of the range unit migration of all targets on the same range gate. At the same time, since the signal is in the range time domain, it is easy to correct for the range cell migration along the slope of the target. The range-Doppler algorithm (RD) is the classical algorithm for SAR echo signal processing with range migration correction in the range-Doppler domain. Until now, the range-Doppler algorithm is still a widely used algorithm for SAR signal processing.

The algorithm calculates the range migration based on the change of the vertical oblique range between the irradiated target and the radar track, and corrects the range unit migration, so the range Doppler algorithm has high accuracy when the synthetic aperture is not too wide, the mapping band is not too wide and the oblique angle of view is not too large. Therefore, the range Doppler algorithm is one of the most classic algorithms in airborne SAR signal processing.

The range Doppler algorithm can still be an accurate and efficient signal processing method in the FMCW SAR imaging mode with forward and side view or small oblique view, general mapping bandwidth and general synthetic aperture width. However, due to the large differences in the signal time domain representation and spectral model characteristics, the FMCW SAR echo signal processing in the range Doppler domain is based on the basic ideas of the traditional range Doppler algorithm, but the specific operation steps and filtering functions are different from those of the traditional algorithm [5]. In this section, the corresponding range Doppler algorithm is derived and described for the FMCW SAR signal characteristics.

For the signal model and spectral characteristics of FMCW SAR, the flow of the range Doppler algorithm is shown in Fig. 2.5, where the key derivation process and operation details are described below.

(1) Removing the residual video phase (RVP) term of the echo signal, which is a necessary operation for all FMCW SAR focusing, is usually performed in the first step of signal processing. Removing the RVP term usually requires three steps, namely, range-to-Fourier transform, phase multiplication, and range-to-Fourier inversion.
(2) The azimuthal Fourier transform of the signal transforms the echo to the two-dimensional frequency domain. The range-Doppler algorithm ignores the second and higher order range-azimuth coupling terms, so the Taylor series expansion can be performed on the root term of the two-dimensional spectral phase of the FMCW SAR echo signal and ignores its higher than second range-azimuth coupling terms.

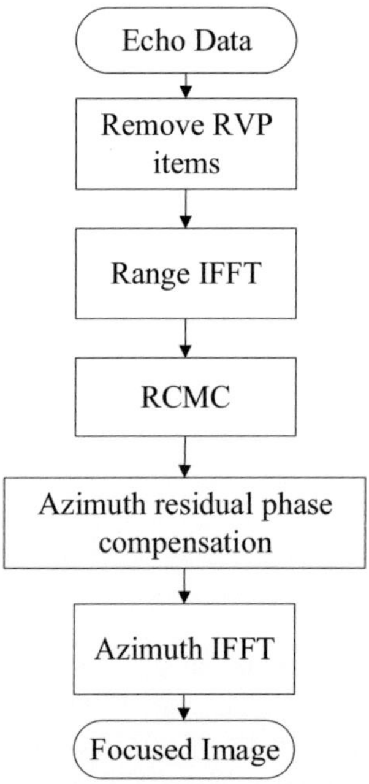

Fig. 2.5 FMCW SAR range doppler imaging algorithm process

(3) The range-to-direction inversion completes the signal range compression. It should be noted that the effects of the range-azimuth coupling terms of second order and higher are neglected in the derivation of the signal expression here. It can be seen that the target trajectory during the radar irradiation time will pass through different range units, so it is called range migration, and the range migration phenomenon of the target becomes more significant with the increase of the target range. At the same time, the migratory trajectory of the target located at the same range gate is a function of the azimuthal Doppler frequency, and the range migration of the target becomes more significant as the absolute value of the Doppler frequency becomes larger. If this migration is not corrected, or if the correction is inaccurate, the energy of the same target will be dispersed in different range units, which will reduce the accuracy of azimuthal compression and lead to the widening of the main flap and the elevation of the side flap in the range and azimuthal directions, and seriously affect the target focus quality.

(4) Perform range cell migration correction. The range unit migration of the point target echoes located at different range gates increases with the vertical slope range from the radar platform track, and is affected by the change of Doppler frequency in azimuth. The goal of range migration correction is to make the target echoes located at the same range gate concentrate at R_0. Therefore, to accurately correct the range cell migration of different range gate echoes, the migration correction factor must be null-varying along both the range and azimuth directions. Assuming that the interpolation operation is accurate, the migration trajectory of the target in the range Doppler domain in the mapped area is only related to the shortest slope range of the target from the radar track after the migration correction is completed. The trajectory of the target echoes in the range Doppler domain located in the mapping area is corrected to a straight line.

(5) After the range migration correction, the signal is in the range Doppler domain, and the azimuthal compression can be achieved precisely and efficiently at this time, i.e., the residual phase compensation is performed by azimuthal matching filtering. At this point, the azimuth compression is achieved by multiplying the azimuth residual phase compensation function along the range direction, and after azimuth compression, the signal is compressed to zero Doppler position. For each point target in the radar exposure area, all phases are compensated except for the linear phase that determines its post-focus position.

(6) After azimuth-matching filtering, the residual phase of the echo signal is accurately compensated, but the signal is still in the range Doppler domain. At this time, the azimuthally matched filtered signal is transformed to the two-dimensional time domain, i.e., the image domain, by inverse azimuthal Fourier transform, to complete the azimuthal focusing.

2. Wave number domain imaging algorithm

The range Doppler algorithm is validated and applied in the echo signal processing of continuous wave SAR, but the variation of the more than quadratic azimuth-range coupling term with the ground target slant range is neglected in the processing, which can provide sufficient accuracy in the case of positive lateral view or small oblique view mode, and the mapping band is not particularly large. However, in large oblique view mode, or with a wide radar mapping band, neglecting the variation of the quadratic azimuth-range coupling term with the ground target slope can lead to scattering of the target far from the center of the scene [6].

The range migration algorithm (RMA) was first derived from seismic wave data processing, where the transfer function model of seismic wave signals is time-varying and has similar characteristics to SAR echo signals [7]. R. H. Stolt has proposed a frequency domain processing method to solve the time-varying nature of this transfer function, and then the RMA uses its key operations to process SAR signals. The Stolt interpolation operation completely overcomes the dependence of range migration and secondary range compression (SRC) on slope range in SAR signals, and thus the wave number domain algorithm has the capability of processing SAR echo data with wide mapping bands and large slopes. The wave number domain algorithm therefore has the ability to handle wide mapping bands and large skew SAR echo data [8]. It should be noted that the assumption that the equivalent radar velocity does not vary with slope during signal processing limits its ability to handle very wide mapping band radar echo data. This can be solved by dividing the broadband radar data into multiple strips in the range direction for processing.

The exact focus flow of the FMCW SAR wave number domain algorithm is shown in Fig. 2.6.

2.2.3 Airborne Millimeter Wave FMCW SAR Measured Data Processing

The previous research on the imaging algorithm of airborne millimeter wave FMCW SAR assumes that the motion state of the carrier is ideal, that is, the carrier moves in a straight line parallel to the ground at a fixed direction and a constant speed, for the convenience of analysis. The actual movement of the carrier cannot be an ideal state. Especially for SAR platforms flying at medium and low altitudes, due to the influence of unstable airflow, the motionless nature of the platform is significant, and the platform inevitably deviates from the ideal uniform linear motion state, resulting in motion errors. This will directly cause phase errors and errors in the radar echo signal. If motion compensation is not carried out, the echo will be greatly distorted by unstable factors, resulting in a decrease in imaging quality. It cannot even be imaged. Therefore, when using the imaging algorithm mentioned in Sect. 2.2.2, it is necessary to compensate for the errors introduced by the carrier motion. The common motion compensation algorithms are inertial navigation based motion compensation and echo based motion compensation. This section presents the imaging results of airborne Ka FMCW SAR processed using wavenumber domain imaging algorithms after two types of motion compensation.

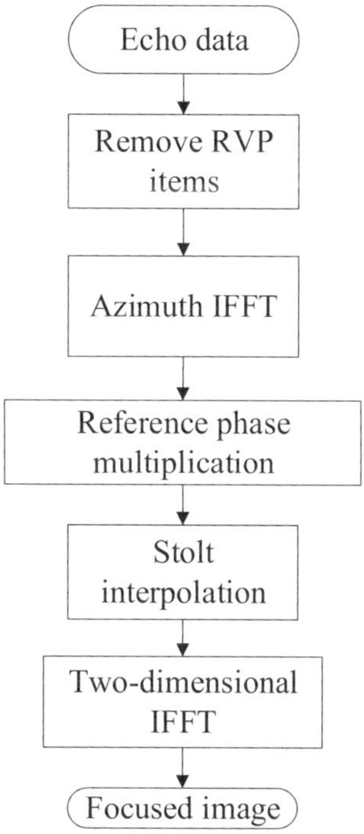

Fig. 2.6 Imaging algorithm process signal processing process of wavenumber domain algorithm

Figure 2.7 is the Ka FMCW SAR image tested on the "Yunwu" as the carrying platform. The carrying system is the light and small Ka FMCW SAR developed by the Key Laboratory of Millimeter Wave Imaging Technology of Shanghai Academy of Spaceflight Technology. The test site is Zhengzhou, Henan Province, and the time is June 2019. The abscissa is the azimuth and the ordinate is the range.

Figure 2.8 The two-dimensional resolution is 0.35 m, the azimuth width is about 8 km, and the range width is 2.5 km. The image covers the southern suburbs of Shangjie District, Zhengzhou City. The scene terrain covers typical land landforms such as urban areas, farmland, hills, etc. Figures 2.8 and 2.9 are typical targets in Fig. 2.7, showing high-voltage lines in large scenes and high-voltage towers in small

Fig. 2.7 Airborne Ka FMCW SAR stripe mode image

Fig. 2.8 High voltage line image

Fig. 2.9 SAR image of iron tower

scenes, respectively. Both images demonstrate the ability of millimeter waves to describe target details and better visual interpretability compared to low-frequency SAR images. It can be seen from Fig. 2.9 that the Ka band SAR image has obvious shadow characteristics, and shadow based target detection can play an important role in static target recognition and moving target detection. See Chap. 7 for more applications of this feature.

2.3 Airborne FMCW SAR Sliding Spotlight Imaging

In the conventional strip mode, the angle between the antenna beam centerline and the mounted platform is fixed, and a continuous imaging scene can be obtained, but its azimuthal resolution is limited by the azimuthal equivalent size of the antenna, which can only reach half of the antenna azimuthal equivalent size under ideal conditions, and the azimuthal equivalent size of the antenna cannot be infinitely miniaturized due to Fresnel's law. In cluster mode, the radar antenna rotates so that the electromagnetic waves irradiate the same scene on the ground, thus obtaining a longer synthetic aperture, and therefore a higher azimuthal bandwidth and azimuthal resolution. However, the spot beam mode can only irradiate a fixed scene, and its imaging scene size is limited, and it can only image a scene within a beam width with high accuracy. Sliding spotlight is an imaging mode that combines the features of both strip and spot beam modes, with higher resolution than strip mode and larger azimuthal scene width than spot beam mode. The sliding spotlight imaging mode can be adjusted by the antenna rotation angle and speed to achieve the observed scene range and resolution, and can be adjusted by the antenna rotation angle and speed to achieve a compromise between azimuthal scene range and resolution, increasing the flexibility of the SAR system in practice.

In this section, we first analyze the azimuthal characteristics of the sliding spotlight FMCW SAR from the airborne FMCW SAR sliding spotlight imaging model, establish the echo signal model of the sliding spotlight SAR, and compare the azimuthal spectral characteristics of the strip mode, the cluster mode, and the sliding spotlight. After that, the classical two-step method and the improved two-step method with the image domain unmixing operation are introduced. Finally, the process and results of the airborne millimeter wave FMCW SAR measurements are presented.

2.3.1 Airborne FMCW SAR Sliding Spotlight Imaging Model

1. FMCW SAR sliding beam echo signal model

When the FMCW SAR operates in sliding spotlight imaging mode, the antenna beam center rotates backward with the motion of the carrier platform. The ground irradiated area moves forward at a lower speed than the carrier aircraft, making the platform displacement during the time each point in the scene is irradiated larger than the width of the actual antenna beam aperture irradiated on the ground, i.e., longer time accumulation and larger azimuthal bandwidth, thus allowing for improved azimuthal resolution [9–15].

The spatial Geometric modeling of FMCW SAR gliding spotlight imaging mode is shown in Fig. 2.10.

Point O in Fig. 2.10 is the position of the carrier at the time (tr, ta) = (0, 0) and is defined as the origin of the coordinates. The carrier moves from point A to point B in the direction of motion shown in the figure with an ideal uniform linear motion of velocity v. In the process of carrier movement, the antenna beam center rotates with angular velocity ω from $+\theta$ (the front view angle is positive and the rear view angle is negative) to $-\theta$. P_0, P_1, P_2 are the positions of the radar beam center when the carrier is at point O, point A and point B. The length of the line segment OP_0 is equal to the reference slope length Rref, and the line segments OP_0, AP_1 and BP_2 are extended to intersect at The length of the line segment AB L_b is the range of

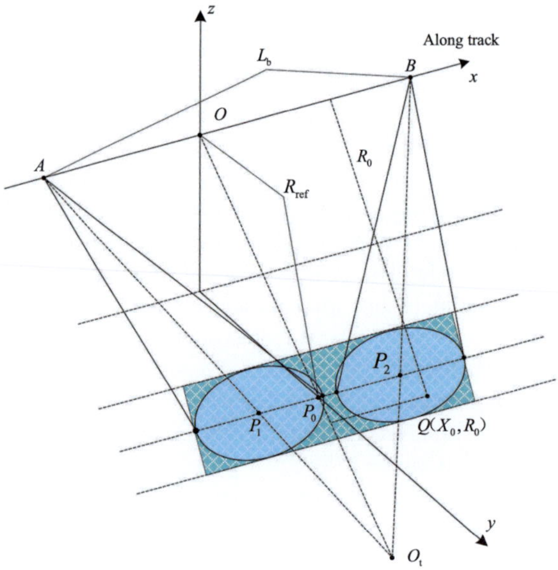

Fig. 2.10 Space geometric modeling of FMCW SAR sliding spotlight imaging mode

the carrier motion in a complete sliding spotlight imaging, and the length of the line segment P_1P_2 Ls is the range of the antenna beam center scanning on the ground, which also corresponds to the range of full resolution imaging in sliding spotlight imaging mode. The shaded area in Fig. 2.10 shows the total observation range of the sliding spotlight imaging.

L_a is the azimuthal width of the antenna beam illuminated on the ground and its value is

$$L_a \approx R_{ref}\theta_a = \frac{\lambda R_{ref}}{D_a} \tag{2.32}$$

where θ_a is the azimuthal angle of the antenna beam; D_a is the azimuthal equivalent width of the antenna. Define the sliding coefficient $p = L_s / L_b$, $0 < p < 1$, then the velocity of the antenna beam centerline scanning on the ground is pv, and the azimuthal window function W_a for the sliding beam pattern imaging is

$$W_a(t_a) = \mathrm{rect}\left(\frac{X_0 - pvt_a}{L_a}\right) \tag{2.33}$$

Combining Eqs. (2.32) and (2.33), the dechirp received echo signal of the point target $Q(X_0, R_0)$ in the FMCW SAR sliding spotlight imaging mode is

$$
\begin{aligned}
s_{\rm d}'(t_r, t_a) = {} & \mathrm{rect}[\tfrac{t_r}{T_{\rm AD}}]W_a(t_a)\times \\
& \exp\left\{-\mathrm{j}\tfrac{4\pi f_{\rm c}}{c}\left\{\sqrt{R_0^2 + [v(t_r+t_a) - X_0]^2} - R_{\rm ref}\right\}\right\}\times \\
& \exp\left\{-\mathrm{j}\tfrac{4\pi K_{\rm r} t_r}{c}\left\{\sqrt{R_0^2 + [v(t_r+t_a) - X_0]^2} - R_{\rm ref}\right\}\right\}\times \\
& \exp\left\{\mathrm{j}\tfrac{4\pi K_{\rm r}}{c}\left\{\sqrt{R_0^2 + [v(t_r+t_a) - X_0]^2} - R_{\rm ref}\right\}^2\right\}
\end{aligned} \tag{2.34}
$$

Neglecting the effect of the Fourier transform on the azimuthal window function of the sliding cluster pattern, the fast time walk term is removed using the method shown in Eq. (2.31) to obtain

$$
\begin{aligned}
s_1(t_r, t_a) = {} & \mathrm{rect}[\tfrac{t_r}{T_{\rm AD}}]W_a(t_a)\times \\
& \exp\left\{-\mathrm{j}\tfrac{4\pi f_{\rm c}}{c}\left[\sqrt{R_0^2 + (vt_a - X_0)^2} - R_{\rm ref}\right]\right\}\times \\
& \exp\left\{-\mathrm{j}\tfrac{4\pi K_{\rm r} t_r}{c}[\sqrt{R_0^2 + (vt_a - X_0)^2} - R_{\rm ref}]\right\}\times \\
& \exp\left\{\mathrm{j}\tfrac{4\pi K_{\rm r}}{c}[\sqrt{R_0^2 + (vt_a - X_0)^2} - R_{\rm ref}]^2\right\}
\end{aligned} \tag{2.35}
$$

Equation (2.35) contains three phase terms: the first phase term contains mainly azimuthal information, and the root term represents the Doppler frequency introduced by the slant range variation, and the azimuthal Doppler bandwidth enables the SAR system to achieve azimuthal high resolution; the second phase term contains mainly range information, and the dechirp reception makes the second phase term a single frequency signal about tr, but due to the presence of ta in the root term, there is a range-azimuthal coupling in the second term. The third phase term is a residual video phase term introduced by the de-frequency reception, which needs to be removed.

The RVP term can be removed in the range frequency domain under the condition that the effect of the Fourier transform on the range-to-window function is neglected:

$$
\begin{aligned}
s_1'(t_r, t_a) &= \mathrm{IFFT_r}\left\{\mathrm{FFT_r}\{s_1(t_r, t_a)\}\times\exp\left\{-\mathrm{j}\pi\tfrac{f_r^2}{K_{\rm r}}\right\}\right\} \\
&= \mathrm{rect}[\tfrac{t_r}{T_{\rm AD}}]W_a(t_a)\times \\
&\quad \exp\left\{-\mathrm{j}\tfrac{4\pi(K_{\rm r}t_r + f_{\rm c})}{c}(\sqrt{R_0^2 + (vt_a - X_0)^2} - R_{\rm ref})\right\}
\end{aligned} \tag{2.36}
$$

where FFTr is the range-to-Fourier transform; IFFTr is the range-to-Fourier inverse transform. The range-to-frequency exists as $K_r\ t_r + f_c \rightarrow f_r$, so the above equation can be written in the form of Eq. (2.35)

$$
\begin{aligned}
s_1'(f_r, t_a) = {} & \mathrm{rect}[\tfrac{f_r - f_c}{K_r T_{AD}}] W_a(t_a) \times \\
& \exp\left\{-\mathrm{j}\tfrac{4\pi f_r}{c}\left(\sqrt{R_0^2 + (vt_a - X_0)^2} - R_{\mathrm{ref}}\right)\right\}
\end{aligned}
\tag{2.37}
$$

From Eq. (2.37), it can be seen that the SAR signal received by de-modulation can be regarded as being in the "pseudo-range frequency domain", and the azimuthal FFT operation can be performed to obtain the azimuthal frequency domain signal. However, the azimuthal Fourier transform of Eq. (2.37) is difficult to derive directly. It is necessary to apply the principle of stationary phase (POSP) to obtain an approximate expression. Let's denote the phase part of Eq. (2.37) as $\varphi(t_a)$.

Make

$$
\frac{\mathrm{d}\varphi(t_a)}{\mathrm{d}t_a} = 0 \tag{2.38}
$$

$$
t_a = -\frac{f_a c R_0}{2v^2\sqrt{f_r^2 - (\frac{f_a c}{2v})^2}} + \frac{X_0}{v} \tag{2.39}
$$

Substituting Eq. (2.39) into Eq. (2.37), we get

$$
\begin{aligned}
s_1'(f_r, f_a) = {} & \mathrm{rect}[\tfrac{f_r - f_c}{K_r T_{AD}}] W_a'(-\tfrac{f_a}{K_a} + \tfrac{X_0}{v}) \times \\
& \exp\left\{-\mathrm{j}\tfrac{4\pi R_0}{c}\sqrt{f_r^2 - (\tfrac{c f_a}{2v})^2}\right\} \times \\
& \exp\{\mathrm{j}\tfrac{4\pi R_{\mathrm{ref}}}{c} f_r\} \exp\{-\mathrm{j}2\pi f_a \tfrac{X_0}{v}\}
\end{aligned}
\tag{2.40}
$$

Equation (2.40) is the expression of the two-dimensional spectrum of the FMCW SAR echo signal with the fast time walk term corrected for the ideal point target $Q(X_0, R_0)$ and the RVP term removed, where the azimuthal modulation frequency K_a is related to the two-dimensional frequency variables f_r and f_a:

$$
K_a = -\frac{2v^2 f_r}{cR_0}\sqrt{1 - (\frac{f_a c}{2v f_r})^2} \tag{2.41}
$$

Figure 2.11 shows the magnitude of the root term in Eq. (2.41) for a typical Ka-band SAR system.

As can be seen in Fig. 2.11, the root term of Eq. (2.41) is infinitely close to 1 when operating a typical Ka-band SAR system, so the azimuthal linear tuning frequency Ka can be approximated as

$$
K_a = -\frac{2v^2 f_r}{cR_0} \tag{2.42}
$$

There exists $f_r \to K_r t_r + f_c$, so the azimuthal linear modulation frequency Ka is mainly influenced by the range-oriented time variable t_r and the shortest slope

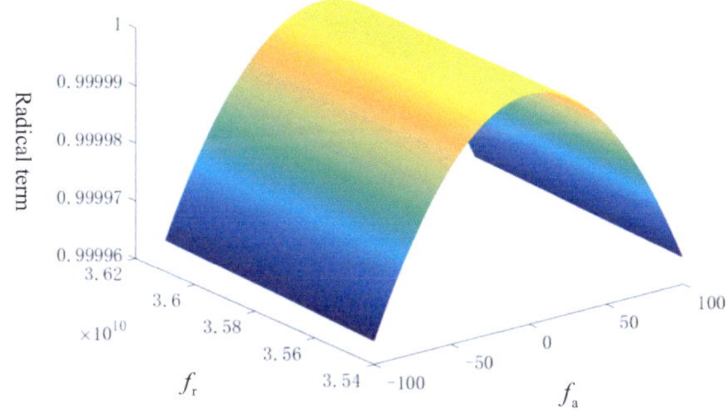

Fig. 2.11 Typical parameter simulation of root term

range R_0 of the point target. In the Ka band and higher, the relative bandwidth of the signal is generally small, with $f_r \approx f_c$, so the azimuthal tuning frequency is generally approximated by $-2v^2/(\lambda R_{ref})$.

2. Comparison of FMCW SAR azimuthal characteristics

The fundamental difference between these three imaging modes is the azimuthal rotation of the antenna: the antenna beam angle is fixed in strip mode imaging, the antenna beam angle is rotated to point to a point in the center of the scene in cluster mode imaging, and the antenna beam angle points to a point below the ground in sliding spotlight imaging. The rotation of the antenna beam in the echo is mainly the azimuth frequency course and azimuth resolution of the difference.

Figure 2.12 shows the operation of the SAR system, where the carrier platform is moving in azimuthal direction with an ideal uniform linear motion at velocity v. Q_1, Q_2, and Q_3 are three ideal point targets uniformly distributed at the scene reference slope R_0.

1) Azimuth frequency course

Azimuthal frequency history is the instantaneous azimuthal frequency of each point in the imaging scene as a function of time. The following is an example of the

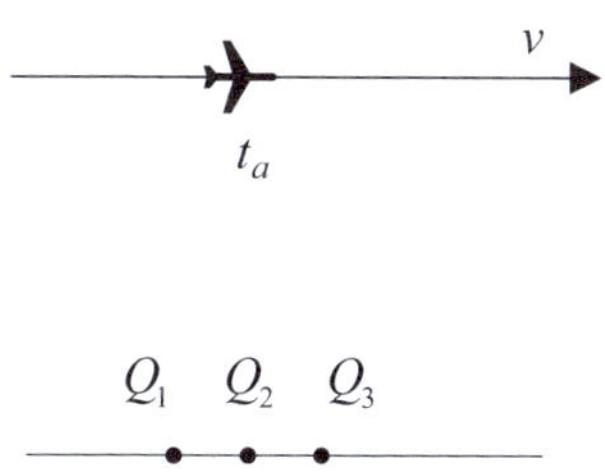

Fig. 2.12 Schematic diagram of SAR system operation

azimuthal frequency history of each of the three imaging modes, using the front-side view.

(1) Strip pattern.

Figure 2.13 shows the diagram of SAR strip mode imaging slope plane, the carrier aircraft is moving at a uniform speed v in a straight line, and the shortest slope range from the carrier platform to the target is R_0.

In Fig. 2.13, L_a is the synthetic aperture length; L_s is the azimuthal width of the antenna beam angle hitting the ground; L_w is the imaging scene width. When the strip mode works, with the flight of the carrier aircraft, Q_1, Q_2, Q_3 points enter the irradiation range of the radar beam angle in turn, and since the antenna is fixed, each point in the scene has the same azimuthal frequency course, so the Doppler frequency course range B_{strip} of the strip mode is equal to the bandwidth B_a of any point in the scene:

$$B_{\text{strip}} = B_{\text{a}} = \frac{2v}{D_{\text{a}}} \tag{2.43}$$

Figure 2.14 shows the strip mode time–frequency diagram.

(2) Spotlight mode.

Figure 2.15 shows a schematic diagram of the oblique range plane of the SAR focused beam mode imaging.

The center of the antenna beam always points to the center of the scene when imaging in cluster mode, and the points Q_1, Q_2, and Q_3 at different azimuthal positions are always in the range of the antenna beam, so they have the same time accumulation. The difference between the points is that the azimuthal positions are different and their Doppler frequency histories are slightly different, being a set of straight lines parallel to each other along the frequency axis, and the start and end times of each line are the same. Figure 2.16 shows the time–frequency diagram of the cluster mode.

As can be seen from Fig. 2.16, the total Doppler bandwidth of the cluster mode is divided into two parts, one is the Doppler bandwidth of the point target in synthetic

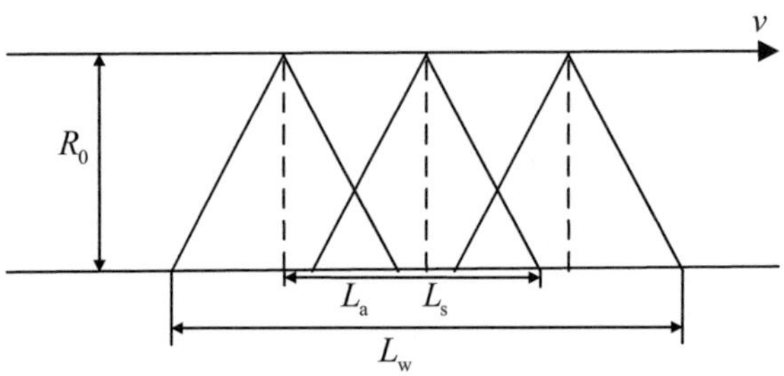

Fig. 2.13 Strip pattern skew plane

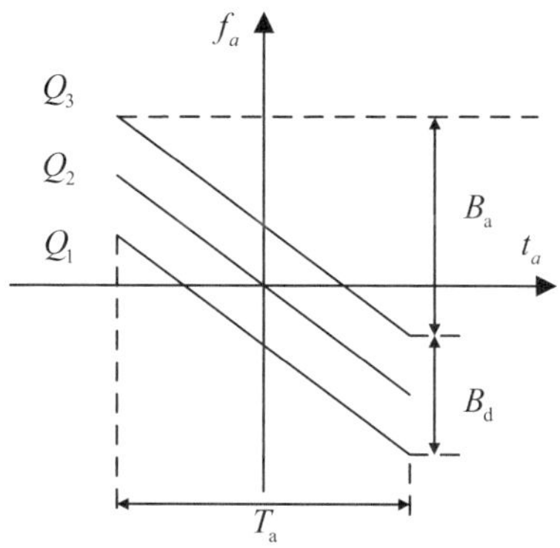

Fig. 2.14 Time frequency diagram of strip mode

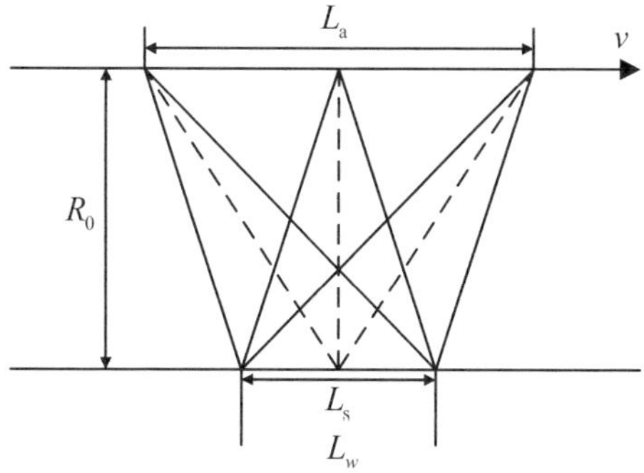

Fig. 2.15 Skew plane of spotlight mode

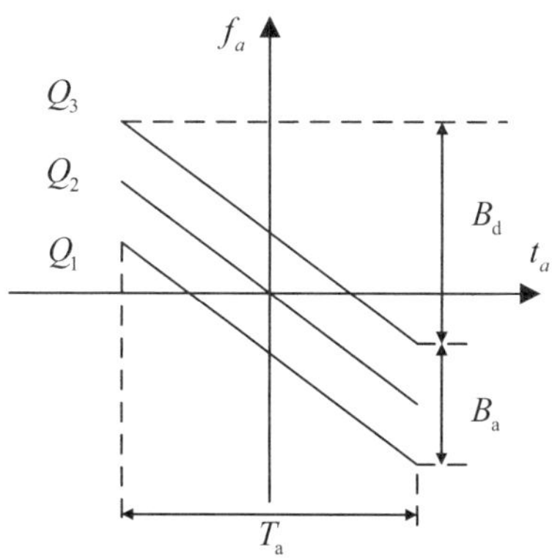

Fig. 2.16 Time frequency diagram of spotlight mode

aperture time B_a and the other is the Doppler bandwidth associated with the azimuthal imaging bandwidth B_d, and the Doppler history of the cluster mode is the sum of these two parts:

$$\begin{aligned} B_{\text{spot}} &= B_{\text{a}} + B_{\text{d}} \\ &= \frac{4v\sin\theta}{\lambda} + \frac{2vL_{\text{w}}}{\lambda R_0} \end{aligned} \tag{2.44}$$

(3) Sliding spotlight mode.

Figure 2.17 shows a schematic diagram of the SAR sliding spotlight mode. imaging slope range plane.

As can be seen in Fig. 2.14, the sliding beam pattern antenna beam rotates from θ to −θ, the beam centerline always points to a point below the ground, i.e. the dashed line extension in Fig. 2.17 will intersect at the same point O_t.

Figure 2.18 shows the time–frequency relationship of the sliding beam pattern.

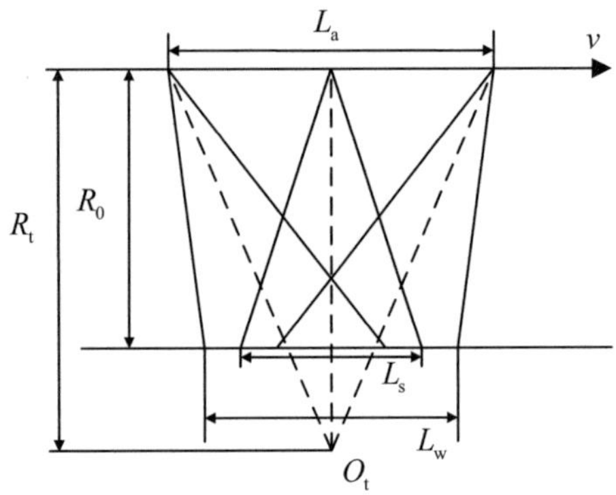

Fig. 2.17 Sliding spotlight mode diagonal plane

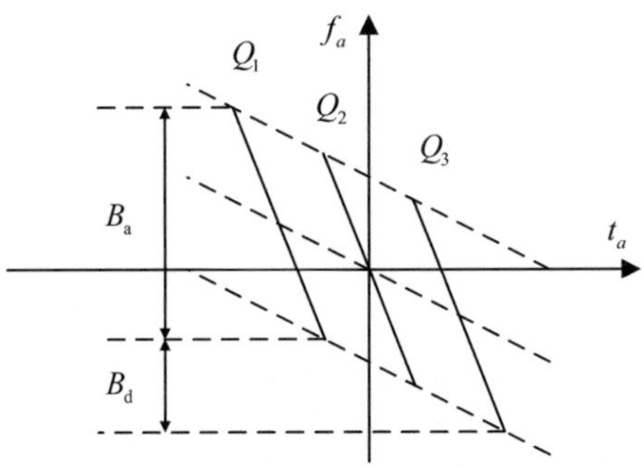

Fig. 2.18 Time frequency diagram of sliding bunching mode

The Doppler history of the sliding beam aggregation mode is also composed of two parts B_a, B_d

$$\begin{aligned} B_{\mathrm{slide}} &= B_{\mathrm{a}} + B_{\mathrm{d}} \\ &= \frac{2v}{D_{\mathrm{a}}} + K_{\mathrm{t}} T_{\mathrm{a}} \end{aligned} \tag{2.45}$$

$$K_{\mathrm{t}} = \frac{2v^2}{\lambda R_{\mathrm{t}}} \tag{2.46}$$

Usually $B_{strip} < B_{slide} < B_{spot}$. The Doppler frequency histories of the three imaging modes show that the strip mode and the cluster mode are special cases of the sliding cluster mode. For a sliding factor $p = 0$, the B_d introduced by the antenna rotation in Eq. (2.43) is 0, the equivalent slope range R_t is ∞, and $B_{slide} = B_{strip}$, and the sliding spotlight becomes a strip mode; for a sliding factor $p = 1$, the sliding spotlight becomes a beam mode, and $B_{slide} = B_{spot}$. All the typical sliding spotlights with $0 < p < 1$ are discussed in the book.

2) Azimuthal resolution

The azimuthal bandwidth of the SAR echo signal is B_a, and the corresponding azimuthal resolution ρ_a of the system is

$$\rho_{\mathrm{a}} = \frac{v}{B_{\mathrm{a}}} \tag{2.47}$$

Equation (2.47) holds for each imaging mode, where B_a varies with the SAR system parameters and imaging mode.

(1) Strip pattern.

The Doppler bandwidth of the strip mode is shown in Eq. (2.43), and the corresponding azimuthal resolution is

$$\rho_{\mathrm{a}} = \frac{D_{\mathrm{a}}}{2} \tag{2.48}$$

The azimuthal resolution of strip-mode SAR images is only related to the azimuthal actual size of the antenna, which overcomes the limitation that the resolution of conventional aperture radar is inversely proportional to the antenna size and facilitates the miniaturization design of radar systems. However, Eq. (2.48) must satisfy the Fresnel approximation, i.e., $R_0 \ll L_a$, and if D_a decreases infinitely and the synthetic aperture length $L_a \to \infty$, the higher-order phase term in the Doppler frequency of the echo signal is no longer negligible, and the azimuthal resolution no longer increases linearly with the increase of the antenna beam angle, and Eq. (2.48) is no longer applicable.

(2) Spotlight mode.

The synthetic aperture time for SAR focused beam mode imaging is

$$T_{\mathrm{a}} = \frac{2R_0 \sin\theta}{v} \tag{2.49}$$

After substitution, we can obtain the resolution of the cluster beam pattern as

$$\rho_{\mathrm{a}} = \frac{\lambda}{4\sin\theta} \tag{2.50}$$

The above equation shows that the azimuthal resolution of the SAR convergence mode imaging is only related to the carrier wavelength λ and the angle of antenna rotation θ, which breaks the limitation of the radar antenna size on the imaging resolution in the strip mode.

(3) Sliding spotlight model.

The synthetic aperture time for SAR sliding spotlight mode imaging is

$$T_{\mathrm{a}} = \frac{\lambda R_0}{D_{\mathrm{a}} v} \frac{R_{\mathrm{t}}}{R_{\mathrm{t}} - R_0} \tag{2.51}$$

The resolution of the sliding spotlight mode after substitution is

$$\rho_{\mathrm{a}} = \frac{v}{B_{\mathrm{a}}} = \frac{v}{\frac{2v^2}{\lambda R_0} T_{\mathrm{a}}} = \frac{D_{\mathrm{a}}}{2} \frac{R_{\mathrm{t}} - R_0}{R_{\mathrm{t}}} \tag{2.52}$$

As can be seen from Eq. (2.52), the azimuthal resolution of sliding spotlight mode is related to the azimuthal size of the antenna and the equivalent slope range R_t caused by the rotation of the antenna beam. The resolution of the sliding beam pattern is $(R_t - R_0)/R_t$ times that of the corresponding strip pattern.

$$R_{\mathrm{t}} = \frac{v}{\omega} \tag{2.53}$$

The equivalent slope range R_t is equal to the ratio of the carrier velocity v to the rotational angular velocity ω of the antenna beam. The azimuthal resolution of $R_t > R_0$ in Eq. (2.52) is not infinitesimal at R_t infinitely close to R_0, but is limited by the scanning range of the antenna, which is approximated by the clustering mode as λ/(4 sinθ).

3) Simulation experiment

Simulation experiments are conducted here to compare the amplitude and resolution of the three imaging modes based on the system parameters shown in Table 2.1.

Figure 2.19a shows the contour map of the 32.fold interpolation of the point target in strip mode, and Fig. 2.16b shows the azimuthal profile of the point target, with the red horizontal line corresponding to the 3 dB position in the figure.

Table 2.1 Imaging parameters for three modes

Parameter	Stripe	Spotlight	Sliding spotlight
f_c/GHz	35.75	35.75	35.75
PRF/Hz	1000	1000	1000
B_r/MHz	700	700	700
V/(m·s^{-1})	50	50	50
Angle/(°)	\	±9	±9
R_s/km	2.5	2.5	2.5
Beam rotation speed/(°·s^{-1})	\	1.7	0.3
Antenna/m	0.35	0.35	0.35

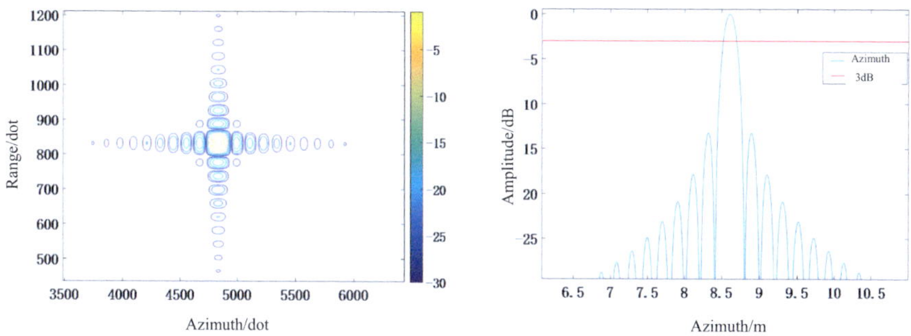

(a) Point target **(b) Azimuth profile**

Fig. 2.19 Strip mode

Figure 2.20a shows the contour map of the point target with 32.fold interpolation for the cluster mode, and Fig. 2.20b shows the azimuthal profile of the point target.

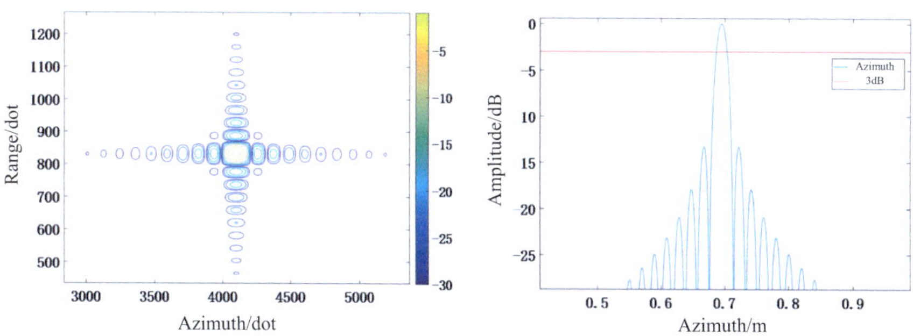

(a) Point target **(b) Azimuth profile**

Fig. 2.20 Spotlight mode

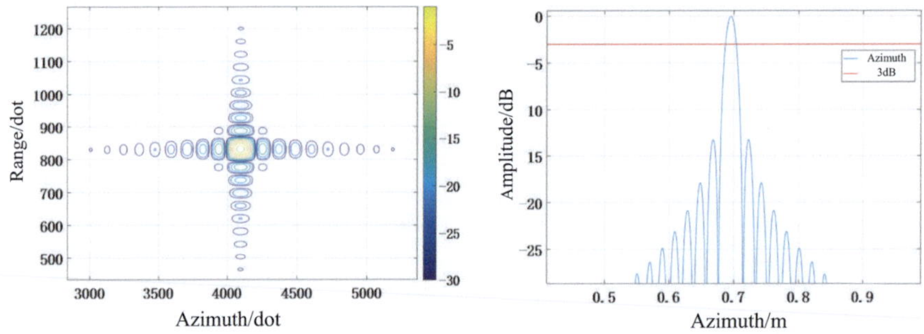

(a) Point target **(b) Azimuth profile**

Fig. 2.21 Sliding spotlight mode

Table 2.2 Results of three modes

Parameter	Stripe	Spotlight	Sliding spotlight
PSLR/dB	−13.26	−13.28	−13.33
ISLR/dB	−10.03	−10.04	−10.24
Resolution/m	0.176	0.017	0.143
Ideal resolution/m	0.175	0.016	0.140

Figure 2.21a shows the contour map of the point target with 32.fold interpolation for the sliding cluster model, and Fig. 2.21b shows the azimuthal profile of the point target.

Table 2.2 shows the simulation results for the three imaging modes, comparing the azimuthal peak partials ratio (PSLR), azimuthal integrated partials ratio (ISLR), azimuthal resolution, and the ideal resolution for the corresponding conditions.

As can be seen from Table 2.2, the 3 dB width of the ideal point target is basically the same as the resolution calculated by Eq. Comparing the theoretical equations and simulation results of the clustering mode and the sliding clustering mode, it can be seen that the faster the rotation speed is, the higher the resolution is at the same rotation angle.

2.3.2 Airborne FMCW SAR Sliding Spotlight Mode Imaging Algorithm

The airborne FMCW SAR sliding spotlight imaging mode relies on the azimuthal rotation of the antenna to achieve high azimuthal resolution, and the resulting azimuthal spectral aliasing makes the traditional strip SAR imaging algorithm no longer applicable. One way to solve the azimuthal aliasing problem is to design the system with sufficient pulse repetition frequency (PRF) for the desired operation.

However, a high PRF leads to increased range ambiguity and increased data volume in the acquisition and downlink. Therefore, high PRF not only limits the mapping bandwidth and data acquisition capability of SAR, but also complicates the design of SAR systems. The traditional methods for processing azimuthal aliasing signals are the subaperture method and the two-step method.

The sub-aperture imaging algorithm divides a complete synthetic aperture time into several sub-apertures. Since the PRF of a sub-aperture is much larger than the azimuthal bandwidth of the signal within the sub-aperture, no azimuthal spectral overlap occurs during signal processing, and the imaging algorithm can be used to focus the imaging on each sub-aperture, and then finally the sub-maps are coherently stitched to obtain the SAR image of the full aperture. The subaperture algorithm has low memory requirements, requires a certain percentage of overlap between the subaperture data, and the post-imaging stitching is more complicated, with low imaging efficiency, which is difficult to meet the demand of one imaging. Figure 2.22 shows the obvious stitching introduced by the subaperture method, with geometric misalignment of the scene target and inconsistent energy intensity at the junction.

Therefore, a two-step method is usually used for sliding spotlight imaging, in which the azimuthal unmixing operation is performed first and the traditional migration correction algorithm is applied in the second step. The two-step method can unify the azimuthal points and requires fewer points for azimuthal expansion, which is relatively more efficient. This section introduces a two-step method based on the RMA algorithm, and improves the traditional two-step method for the possible image domain overlap problem.

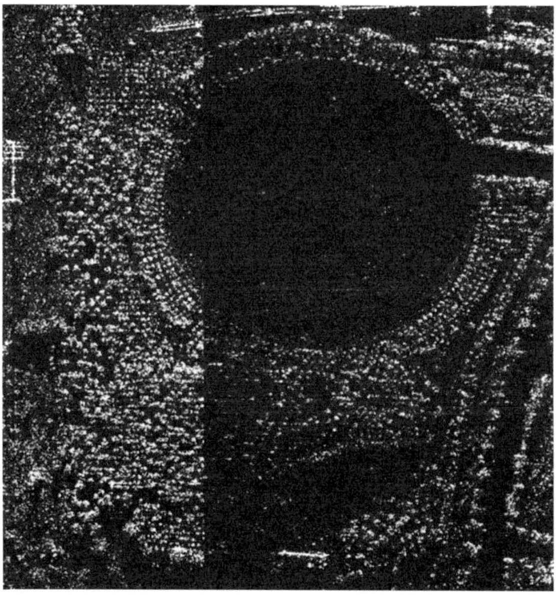

Fig. 2.22 Splice generated by sub aperture method

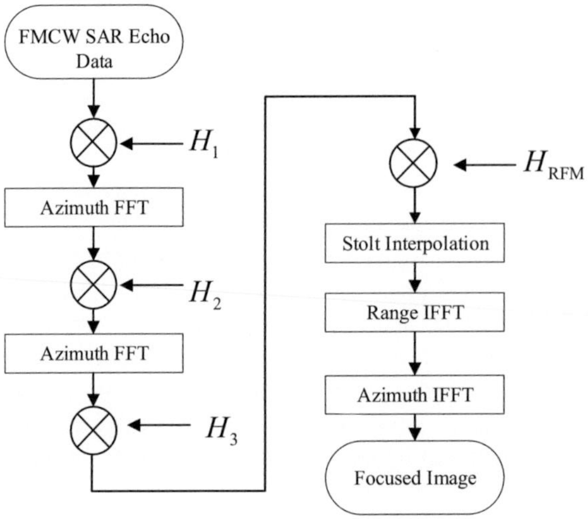

Fig. 2.23 Two step method

1. Two-step method

For the echo signal of FMCW SAR sliding beam pattern, the azimuthal FFT alone cannot obtain the signal in the shape of Eq. (2.55). Because B_{slide} is usually larger than PRF, the FFT will produce azimuthal aliasing, which will not allow for the subsequent range-variance correction (RCMC) operation, and RCMC of the signal with azimuthal aliasing will definitely produce ghosting and reduce the resolution of the main signal. The main workflow of the two-step method is shown in Fig. 2.23.

The time–frequency relationship of the signal $s'_1(f_r, t_a)$ in the azimuthal time domain is shown in Fig. 2.24a, where the Doppler frequency course introduced at the center of the antenna beam is removed by introducing a linear FM term H_1:

$$H_1 = \exp\{j\pi K_t t_a^2\} \tag{2.54}$$

$$s_2(f_r, t_a) = s_1'(f_r, t_a) H_1 \tag{2.55}$$

The time–frequency diagram of $s_2(f_r, t_a)$ is shown in Fig. 2.24b. After eliminating the Doppler frequency introduced by the rotation of the antenna beam center, only the Doppler bandwidth of the strip mode with the same system parameters remains for the Doppler history. The azimuthal resampling is performed after the azimuthal de-modulation, and the new sampling frequency Fn is used for the azimuthal axis, and the time variable after the resampling is t'_a. In order to ensure the effectiveness of the interpolation resampling operation, the resampling frequency F_n needs to satisfy the Nyquist sampling law compared with B_h, and the oversampling rate is generally chosen to be 1.2~1.5.

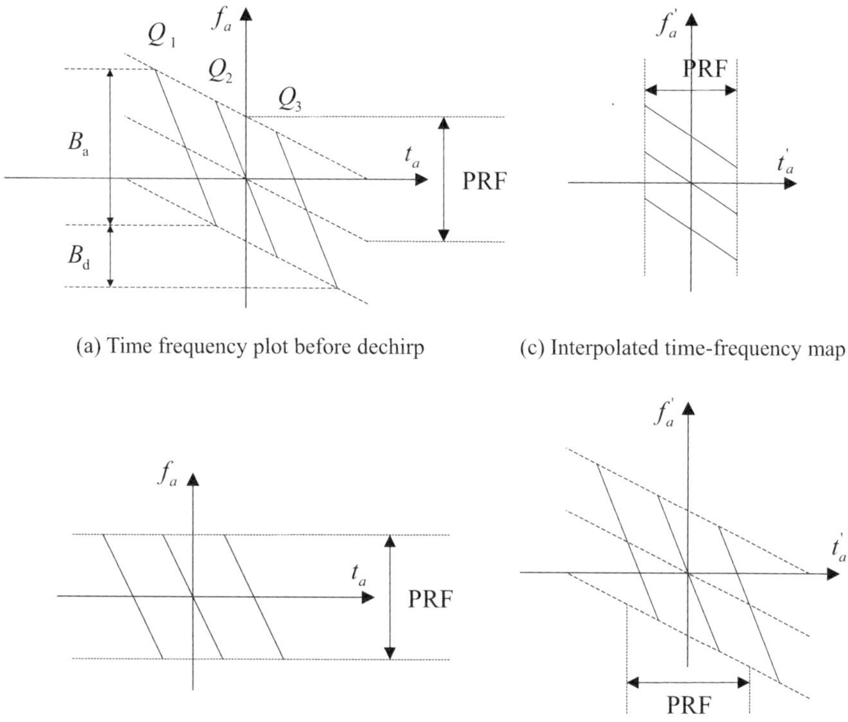

(a) Time frequency plot before dechirp (c) Interpolated time-frequency map

(b) Time frequency map after dechirp (d) After azimuth de aliasing

Fig. 2.24 Time frequency relationship diagram

$$H_2 = \exp(\mathrm{j}\pi K_t t_a'^2) \tag{2.56}$$

$$s_3(f_a', t_a') = \mathrm{FFT_a}(s_2(f_a, t_a))H_2 \tag{2.57}$$

The azimuthal complement of S_2 is followed by an azimuthal FFT and multiplication by H_2. It should be noted that this step is not a time–frequency conversion, but a sinc interpolation. The total number of azimuthal points after resampling is

$$N_n = \frac{\mathrm{PRF}}{K_t} F_n \tag{2.58}$$

Figure 2.24 shows the azimuthal time–frequency diagrams before the demodulation operation, after the de-modulation operation, after the interpolation, and after the azimuthal demixing.

The frequency components introduced by the rotation of the antenna beam center that were removed during the azimuthal de-modulation operation need to be compensated into the signal after the resampling is completed:

After the resampling is completed, the frequency component introduced by the rotation of the antenna beam center, which was removed from the azimuth declination operation, needs to be compensated into the signal

$$H_3 = \exp\left\{ j\pi \frac{f_a'}{K_t} \right\} \tag{2.59}$$

$$\begin{aligned} S_3(f_r, f_a') &= \mathrm{FFT_a}(s_3(f_r, t_a'))H_3 \\ &= W_a(f_a')\exp[-j\Phi(f_r, f_a)] \end{aligned} \tag{2.60}$$

S_3 can be interpreted as an FFT operation without azimuthal aliasing, which results in a true FMCW SAR sliding beam pattern in the two-dimensional frequency domain. After the azimuthal decluttering operation, the focus operation can be completed by applying the kernel migration correction algorithm.

2. A two-step approach to improvement

The RCMC corrected two-dimensional frequency domain signal is subjected to a range-to-IFFT operation to obtain

$$S_4(t_r', f_a') = W_a(f_a')p_r\left(t_r' - 2\frac{R_0 - R_{\mathrm{ref}}}{c}\right)\exp(-j2\pi f_a' t_0) \tag{2.61}$$

From Fig. 2.24d, it can be seen that the signal to unmix the azimuthal frequency domain is still limited by the PRF and requires a large PRF, which limits the degree of freedom in the wave position design and is not conducive to the achievement of the overall SAR performance index. The azimuthal frequency domain can be decoupled from the azimuthal time domain by referring to the method of decoupling the azimuthal spectrum, but another interpolation operation is performed, and the amount of operations is large, so the imaging method of azimuthal frequency domain de-modulation is chosen, considering that there is no longer any azimuthal frequency domain decoupling at this time. The specific algorithm flow chart is shown in Fig. 2.25.

The time–frequency relationship diagram and the time–frequency relationship after phase multiplication are shown in Fig. 2.26a and b. Figure 2.26 shows the time–frequency relationship diagram after three deblurring operations in the process of resolving image domain aliasing.

The time frequency relationship is shown in Fig. 2.26c, which shows that the azimuthal points are single frequency signals:

$$H_5 = \exp(-j\pi K_{a2} t_a'^2) \tag{2.62}$$

$$\begin{aligned} S_6(t_r', f_a') &= \mathrm{IFFT_a}(S_5 \times H_5) \\ &= p_r\left(t_r' - \frac{2(R_t - R_{\mathrm{ref}})}{c}\right)p_a(f_a' + K_{a2}t_0)\ \exp\left(-j\pi\frac{f_a'^2}{K_{a2}}\right) \end{aligned} \tag{2.63}$$

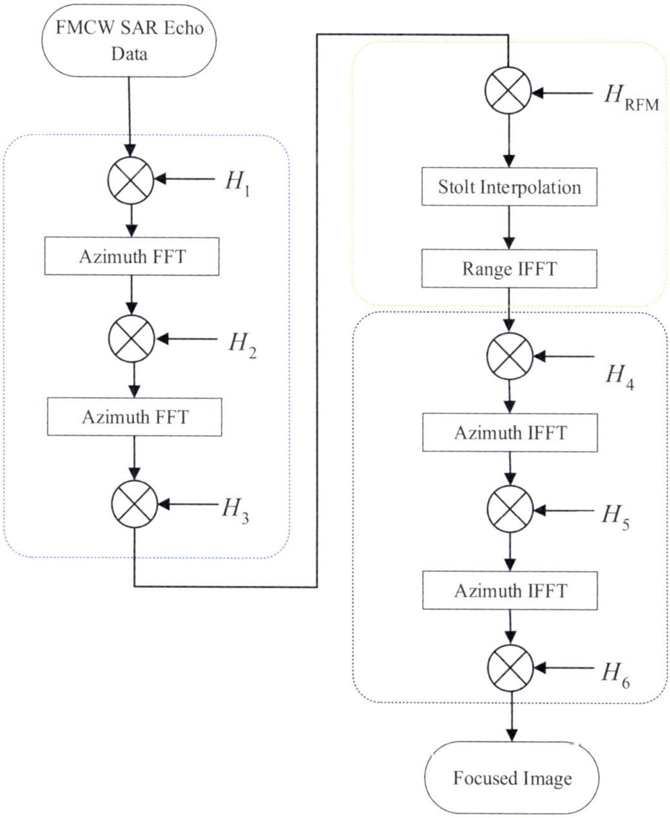

Fig. 2.25 The flowchart of the improved two-step method

S_6 is already a well-focused image. If a highly phase-preserving image is required for subsequent data applications such as interferometry, the last phase term can be compensated for:

$$H_6 = \exp(\mathrm{j}\pi \frac{f_a'^2}{K_{a2}}) \tag{2.64}$$

$$S_7(t_r', f_a') = p_r(t_r' - \frac{2(R_t - R_{ref})}{c}) p_a(f_a' + K_{a2} t_0) \tag{2.65}$$

The two-step method can effectively solve the azimuthal spectral aliasing problem, but will produce image domain aliasing. To overcome the image domain aliasing, the pulse repetition frequency needs to be increased, which leads to an increase in the radar received echo data and puts pressure on the processing. The improved two-step imaging algorithm is based on the traditional two-step method to remove the azimuthal spectral overlap, and the azimuthal frequency domain imaging is chosen

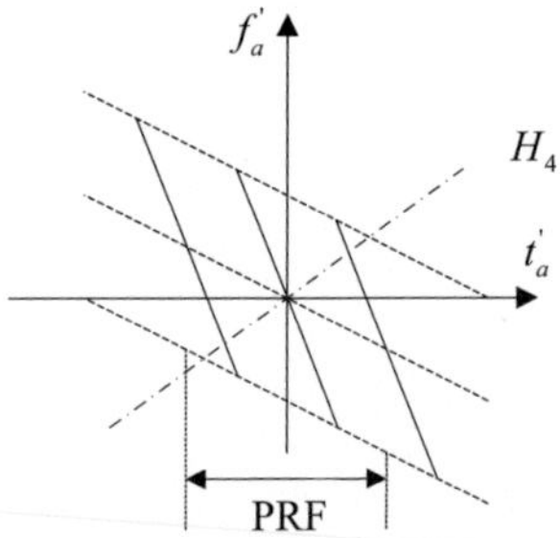

(a) Frequency domain declination

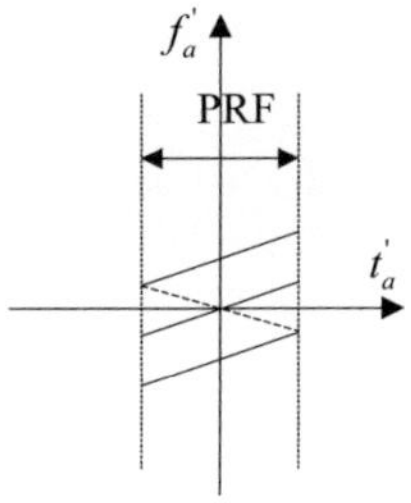

(b) Azimuth frequency domain declination

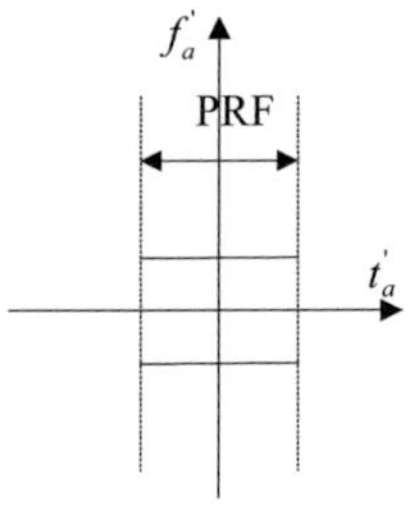

a

(c) Quadratic offset

Fig. 2.26 Removing image domain aliasing

to complete the focus by the azimuthal frequency domain de-modulation operation, which is not limited by the pulse repetition frequency and no longer generates image domain overlap.

2. Simulation experiments

In order to verify the effectiveness of the described algorithm, simulation experiments are carried out in MATLAB. First, the echo data are generated according to the derived echo model, and the simulation parameters are based on the light and small Ka FMCW SAR system developed in the laboratory. The system parameters are shown in Table 2.3.

Table 2.3 FMCW SAR system parameters

Parameter	Value	Parameter	Value
fc/GHz	35.75	Beam rotation angle/(°)	±9
PRF/Hz	1000	Rs/km	2.5
Br/MHz	700	Beam rotation speed/(°·s^{-1})	0.33
V/(m·s^{-1})	50		

Six equally spaced point targets are arranged in the imaging scene as shown in Fig. 2.27. Points 1 and 6 are located at the edge of the scene and are non-full aperture imaging points.

Figure 2.28 shows the resultant profile of the sliding cluster imaging simulation experiment using the conventional two-step method.

Figure 2.29 shows the resultant profiles of the simulation experiments using the modified two-step method.

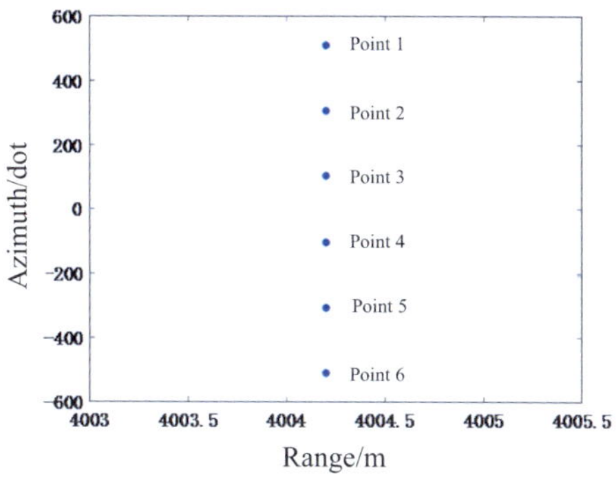

Fig. 2.27 Layout diagram

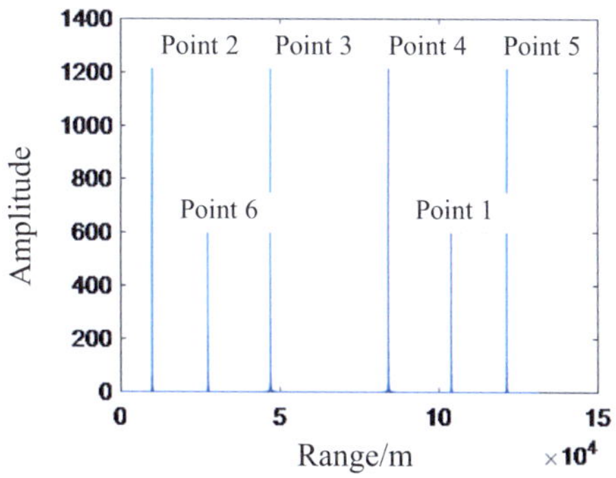

Fig. 2.28 Traditional two-step method

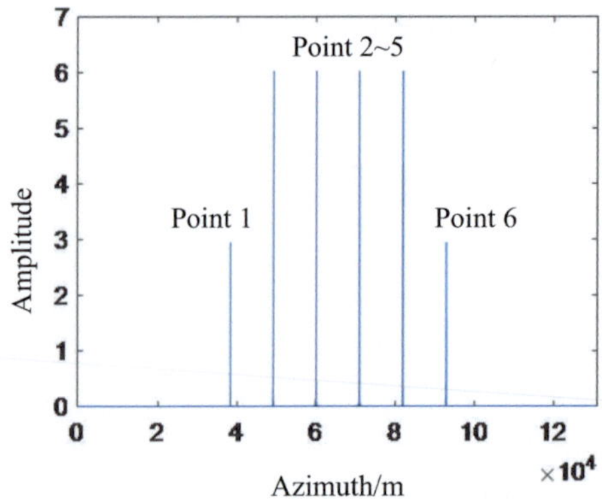

Fig. 2.29 Improved two-step method

The echo signal generated from the six point targets laid out in Fig. 2.29 was imaged, and the results of the 32.fold sinc interpolation of the image points are shown in Fig. 2.30.

As can be seen in Fig. 2.30, the six image points obtained from the imaging process are well focused. The range, azimuthal PSLR, ISLR, and resolution of these 6 image points are shown in Table 2.4.

It can be seen from Fig. 2.30 that the position of the image points in the simulation of the improved two-step method is the same as the position of the points in the distribution map, and there is no longer any overlap, which proves that the improved two-step method is effective in achieving the focus operation in the azimuthal frequency domain.

As can be seen in Fig. 2.30 and Table 2.4, the algorithm proposed in this section focuses well from the edge points to the full-aperture imaging points, and the range and azimuthal peak partials and integral partials are in agreement with the theoretical values, where points 1 and 6 are non-full-aperture imaging at the edge, and the intensity of the image points and the 3 dB width after 32.fold interpolation are only half of the full-aperture imaging points. The azimuthal 32.fold interpolation of the FMCW SAR strip mode with the same parameters. The theoretical ratio of the 3 dB width strip pattern for the sliding spotlight is 0.4491, which is consistent with the azimuthal resolution improvement factor $A = 119/264 = 0.45$ calculated from the simulation results.

2.3.3 Airborne Millimeter Wave FMCW SAR Sliding Cluster SAR Real-World Data Processing

1. Process flow

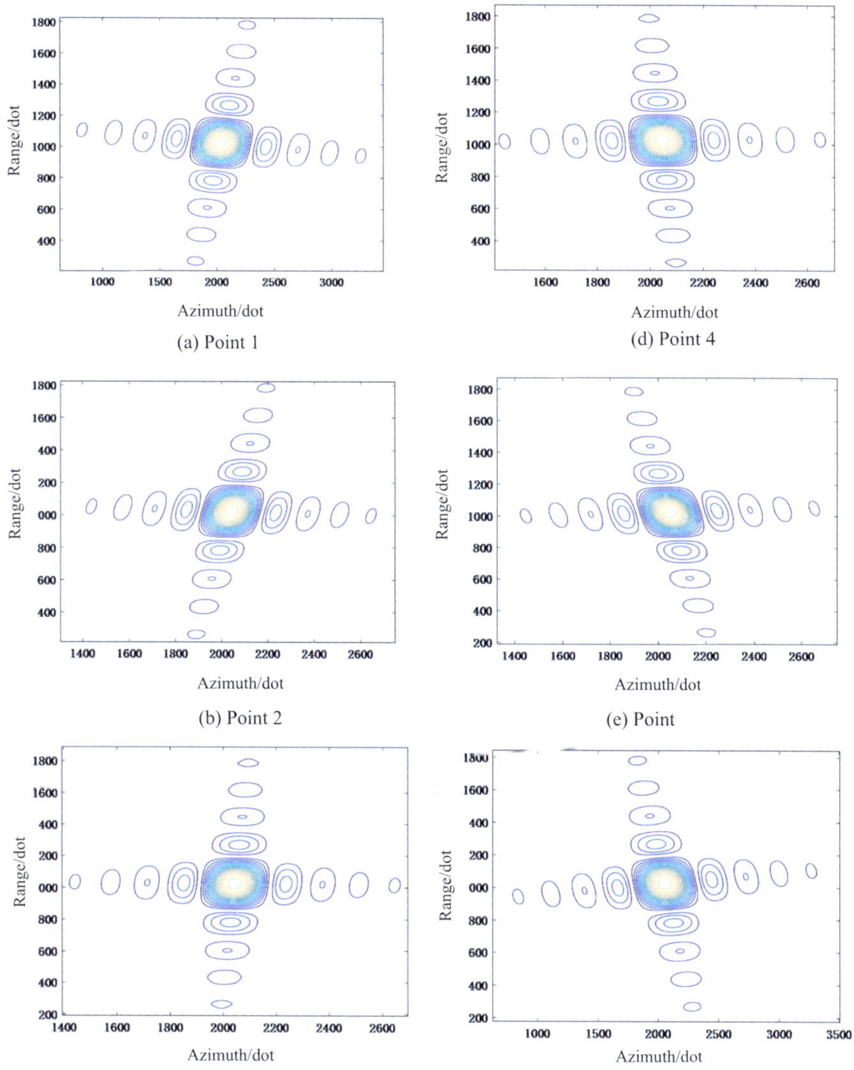

(a) Point 1

(b) Point 2

(c) Point 3

(d) Point 4

(e) Point

(f) Point 6

Fig. 2.30 Simulation results of 32 times interpolation for point targets

Here, we present a scheme and results of the slip aggregation data processing using inertial guidance compensation and multiplexed phase gradient autofocus (PGA) for motion compensation, and a modified two-step method for migration correction.

Figure 2.31 shows the flow chart of the designed algorithm.

In Fig. 2.31, PGA: phase gradient autofocus.

The processing flow of airborne FMCW SAR sliding spotlight mode imaging is divided into seven main steps.

Table 2.4 Simulation results

	Index	Point 1	Point 2	Point 3	Point 4	Point 5	Point 6
Range	PSLR/dB	−13.27	−13.26	−13.21	−13.43	−13.26	−13.26
	ISLR/dB	−10.42	−10.52	−10.52	−10.15	−10.52	−10.41
	RES/m	0.244	0.244	0.248	0.246	0.243	0.244
Azimuth	PSLR/dB	−13.26	−13.30	−13.26	−13.27	−13.30	−13.25
	ISLR/dB	−10.09	−10.32	−10.04	−10.04	−10.32	−10.08
	RES/m	0.423	0.210	0.207	0.214	0.215	0.426

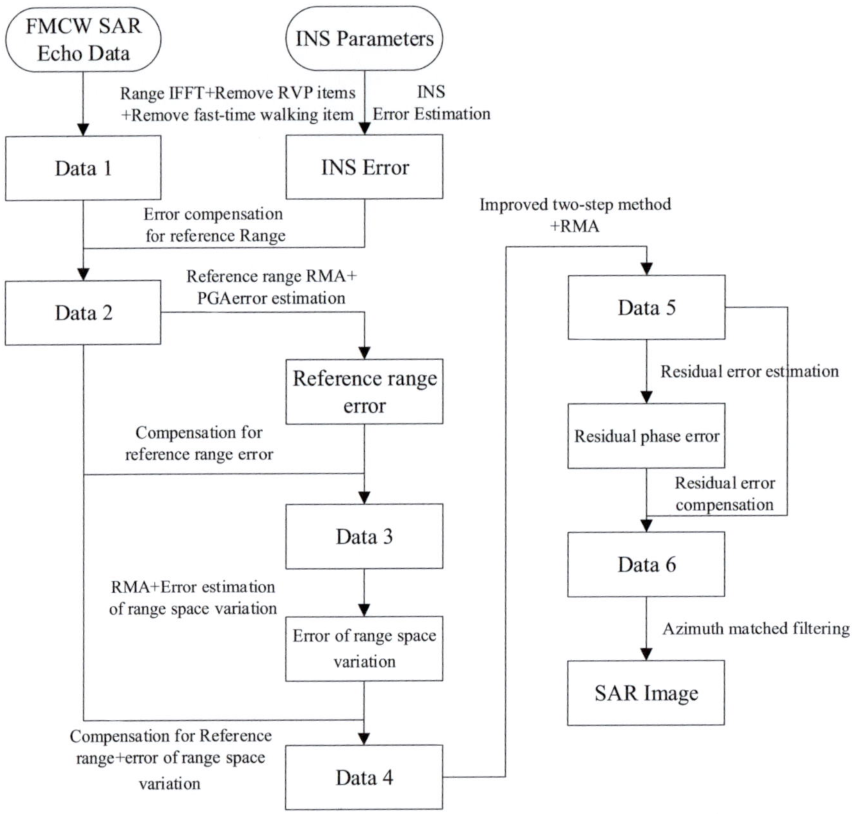

Fig. 2.31 Airborne FMCW SAR sliding spotlight mode focusing processing flow

(1) Pretreatment.

The FMCW SAR sliding spotlight mode echo data range to the IFFT enters the migrating line domain, completing the fast time walk correction and RVP correction.

(2) Inertial guidance error compensation.

The amount of error introduced by the carrier motion error is the same for each range gate at the moment corresponding to each azimuthal sampling point, so the amount of error compensation of the inertial guide for the slant range is calculated with the reference position.

(3) PGA reference position error compensation.

The PGA performs phase extraction without considering the range null variation, and adopts the approach of compensating the reference position error first and then the range null variation error. The 512 or 1024 range gates near the reference position are selected as input, the azimuthal direction is divided into several sub-blocks, the phase error within each sub-block is estimated, and then the phase error of each sub-block is stitched together to produce the reference position phase error curve of the whole block of echo data.

(4) PGA range null variation error compensation.

The change of the proximal and distal slope range of the range-oriented mapping band in the imaging scene area is inconsistent, and there are differences in the required phase error compensation. The reference error compensated azimuthal sub-blocks are divided into sub-blocks along the range direction, and the phase error information of each range direction sub-block is extracted, then a curve of equal length to the range direction sampling points is fitted, and the phase error of this curve is used to compensate the error of each range gate.

(5) De-blending operation and migration correction.

Following the method discussed in Sect. 2.3.2, an improved two-step method and the RMA algorithm are used to perform azimuthal de-mixing and migration correction on the echoes after the range null-variation error correction is completed.

(6) PGA residual error compensation.

The inertial guidance compensation and the two PGA motion error compensation complete the correction of the main error amount, and the compensated data still have some residual error phases. Theoretically, the phase error that can achieve good pulse compression needs to be within $\pi/4$, and the residual error compensation in step (6) can adjust the number of runs of step (6) according to the size of the actual residual error phase.

(7) Azimuthal matched filtering.

The azimuthal matching filtering operation is performed to complete the azimuthal pulse compression and obtain a well-focused FMCW SAR sliding beam image.

If the image quality does not meet the requirements after analyzing the image resolution, integral partials ratio, peak partials ratio, etc., it is possible to further improve the energy focus by performing PGA operation on the image domain within the system parameters.

2. Processing results

The algorithm described in the previous section is used to process the airborne experimental data of the Ka band FMCW SAR system. Figure 2.34 shows the measured data image of Ka band FMCW SAR gliding spotlight imaging mode with "Yunwu" as the carrying platform. The abscissa is in the azimuth direction, the ordinate is in the range direction, and the scene area is Shangjie Airport in Zhengzhou, Henan Province.

Figure 2.33 are enlarged images of typical targets in Fig. 2.32. Figure 2.33a shows an enlarged view of the corner reflector in the red rectangular box of Fig. 2.32, and Fig. 2.33b shows an enlarged view of the Y-5 aircraft in the yellow rectangular box of Fig. 2.32. From Fig. 2.33b, it can be seen that the contour of the aircraft is relatively clear and easy to interpret.

Figure 2.34 show the amplitude values of the corner reflector in logarithmic form in Fig. 2.33a. Figures 2.35 and 2.36 show the distance and azimuth profiles of the corner reflection, respectively. The calculated two-dimensional peak sidelobe ratio (PSLR), integrated sidelobe ratio (ISLR), and resolution are shown in Table 2.5.

Select the corner reflector at the center of the corner reflection Contour line in Fig. 2.33 for analysis. From Fig. 2.36, it can be seen that the distance side lobe of the

Fig. 2.32 Ka FMCW SAR sliding spotlight image

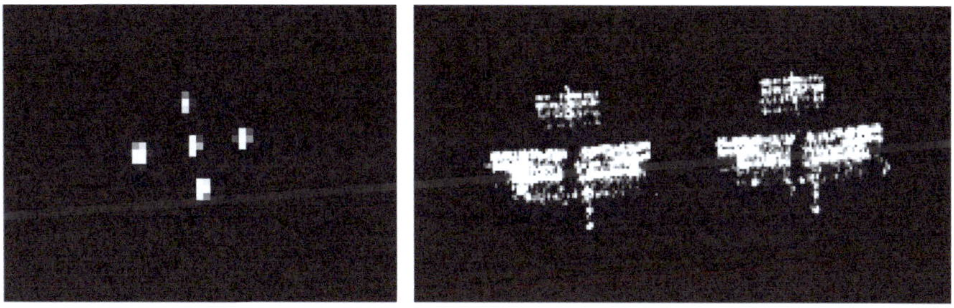

(a) corner reflector (b) "Yunwu" Aircraft

Fig. 2.33 Typical objectives

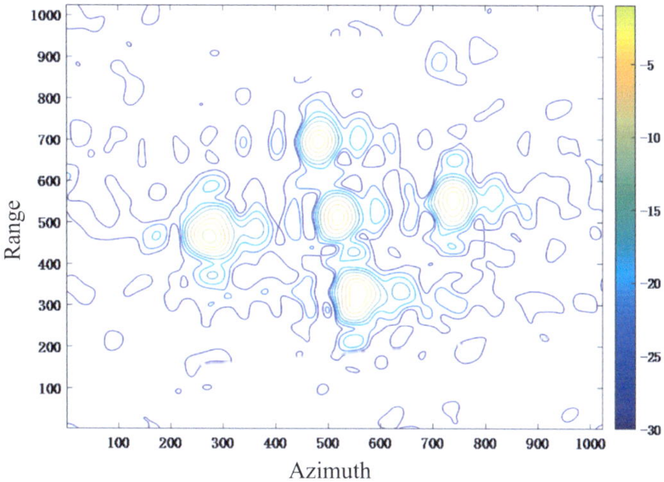

Fig. 2.34 Inverse contour line of angle

corner reflector is approximately −18 dB, and the azimuth side lobe is approximately −17 dB, both of which are better than the criterion of good focusing −13 dB.

Table 2.5 shows the integrated side lobe ratio, peak side lobe ratio and resolution of the inverse two dimensions of the central Position angle.

Due to the windowing process used in the actual airborne imaging processing, the two-dimensional PSLR and ISLR of the final angular backscatter analysis are better than the theoretical values. Figure 2.37a and b show the measured data images of sliding bunching mode and stripe mode in the same scene. Figure 2.37a has a resolution of 0.25 m and the scene targets two sets of metal frame structured greenhouses. Figure 2.37b has a resolution of 0.35 m. It can be clearly seen that the image quality of the sliding mode is better than that of SAR images in the same system with stripe mode, with higher resolution and better image readability.

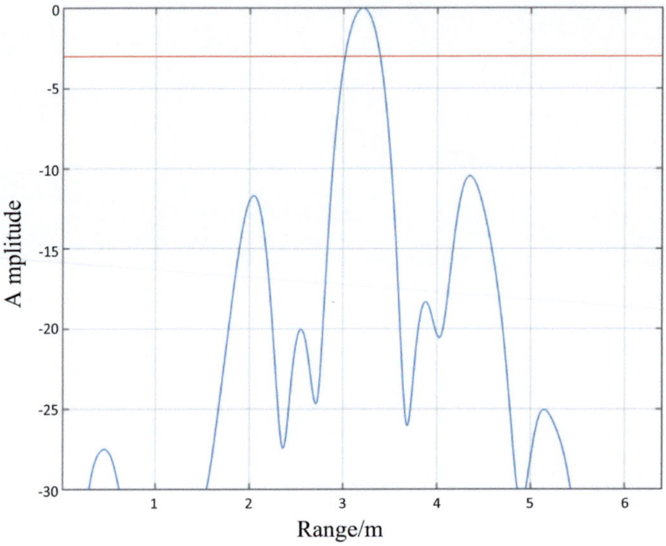

Fig. 2.35 Range profile

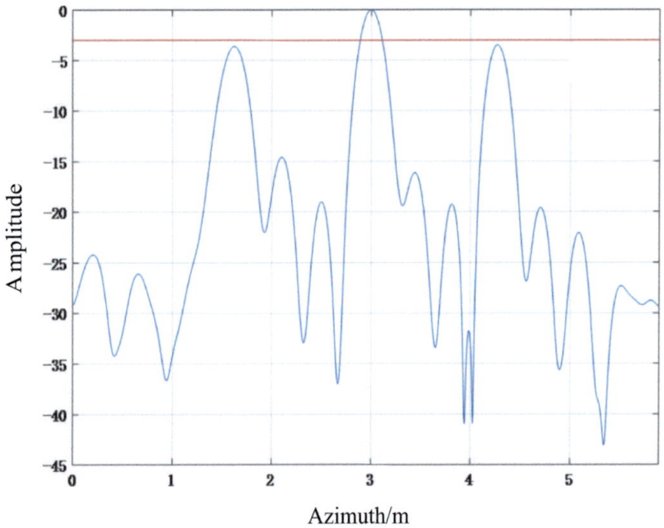

Fig. 2.36 Azimuth profile

Table 2.5 Analysis of corner reflectors

	Range	Azimuth
PSLR/dB	−18.45	−16.12
ISLR/dB	−12.84	−11.59
RES/m	0.355	0.249

(a) Sliding Spotlight mode

(b) Strip mode

Fig. 2.37 Sliding and banding modes

2.4 W-Band Small Unmanned Aerial Vehicle (UAV) Synthetic Aperture Radar (SAR) High-Resolution Imaging

Due to its small size, light weight, and susceptibility to aerodynamic disturbances, W-band UAV SAR suffers from significant motion errors [16]. Additionally, W-band SAR operates at very high carrier frequencies and short wavelengths, making the focusing effect of SAR images more sensitive to motion errors [17–19]. Therefore, high-resolution imaging algorithms require more stringent accuracy in estimating motion errors. However, based on the limited accuracy of the inertial measurement unit (IMU) navigation, it is not possible to achieve high-precision imaging solely based on inertial information. Furthermore, due to the limited transmit power of the

W-band radar system, SAR has a short slant range, resulting in pronounced scene-induced spatial variations and more severe range phase errors. Thus, accurately estimating and compensating for the motion errors of the aircraft's platform carrying the SAR antenna phase center (APC) is a key technology for achieving high-resolution imaging in W-band UAV SAR [20–26].

2.4.1 *Efficient Real-Time Imaging Algorithm for Small UAV-Based W-Band SAR*

Real-time imaging, constrained by time and memory resources, requires reducing computational complexity while maintaining imaging accuracy. Therefore, unnecessary computational steps should be eliminated as much as possible when implementing real-time imaging. In this section, by combining the flight test parameters of the system, an optimized real-time imaging algorithm based on inertial navigation and motion error estimation is proposed.

1. Analysis of imaging errors

The flight test parameters of the W-UAV-SAR system are shown in Table 2.6. Next, considering the system parameters, the phase error caused by intra-pulse continuous motion and the envelope error caused by range migration during the imaging processing are analyzed to simplify the processing reasonably during real-time imaging.

(1) Analysis of intra-pulse continuous motion error

In FMCW SAR, the signal is continuously transmitted with a very high duty cycle, reaching or approaching 100%. Compared to narrow-pulse transmission, intra-pulse radar motion introduces additional errors, and the stop-and-go assumption no longer holds. Unlike the frequency function in pulse-based systems, the frequency function in FMCW systems can be expressed as

Table 2.6 W-UAV-SAR system initial flight test parameters

Parameter	Value	Parameter	Value
Transmitting waveform	LFM	Beam width in azimuth direction	1.6°
Receiving mode	Dechirp reception	Beam width in the range direction	8°
Signal bandwidth	2 GHz	Center slant distance	684 m
PRF	1000	Wavelength	0.0032 m
Range resolution	0.075 m	Height	300 m

$$f = [f_0 + K_r(\tau - 2R/c)] \tag{2.66}$$

Compared to pulse-based systems, this frequency difference corresponds to a time drift in the filter input, which is equivalent to a range migration in the filter output.

$$\Delta R \approx \frac{f_0}{K_r} \cdot v \cdot \sin\theta = \frac{f_0}{K_r} \cdot \frac{v^2 t}{\sqrt{R_0^2 + v^2 t^2}} \tag{2.67}$$

Where: θ is the azimuth angle between the radar and the target, $\theta \in [\theta_{\min}, \theta_{\max}]$; $\theta_{\min}$ and $\theta_{\max}$ are the azimuth angles corresponding to the rear and front edges of the radar azimuth beam, respectively, when illuminating the target. Therefore, the maximum range migration caused by intra-pulse motion is

$$\Delta R_{\max} \approx \frac{f_0}{K_r} \cdot v \cdot (\sin\theta_{\max} - \sin\theta_{\min}) \tag{2.68}$$

Based on the system parameters in Table 2.6, the maximum range migration caused by intra-pulse motion is calculated to be 0.02 m, which is smaller than half of its slant range resolution of 0.075 m. Even without compensation, the impact on imaging is negligible.

(2) Analysis of envelope error introduced by range migration

The schematic diagram of range migration (RCM) in the positive side view is shown in Fig. 2.38. RCM refers to the change in distance between a point target P in the observed scene and the SAR payload on the flight platform. The azimuth beam width of the SAR payload antenna is. When the flight platform reaches point A, the front edge of the azimuth beam of the SAR payload scans the point target P, and when the flight platform reaches point B, the rear edge of the azimuth beam of the SAR payload moves away from the point target P. The shortest distance between point target P and the flight platform is called the closest distance. RCM is usually expressed as the difference between the slant range when the SAR payload antenna beam scans the point target at the edge and the closest distance.

$$R_q = R_e - R_B = R_B \sec\frac{\theta_{BW}}{2} - R_B \tag{2.69}$$

Based on the system parameters shown in Table 2.6, the maximum range migration caused by intra-pulse motion is calculated to be m. Due to distance downsampling in the simplified algorithm, the distance unit is 0.15 m. The range migration is less than half of the distance unit. Therefore, the envelope error caused by range migration can be neglected.

2. Efficient Real-Time Imaging Algorithm Based on Inertial Navigation and Motion Error Estimation

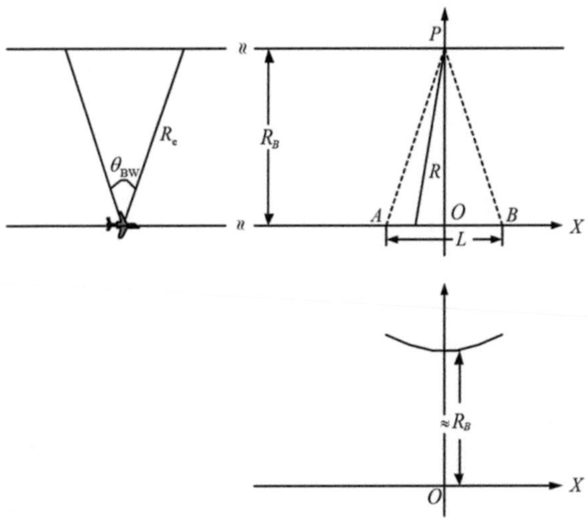

Fig. 2.38 Schematic diagram of range migration in front and side view

Due to the short real-time imaging time, large computation requirements, and limited resources, it is necessary to reduce computational complexity as much as possible while maintaining imaging accuracy. Therefore, unnecessary computational steps should be eliminated as much as possible when implementing real-time imaging. Real-time imaging also needs to be implemented in a pipelined design, where partial data is processed until the full aperture data is accumulated for azimuth compression. For longer imaging processes, imaging is performed by processing each pair of full aperture data separately. To ensure that all points are fully resolved, only one full aperture image size is retained in each frame, and adjacent frames are stitched together by overlapping half of the images. At this stage, the processing of the overlapping data can be preserved for the imaging process of the next frame.

Next, the analysis of the real-time imaging algorithm for W-band UAV SAR system is conducted based on the system parameters shown in Table 2.6. The received signals need to undergo dechirping and removal of the residual phase (RVP) term to eliminate its impact on azimuth focusing. For dechirped received signals, a dechirp-to-baseband conversion approach is designed to better integrate with the imaging algorithm. By calculation, the ground swath width corresponding to the beamwidth of 8° in the range direction is 222 m, and in the slant range plane, it is 300 m. The 200 m swath width corresponds to 2666 range samples with a spacing of 0.075 m. Considering that the beamwidth corresponds to 4000 scene points, a maximum of 4096 range samples are sufficient for imaging. The optimal azimuth resolution is 0.11 m * 1.2 = 0.13 m. The maximum azimuth bandwidth is less than 300 Hz, considering a PRF of 1000, azimuth data extraction can be performed once. The number of full aperture data points is within 2000. Considering azimuth extraction by a factor of 2, there are only 1000 points.

Based on the analysis, the designed system, limited by the beamwidth, only requires a maximum of 4096 range samples to meet the imaging swath width requirement, and 2048 azimuth samples are sufficient to meet the full-resolution imaging requirement in the azimuth direction. Therefore, the flowchart of the designed real-time imaging algorithm is shown in Fig. 2.39.

The sub-aperture echo data is processed in one step. First, the RVP term is used for correction to remove the azimuth phase modulation caused by dechirping reception.

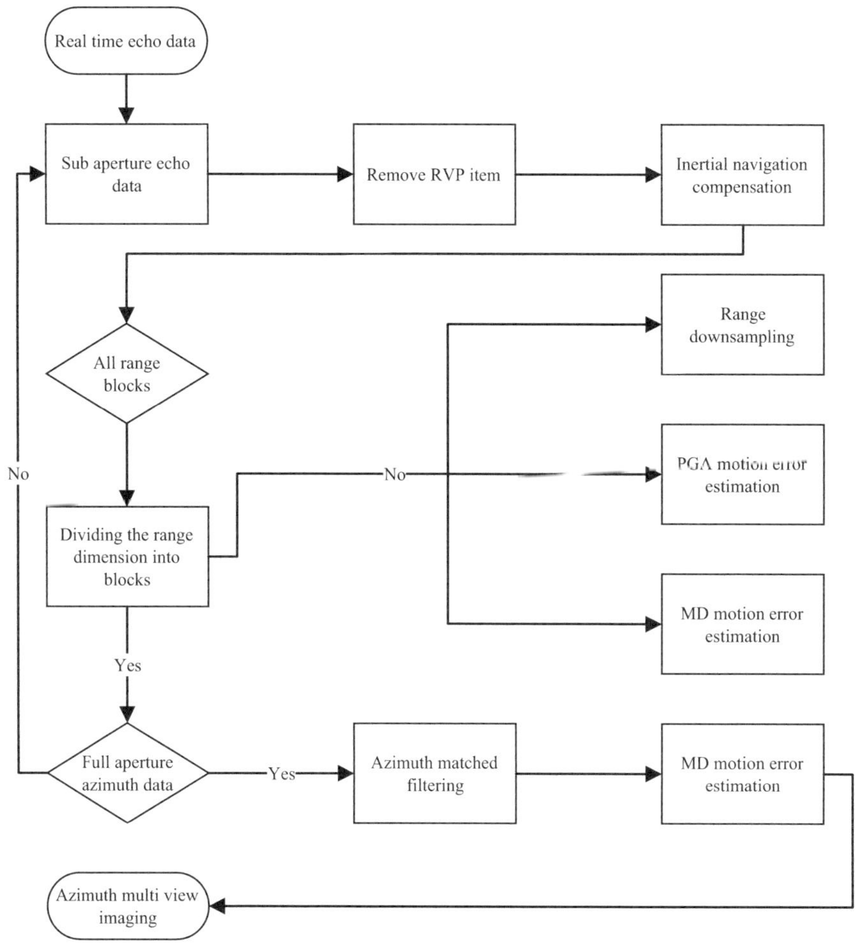

Fig. 2.39 Flowchart of the real-time imaging algorithm

This step can be applied to each recorded echo, and there is no need to accumulate a certain number of sub-aperture echo data pulses. Then, motion error estimation is performed on the range direction in blocks. This is because, due to the close observation distance, the range spatial variation of motion errors cannot be ignored, and motion error estimation in the range direction variation needs to be achieved by fitting multiple range blocks. The motion error estimation of each sub-block is achieved through a combination of PGA algorithm and MD algorithm. When the data accumulates to 2048 points, which corresponds to the length of two full apertures in the azimuth direction, a full-resolution image of the full aperture size can be obtained through matched filtering. Motion error compensation is applied to the data, and imaging is performed using azimuth matched filtering. Finally, multi-look processing is performed in the azimuth direction.

3. Simulation Experiment

To verify the effectiveness of the algorithm, this section conducts point target simulation experiments for the efficient real-time imaging algorithm based on inertial navigation and motion error estimation. The system simulation parameters are shown in Table 2.7.

The simulated XYZ-axis motion error input is shown in Fig. 2.40.

When motion errors are introduced without motion compensation using the Range-Doppler (RD) algorithm, the 2D compressed results for point target simulation are shown in Fig. 2.41.

The 2D compressed results show severe defocusing.

When motion errors are introduced and the efficient real-time imaging algorithm based on inertial navigation and motion error estimation is used, the 2D compressed results for point targets are shown in Fig. 2.42.

Comparing Figs. 2.41 and 2.42, it can be seen that the simplified real-time imaging algorithm based on inertial navigation and motion error estimation can compensate for motion errors and significantly improve the imaging of point targets affected by motion errors. This demonstrates the effectiveness of the algorithm presented in Chap. 3.

The two-dimensional profile analysis of the point target imaging results is shown in Table 2.8.

From the above results, it can be observed that the peak sidelobe ratio and integrated sidelobe ratio of the point targets in two dimensions are close to the theoretical values. The simulation results validate the effectiveness of the proposed method.

Table 2.7 Simulation parameters

Parameter	Value	Parameter	Value
Center frequency	94 GHz	Center slant distance	684 m
Transmission signal bandwidth	2 GHz	Antenna azimuth beam width	1.6°
PRF	1000	Scene width	222 m
v	10 m/s	Sampling frequency	40 M

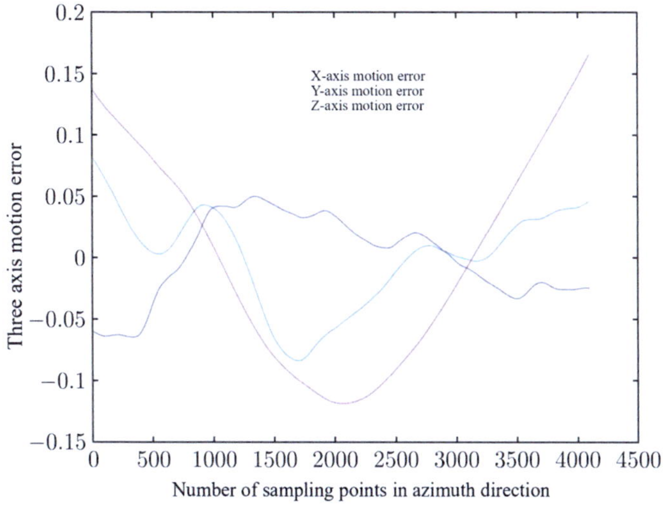

Fig. 2.40 XYZ three axis motion error simulation input

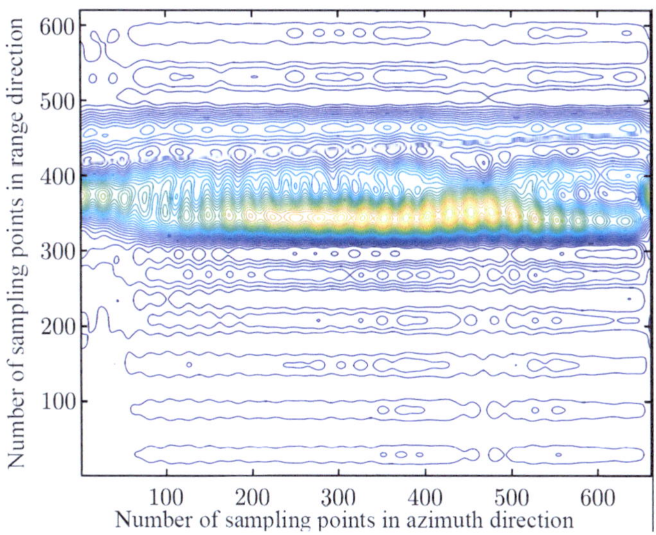

Fig. 2.41 2D compressed results for point targets without motion compensation

2.4.2 *Design of Real-Time Imaging Processing Based on GPU*

According to the stand-alone functional requirements, the digital imaging processing stand-alone design can be decomposed into monitoring and timing functional units,

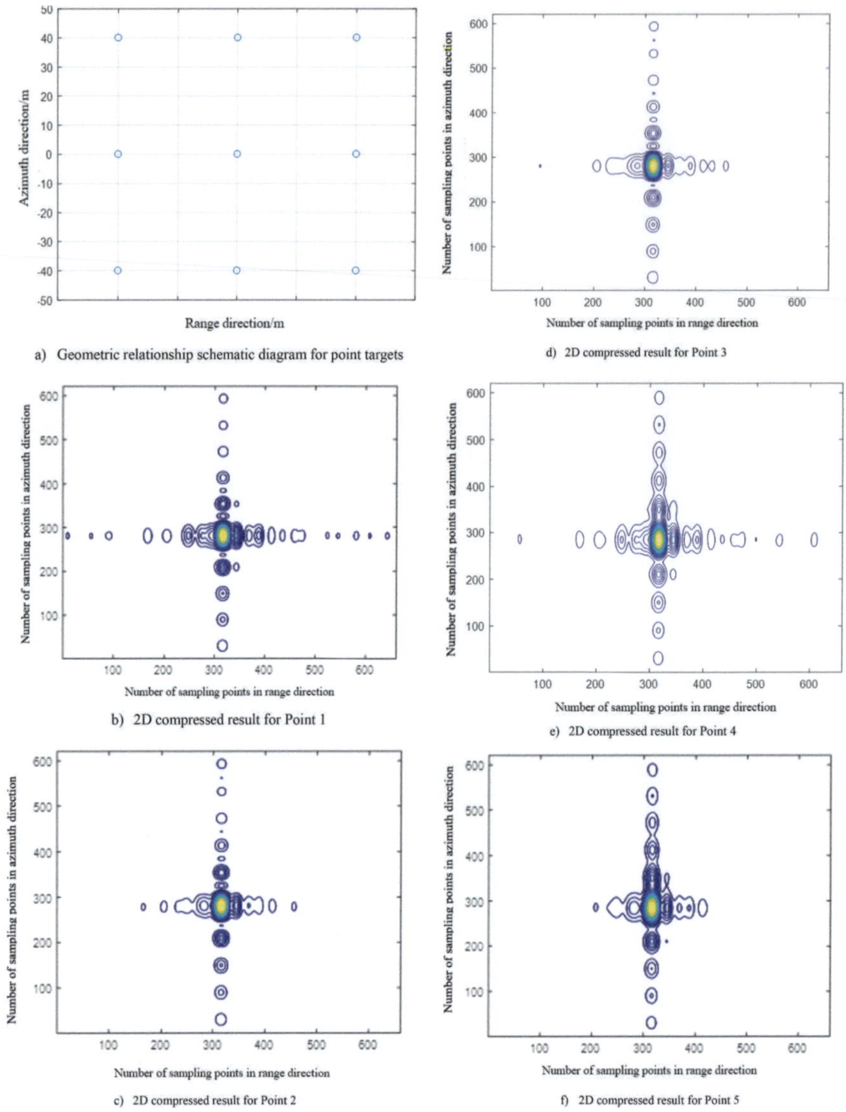

Fig. 2.42 2D compressed results for point targets

digital receiving functional units, real-time imaging processing functional units, data storage functional units and other auxiliary functional units.

Figure 2.43 is a block diagram of a digital stand-alone system based on field programmable gate array (FPGA) + general-purpose graphics processing unit (GPU). FPGA + GPU is used as the main control and processing chip to complete the functional module design of the above-mentioned digital stand-alone division.

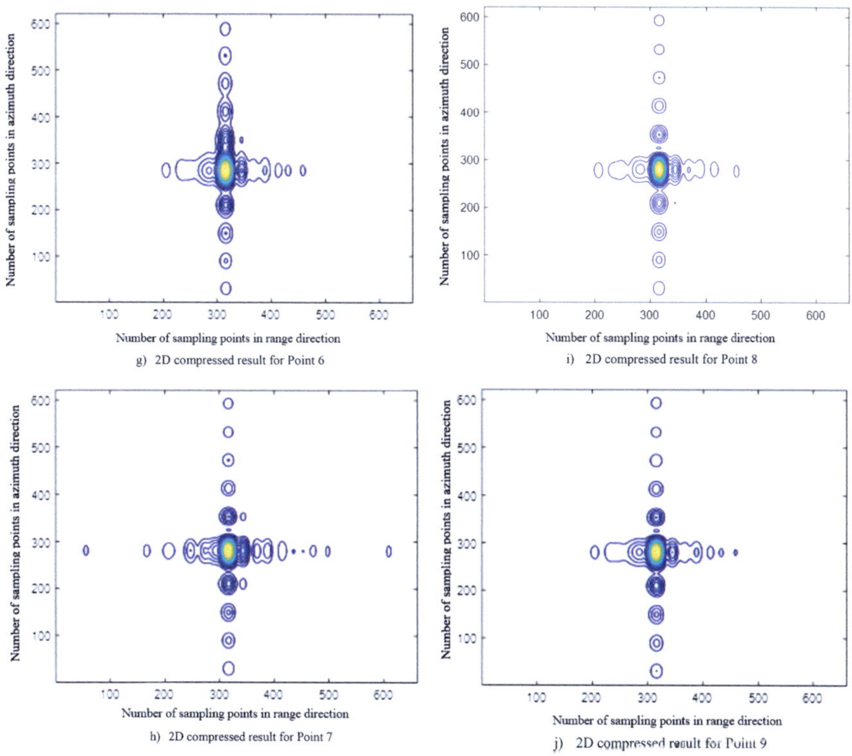

g) 2D compressed result for Point 6

i) 2D compressed result for Point 8

h) 2D compressed result for Point 7

j) 2D compressed result for Point 9

Fig. 2.42 (continued)

Table 2.8 Two-dimensional profile analysis of point target imaging results

Parameter	PSLR/dB	ISLR/dB
1st point	−12.22	−9.05
2nd point	−14.72	−11.44
3rd point	−15.54	−10.92
4th point	−12.06	−9.07
5th point	−14.84	−11.01
5th point	−15.57	−10.89
7th point	−12.04	−9.36
8th point	−14.74	−11.42
9th point	−15.81	−10.99

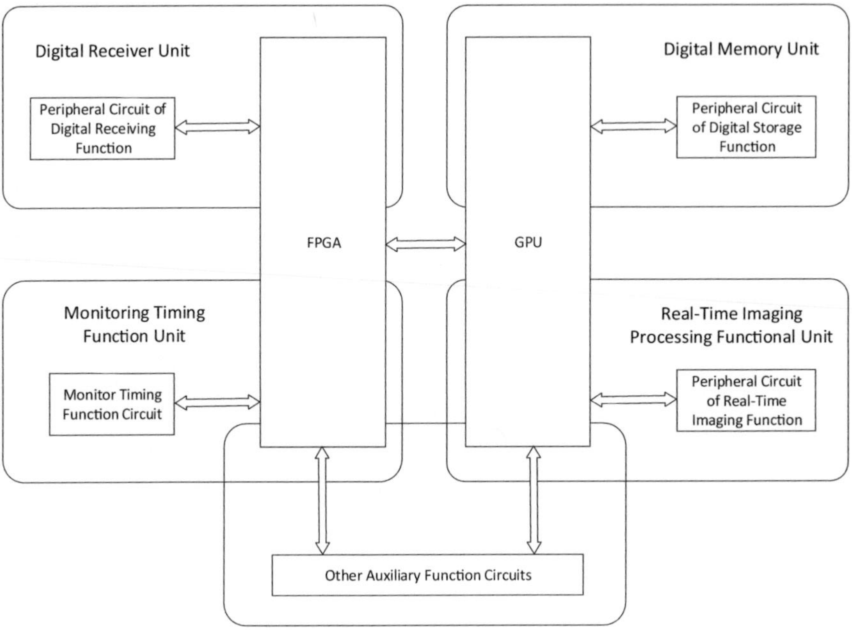

Fig. 2.43 Function realization block diagram of digital imaging processing module based on FPGA + GPU

Among them, the GPU module is Jetson TX2 produced by NVIDIA Corporation. The GPU module is responsible for executing signal processing algorithms and intelligent processing algorithms, and as the main control computer, completes the business process control, display, control, and storage functions. The GPU module is interconnected with the FPGA through the X4 PCIE interface connector REF-186138-01. The GPU can directly fix a standard SATA (serial ATA) solid-state hard drive and connect to the system through the SATA cable.

The GPU processing module uses the standard Jetson Module provided by NVIDIA. The interface between the Module and the FPGA board adopts the standard interface (400-pin, 8 × 50) of NVIDIA and SAMTEC.

The Jetson TX2 features an NVIDIA Pascal GPU with 256 Unified Computing Architecture (CUDA) capable cores. The central processing unit (CPU) complex consists of two clusters of ARM v8 64-bit CPUs connected by a high-performance coherent interconnect. The Denver 2 (dual-core) CPU cluster is optimized for single-thread performance; the second CPU cluster is an ARM Cortex-A57 (quad-core), which is better suited for multi-threaded applications.

The memory subsystem includes a 128-bit memory controller that provides high-bandwidth LPDDR4 support. 8 GBLPDDR4 main memory and 32 GB eMMC flash memory are integrated on the module. Designing from TX1 64-bit to 128-bit is a major performance boost.

Table 2.9 TX2 performance and interface description

	Jetson TX2	Jetson TX1
GPU	NVIDIA Pascal,256 CUDA Cores	NVIDIA Maxwell,256 CUDA Cores
CPU	HMP Dual Denver 2/2 MB L2+ Quad ARM A57/2 MB L2	Quad ARM A57/2 MB L2
Video	4k*2K 60HZ Coding（HEVC) 4K*2K 60HZ Coding（12 bit)	4K*2K 30HZ Coding（HEVC) 4K*2K 60HZ Coding（10 bit)
Memory	8GB 128位LPDDR4 58.3GB/s	4GB 64位LPDDR4 25.6GB/s
Display	2个DSI Interface、2个DP1.2 Interface /HDMI 2.0 Interface/eDP 1.4 Interface	2个DSI Interface、1个DP1.4 Interface /HDMI 2.0 Interface/eDP 1.2 Interface
CSI	2 Channel CSI2 D-PHY 1.2(Channel 2.5 Gbps)	2 Channel CSI2 D-PHY 1.1(Channel 1.5 Gbps)
PCIE	Gen 2 | 1*4 + 1*1或1*1 + 1*2	Gen 2 | 1*4 + 1*1
Storage	32 GB eMMC、SDIO、SATA	16 GB eMMC、SDIO、SATA
Others	CAN、UART、SPI、I2C、I2S、GPIO	UART、SPI、I2C、I2S、GPIO
USB	USB3.0 + USB2.0	USB3.0 + USB2.0
Connect	1 Gigabit Ethernet、802.11 ac WLAN、Bluetooth	1 Gigabit Ethernet、802.11 ac WLAN、Bluetooth
Mechanical	50mm*87mm (400-pin board-to-board connector)	50mm*87mm (400-pin board-to-board connector)

Among them, TX2 performance and interface description are shown in Table 2.9. The physical picture of NVIDIA Jetson TX2 module is shown in Fig. 2.44.

The functional schematic diagram of this module is shown in Fig. 2.45, where the red circles indicate the interfaces used in the design.

1. **Overall workflow of GPU-based software**

Fig. 2.44 NVIDIA Jetson TX2 module physical picture

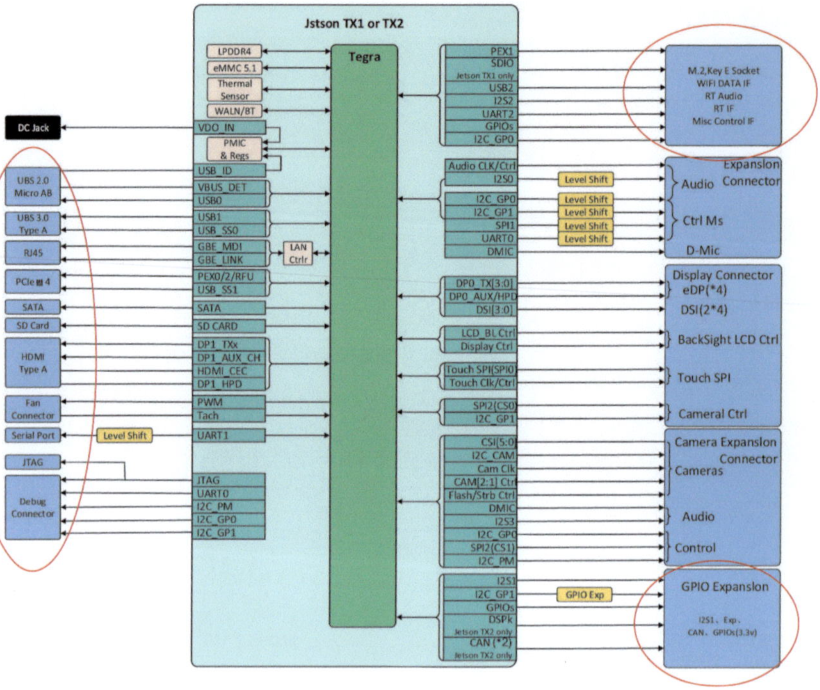

Fig. 2.45 Schematic diagram of NVIDIA Jetson TX2 module functions

The overall workflow of the software is shown in Fig. 2.46.

(1) The system starts up, automatically loads the driver, starts the application program, and completes the parameter configuration. There are two data acquisition modes: original echo data and echo data preprocessed by FFT.
(2) After receiving the start recording command of data transmission, the system starts to collect data, cyclically collects and stores 10 s long data into a file, and returns the remaining space of the data disk.
(3) After receiving the start imaging command of the data transmission, the system will pause the acquisition and start the imaging processing process after the current data file is completely recorded (within 10 s), and then perform acquisition storage and imaging processing after the imaging is completed, and repeat this process.
(4) Receive the data transmission stop recording command, the system stops data collection and waits for the next command.

2. **W-UAV-SAR real-time imaging algorithm GPU implementation process**

The SAR imaging algorithm flow is shown in Fig. 2.47, including echo data preprocessing, Doppler center estimation, range walking correction, first PGA + MD, second PGA + MD, matched filtering, etc. The algorithm involves a large number

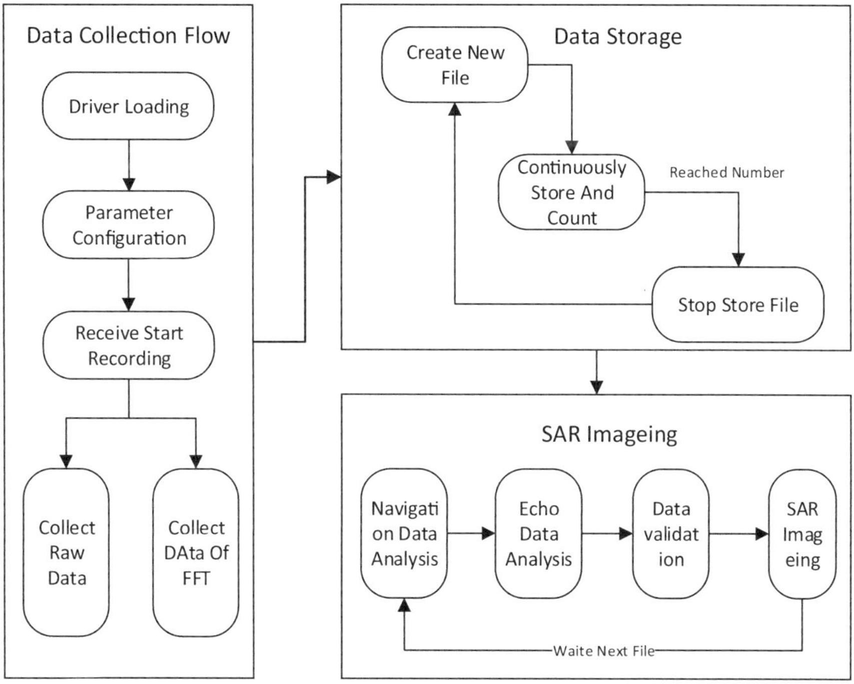

Fig. 2.46 The overall workflow design of the software system

of FFT, matrix, and vector operations, and has the characteristics of highly parallel computing. Therefore, the mixed programming of the central processing unit (CPU) and the graphics processing unit (GPU) is selected to facilitate algorithm performance optimization. In addition, CUDA provides some high-performance computing interfaces: cuFFT, cuCmulf, cuCabsf, etc., which can be called in imaging algorithms.

3. **GPU-based real-time imaging performance analysis**

The complete imaging algorithm takes about 44 min, or 2640 s, to run on a single core and single thread on the TX2 CPU. According to the TX2 CPU and GPU hardware performance calculation, the speedup ratio is up to 20 times, which is the speedup ratio under ideal conditions, and it is almost impossible to achieve in practical applications.

The calculation process with a speedup of 20 times is as follows:

CPU: ARM Cortex-A57 (quad-core) @ 2 GHz + NVIDIA Denver2 (dual-core) @ 2 GHz.

The CPU is based on the Armv8-A architecture, has 6 physical cores, and does not support hyperthreading. It contains 16 single-precision floating-point

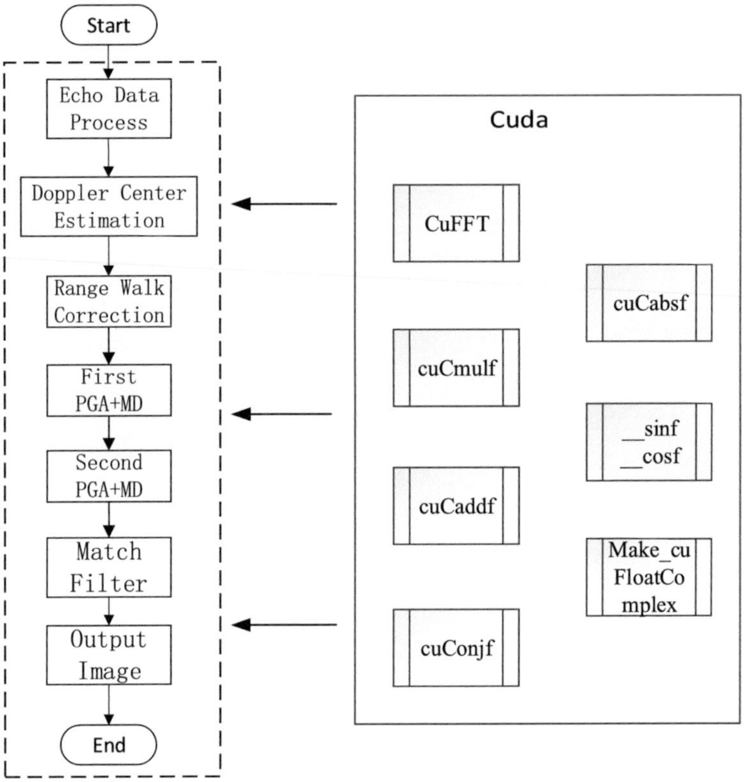

Fig. 2.47 SAR imaging implementation process design

execution units, and the calculation formula of its theoretical single-precision peak floating-point computing power is:

2035 MHz (maximum power) × 16 (number of single-precision floating-point execution units) × 6 (number of physical cores) = 195.36GFlops = 0.19536TFlops.

GPU: 256-core NVIDIA Pascal @ 1300 MHz.

NVIDIA Tegra X2 GPU is based on Pascal architecture, it contains 2 SM units, each SM unit contains 128 CUDA cores, its theoretical single-precision peak floating-point computing power calculation formula is:

1300 MHz (GPU Boost main frequency) × 256 (CUDA core number) × 2 (the number of floating-point calculations that can be processed in a single clock cycle) = 665.6GFlops = 0.6656TFlops.

The CPU program is a single-threaded program, so only one core of the CPU is used. The single-core single-precision peak floating-point computing capability of the CPU is 32.56GFlops (0.19536 TFlops /6).

GPU programs use all cores during execution, so their single-precision peak floating-point computing power is 665.6GFlops (0.6656 TFlops).

Therefore, the ratio of the peak floating-point computing power of GPU to CPU single core is 665.6GFlops/32.56GFlops = 20.44

Therefore, for the single-threaded CPU version of the program, its theoretical acceleration ratio on the GPU is 20:1, and the acceleration ratio cannot break through the limitation of hardware resources, that is, the ratio of the peak floating-point computing capabilities of the GPU to the CPU.

After GPU high-performance computing optimization on TX2, the ideal operation time of the complete algorithm is 2640 s/20 = 132 s. Currently, the optimization effect is 193 s, that is, the acceleration factor is 2640/193 = 13.68. Therefore, there is still some room for improvement, mainly in the optimization of memory access between core operations of various parts, and the optimization of computing has almost reached the maximum level.

The calculation hotspot in the main program includes 6 parts. respectively as shown below. The motion error estimation function based on PGA and MD is called 2 times, which takes 80.5% of the total running time of the algorithm. The distance walk correction function is called 2 times, taking 7.0% of the total time. The distance migration correction function is called 3 times, taking 3.0% of the total time. The azimuth matched filter function is called 1 time, which takes 2.1% of the total time. The envelope compensation function is called 2 times, taking 1.3% of the total time. Preprocessing operations like FFT\IFFT and shift of raw echoes take 2.5% of the total time. The following analyzes each hotspot part separately, and the calculation analysis and time-consuming ratio are shown in Table 2.10.

Motion error estimation function based on PGA and MD

(1) Operation analysis.

This part of the PGA operation accounts for 80.5% of the total running time of the algorithm. Most of the time is spent on PGA-based motion error estimation functions, MD-based motion error estimation functions, and beat functions. They account for 54.1%, 24.8% and 13.5% respectively, and the sum of the three is 92.4%, which accounts for 80.5% $\times$ 92.4% = 74.38% of the total time of the algorithm. The operations of these three functions are analyzed separately in the following.

(2) Motion error estimation function based on PGA.

The most time-consuming calculation in the PGA-based motion error estimation function is the PGA-based motion error estimation subfunction, which accounts for 99% of the time of the PGA-based motion error estimation function. Table 2.11 is the number of times and time-consuming ratio of various operations in the PGA-based motion error estimation sub-function. Since a complex multiplication requires 7 floating point operations, a complex addition requires 3 floating point

Table 2.10 Operation analysis and time-consuming ratio

Number	Function name	Function calls	Proportion (%)	Description
1.	Motion error estimation function based on PGA and MD	2	80.5	Proportion of the total running time of the algorithm
1.1	Motion error estimation function based on PGA (motion error estimation function 1)	52	54.1	Proportion of running time of the parent function (1.)
1.1.1	Motion error estimation subfunction based on PGA	52	99	The proportion of running time of the parent function (1.1)
1.1.1.1	Fftshift	624,000	4.3	The proportion of running time of the parent function (1.1.1)
1.1.1.2	Fft\Ifft	780	4.3	
1.1.1.3	Find extremum function	468,000	14.9	
1.1.1.4	Complex vector Addition\multiplication	~4,680,000	~25.5	
1.1.1.5	Complex addition\multiplication	>4,233,120,000	22.8	
1.1.1.6	Fft\Ifft	780,000	~25.3	
1.1.1.7	Other		2.9	
1.1.2	Multi-view function	52	0.9	
1.1.3	Other		0.1	
1.2	Md-based motion error estimation function	52	24.8	
1.2.1	Md-based motion error estimation subfunction	52	97.2	
1.2.1.1	Isft	104	47.4	
1.2.1.2	Fft\Ifft	104	17.8	
1.2.1.3	Fftshift	416	3.2	
1.2.1.4	Complex vector square root function	104	10.9	
1.2.1.5	Complex vector magnitude sum function	208	11.3	
1.2.1.6	Fft\Ifft	208	5.6	
1.2.1.7	Other		3.8	
1.2.2	Multi-view function	52	1.9	
1.2.3	Fft\Ifft	104	0.5	
1.2.4	Other		0.4	

(continued)

Table 2.10 (continued)

Number	Function name	Function calls	Proportion (%)	Description
1.3	Beat function	52	13.5	
1.3.1	Vector multiplication	159,744	99.9	
1.3.2	Other		0.1	
1.4	Fft\Ifft	104	1.1	
1.5	Other		6.5	
2.	Range cell migration correction	2	7.0	
2.1	Vector multiplication	20,240	98.6	
2.2	Other		1.4	
3.	Range cell migration correction function	3	3.0	
3.1	Isft	3	39.6	
3.2	Fft2\Ifft2	3	22.1	
3.3	Doppler spectrum compensation function	6	8.1	
3.4	Other (vector multiplication)		30.2	
4.	Echo preprocessing (Fft\Ifft, flipping, etc.)	6	2.5	
5	Azimuth matched filter function	1	2.1	
5.1	Isft	3,072	17.2	
5.2	Fft\Ifft	3,072	15.0	
5.3	Zero padding function	2	6.7	
5.4	Other (vector multiplication)		61.1	
6.	Envelope compensation function	2	1.3	
6.1	Fftshift	8	7.2	
6.2	Other (vector multiplication)	18,432	92.8	
7.	Azimuth multi-view processing function	1	0.8	
7.1	Isft	1	19.0	
7.2	Fft\Ifft	1	16.6	
7.3	Zero padding function	1	15.0	
7.4	Multi-view function	3072	8.1	
7.5	Other		41.3	
8.	Isft	3	0.6	
9.	Doppler center estimation function	1	0.6	
10.	Fft\Ifft	3	0.5	
11.	Filter function	1	0.2	

(continued)

Table 2.10 (continued)

Number	Function name	Function calls	Proportion (%)	Description
12.	Zero padding function	1	0.1	
13.	Other		0.8	

Note Sequence number 1 calls the function of sequence number 1.1, sequence number 1.1 calls the function of sequence number 1.1.1, and so on

Table 2.11 The number of times and time-consuming ratio of various operations in the PGA-based motion error estimation sub-function

Function name	Number of calls	Proportion (%)
FFTShift	624,000	4.3
FFT\IFFT	780	4.3
Find extremum function	468,000	14.9
Complex vector addition\multiplication	~4,680,000	~25.1
Complex addition\multiplication	>4,233,120,000	22.8
FFT\IFFT	780,000	~25.3
Other		2.9
Total		100

operations. Therefore, there are about 4,680,000 "complex number vector addition\multiplication" in the table, and the length of the vector is 1024, so the number of floating-point operations is 10240 × 4,680,000, about 48 G floating-point operations. In the table, there are 4,233,120,000 times of "complex number addition \ multiplication", that is, about 42 G times of floating point operations, and the total is 90G times. These two types of operations account for about 48.3% (25.5% + 22.8%) of the time of the PGA-based motion error estimation sub-function. In addition, the time-consuming operations of FFT, IFFT, modulo, and maximum calculation also account for about 48%. Therefore, it is estimated that the total number of floating-point calculations in this part is about 180G times (90G × 2). This is comparable in magnitude to the 280G times of floating-point calculations in the overall imaging algorithm counted by the Intel VTune Profiler tool. Therefore, except for the PGA-based motion error estimation sub-function part, the number of floating-point operations in other parts of the algorithm is about 100G times.

The operations in the PGA-based motion error estimation function are mainly common matrix/vector operations, FFT\IFFT, maximum value reduction, etc., which are very suitable for operations on GPUs. Moreover, this part accounts for the highest proportion of the total algorithm time, so the optimization of GPU high-performance computing is focused on this part.

(3) MD-based motion error estimation function.

Table 2.12 The number of times and time-consuming ratio of various operations in the MD-based motion error estimation sub-function

Function name	Number of calls	Proportion (%)
ISFT	104	47.4
FFT\IFFT	104	17.8
FFTShift	416	3.2
Complex vector square root function	104	10.9
Complex vector magnitude summation	208	11.3
FFT\IFFT	208	5.6
Other		3.8
Total		100

The most time-consuming calculation in the MD-based motion error estimation function is the MD-based motion error estimation sub-function, which accounts for 97.2% of the time of the MD-based motion error estimation function. Table 2.12 shows the number of times and time-consuming ratios of various operations in the MD-based motion error estimation sub-function. It can be seen that the most time-consuming operations are FFT, IFFT and vector shift operations. Regarding the number of vector points for FFT\IFFT, it can be seen from the source code that it is 512 points, and 512×9 points and 512×27 points after zero padding. Similarly, operations such as sqrt\abs\sum are also based on vectors with so many points, so although the number of calls is hundreds of times, it still takes a lot of time.

This part of the operation is mainly FFT\IFFT, followed by common operations and summation of complex vectors\matrixes, etc., which are very suitable for operations on GPUs. Therefore, these operations are optimized for GPU high-performance computing.

(4) Beat function.

In the beat function, 99.9% of the time is spent on large-scale complex vector multiplication. About 159,744 times, these operations are optimized for GPU high-performance computing.

(5) GPU accelerated effects and analysis.

Due to the common operations of matrix/vector, FFT\IFFT, reduction summation, reduction maximum value, etc. in this part of PGA operations, they account for a large proportion of operations. They are very suitable for computing on the GPU, so this part of the algorithm program has been deeply analyzed and rewritten. First, the program is divided into several segments, which can be easily rewritten as code segments of CUDA kernel functions. Then, they were respectively rewritten with CUDA and verified for correctness. Finally, the kernel functions of each segment are integrated and the memory management is unified, so that most of the operations are continuously run on the GPU, which minimizes the copying of data between the device and the host.

The imaging algorithm as a whole runs for 44 min on a single core and single thread on the Jetson TX2 CPU, and the running time of the PGA part of the algorithm is 80.5% × 44 × 60 s = 2125 s, which is the running time before GPU high-performance computing optimization. After optimization, its operation time is about 114 s, and the speedup ratio is 2125/114 = 18.6. Through the analysis of TX2 hardware, the speedup ratio is up to 20 times, which is an ideal speedup ratio, but it is almost impossible to achieve in practice. Therefore, the optimization of this part of the algorithm has made full use of GPU resources to achieve the maximum optimization performance.

2) Range walking correction function and distance migration correction function

(1) Operation analysis.

It can be seen from Table 2.10 that the distance walking correction function accounts for 7% of the total time consumption of the algorithm, that is, the time spent is 2640 s × 7% = 185 s. Among them, 98.6% of the operation time is spent on 20,240 vector multiplications. The distance migration correction function accounts for 3% of the total time consumption of the algorithm, that is, 2640 s × 3% = 79 s. Among them, more than 90% of the time is spent on FFT\IFFT, shift and vector multiplication. These operations are accelerated by GPU high-performance computing.

(2) GPU accelerated effects and analysis.

After GPU high-performance computing acceleration, the distance walking correction function and distance migration correction function take 34.3 s, and the time-consuming before acceleration is 185 s + 79 s = 264 s. Therefore, the speedup ratio is 264/34.3 = 7.7. This is still far from the ideal speedup of 20. This is because the number of calls to this part of the operation is not many, and only frequent calls can improve the speedup ratio. Therefore, there is not much room for further optimization in this part of the calculation.

3) Azimuth matched filter function

(1) Operation analysis.

It can be seen from Table 2.10 that the azimuth matched filter function accounts for 2.1% of the total time consumption of the algorithm, namely 2640 s × 2.1% = 55.4 s. Among them, more than 90% of the time is spent on FFT\IFFT, shift and vector multiplication. Therefore, these operations are accelerated by GPU high-performance computing.

(2) GPU accelerated effects and analysis.

After GPU high-performance computing acceleration, the azimuth matched filter function takes 13.5 s, and it takes 55.4 s before acceleration. Therefore, the speedup ratio is 55.4/13.5 = 4.1, which has a certain gap with 20 times. This is because this part of the operation in the algorithm is only called once, and only frequent calls can improve the speedup ratio, and the concurrency is not high. Therefore, there is not much room for further optimization in this part of the calculation.

4) Envelope compensation function

(1) Operation analysis.

It can be seen from Table 2.10 that the envelope compensation function accounts for 1.3% of the total time consumption of the algorithm, that is, the time spent is 2640 s × 1.3% = 34.3 s. Among them, more than 90% of the time is spent on vector multiplication. Therefore, these operations are accelerated by GPU high-performance computing.

(2) GPU accelerated effects and analysis.

After GPU high-performance computing acceleration, the envelope compensation function takes 13.36 s, and it takes 34.3 s before acceleration. Therefore, the speedup ratio is 34.3/13.36 = 2.58, which is a certain gap from 20 times. Because this part of the operation in the algorithm is only called twice, and only frequent calls can improve the speedup ratio, and the degree of concurrency is not high. Therefore, there is not much room for further optimization in this part of the calculation.

5) Echo data preprocessing

(1) Operation analysis.

Echo data preprocessing mainly includes FFT\IFFT, flipping and other operations. It can be seen from Table 2.10 that it accounts for 2.5% of the total algorithm time consumption, that is, the time spent is 2640 s × 2.5% = 66 s. These operations are accelerated by GPU high-performance computing.

(2) GPU accelerated effects and analysis.

After GPU high-performance computing acceleration, the preprocessing part takes 4.2 s, and the time before acceleration is 66 s. So, the speedup is 66/4.2 = 15.7. The gap with the ideal speedup ratio of 20 is small, and a greater degree of optimization has been achieved.

6) Azimuth multi-look function

(1) Operation analysis.

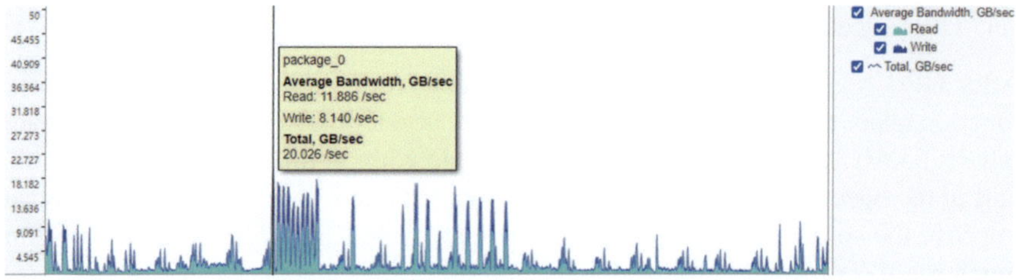

Fig. 2.48 Most of the time when the algorithm is running, the bandwidth usage does not exceed 10 GB/s

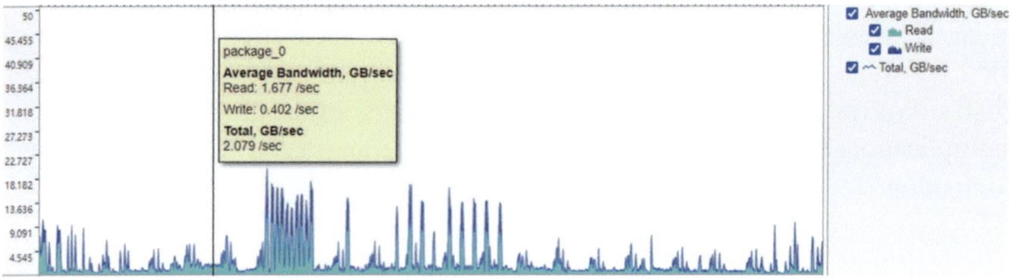

Fig. 2.49 The algorithm uses the maximum read and write bandwidth

It can be seen from Table 2.10 that this part accounts for 0.8% of the total time consumption of the algorithm, that is, the time spent is 2640 s × 0.8% = 21 s. There are many types of operations involved, such as FFT\IFFT, shift, vector operations, etc., so some operations are accelerated by GPU high-performance computing.

(2) GPU accelerated effects and analysis.

After GPU high-performance computing acceleration, this part takes 7 s, and before acceleration takes 21 s, that is, the acceleration ratio is 21/7 = 3. Since this part of the operation is only called once, and the degree of concurrency is not high, it cannot approach the ideal speedup ratio of 20, and there is almost no room for re-optimization.

4. **Bandwidth analysis**

Jetson TX2 has a memory bandwidth of 59.7 GB/s. During the operation of the algorithm, most of the time the bandwidth occupation does not exceed 10 GB/s, and the average occupation is less than 20%, as shown in Fig. 2.48. The monitored read and write bandwidth used is up to 20 GB/s, as shown in Fig. 2.49. However, when running on a single processor, the average processor utilization is above 90%. Therefore, bandwidth is not the bottleneck of algorithm operation, but floating-point computing power is the bottleneck of algorithm operation.

5. **Summary**

Table 2.13 is a summary of the acceleration effect of computing hotspots on the Jetson TX2 platform.

Table 2.13 Summary of accelerated optimization of computing hotspots

Number	Operation	Before-optimization(s)	After-optimization(s)	Ratio
1	Motion error estimation function based on PGA and MD	2125	114	18.6
2	Distance walking correction function and distance migration correction function	264	34.3	7.7
3	Azimuth matched filter function	55.4	13.5	4.1
4	Envelope compensation function	34.3	13.36	2.58
5	Echo data preprocessing	66	4.2	15.7
6	Azimuth multilook function	21	7	3
7	Other	74.3	6.64	–
8	Total	2640	193	13.68

References

1. Liu, Yue. Research on new FMCW SAR system and signal processing method [D]. Beijing: Institute of Electronics, Chinese Academy of Sciences, 2012.
2. Xu Wei. Study of satellite-based TOPSAR models [D]. Beijing: Institute of Electronics, Chinese Academy of Sciences, 2010.
3. Saunders D, Bingham S, Menon G, et al. A single-chip 24GHz SiGe BiCMOS transceiver for low cost FMCW airborne radars[C]// Proc. Aerospace & Electronics Conference (NAECON), 2009: 244–247.
4. Lanari R, Tesa M. Spotlight SAR data focusing based on a two-step processing approach[J]. IEEE Transaction on Geoscience and Remote Sensing, 2001: 363–370.
5. Reigber A, Moreira A. First demonstration of airborne SAR tomography using multibaseline L-band data [J]. IEEE Transaction on Geoscience and Remote Sensing Letters, 2000, 38(5): 2142-2152.
6. Fornaro G, Reale D, Serafino F. Four-dimensional SAR imaging for height estimation and monitoring of single and double scatterers [J]. IEEE Transaction on Geoscience and Remote Sensing, 2009, 47(1):224-237.
7. Wang R, Loffeld O, Nies H, et al. Focus FMCW SAR data using the wavenumber domain algorithm[J]. IEEE Transaction on Geoscience and Remote Sensing, 2010, 48(4):2109-2108.
8. Fornaro G, Lombardini F, Serafino F. Three-dimensional multipass SAR focusing: experiments with long-term spaceborne Data[J]. IEEE Transactions on Geoscience and Remote Sensing, 2005, 43(4):702-714.
9. Jing Guobin. Research on airborne/on-board ultra-high resolution SAR imaging technology [D]. Xi'an: Xi'an University of Electronic Science and Technology, 2018.
10. Li Chunsheng, Yang Wei, Wang Pengbo. A review of satellite-based SAR imaging processing algorithms [J]. Journal of Radar, 2013, 2(1):111-122.
11. Y. Liang. FM continuous wave SAR signal processing [D]. Xi'an: Xi'an University of Electronic Science and Technology, 2009.

12. Liu Hanyan, Song Hongjun, Cheng Zengju. Comparison of strip mode, cluster mode and sliding cluster mode [J]. Journal of the University of Chinese Academy of Sciences, 2011, 28(3):410-417.
13. Zhang Dawei, Wei Fang, Wang Yanfei. Analysis of the effect of Doppler effect on the imaging performance of FMCW-SAR system [J]. Journal of Electronics and Information, 2008, 30(3):55-59.
14. Liang Y, Wang H X, Xing M D, et al. Analysis and imaging of FM continuous wave SAR signals [J]. Journal of Electronics and Information, 2008, 30(5):5-9.
15. Meta A, Hoogeboom P. Development of signal processing algorithms for high resolution airborne millimeter wave FMCW SAR [C]// IEEE International Radar Conference, 2005.
16. ZHANG Hui, WANG Hui, PAN Jiaqi. Imaging and experimental validation of W-band FMCW system ISAR system [J]. Shanghai Space, 2018, 35(6):24–29.
17. Liang Y, Wang HX, Xing MD. Analysis and imaging of FM continuous wave SAR signals [J]. Journal of Electronics and Information, 2008, 30(5):1017-1021.
18. Zhang Dawei. Research on airborne SAR real-time processing technology and FMCW-SAR imaging [D]. Beijing: Institute of Electronics, Chinese Academy of Sciences, 2007.
19. He Xuehui. Real-time signal processing of airborne synthetic aperture radar imaging [D]. Xi'an: Xi'an University of Electronic Science and Technology, 2004.
20. Bao Zheng. Radar imaging technology [M]. Beijing: Electronic Industry Press, 2005: 24-30.
21. Xing, Mengdao, Bao, Zheng. SAR imaging based on motion parameter estimation [J]. Journal of Electronics, 2001, 29(S1):1824-1828.
22. Zhang Yulun, Zhang Tao, Li Tao. Research on airborne oblique SAR motion compensation based on inertial guidance [J]. Fire Control Radar Technology, 2014, 43(1):18-21.
23. Gong W, Xiang Mao S. Motion compensation for radar imaging using inertial guidance and GPS data [J]. Electronic Measurement Technology, 2006, 29(2):11-13.
24. Liu RH, Zhao YX. A low-cost INS-based SAR motion compensation algorithm [J]. Radar Science and Technology, 2009(4):257-261.
25. Zhang L , Wang G , Qiao Z , et al. Azimuth motion compensation with improved subaperture algorithm for airborne SAR imaging [J]. IEEE Journal of Selected Topics in Applied Earth Observations and Remote Sensing, 2016, 10(1):1-10.
26. Pi Yiming, Yang Jianyu, Fu Yusheng. Synthetic aperture radar imaging principles [M]. Beijing: University of Electronic Science and Technology Press, 2007: 115-120

Chapter 3
Pulse Regime Airborne Millimeter Wave Synthetic Aperture Radar

3.1 Overview

Wide swath and high resolution are the eternal pursuit of SAR, and they are also a pair of contradictory quantities. For wide-swath observation requirements, it is difficult to obtain good antenna gain in the entire width due to factors such as the broadening of the main lobe of the antenna pattern and the restriction of range ambiguity on the long-time modulated signal, thereby reducing the signal signal-to-noise ratio [1]. For millimeter wave, due to its short wavelength and low flight altitude when applied to airborne platforms, it is more difficult for SAR designs to obtain large swath at reasonable output power levels. Digital beamforming forming scan-on-receive (DBF-SCORE) technology can use the multiple antenna sub-apertures in range, by setting the weighting coefficient of each sub-aperture, synthesize a narrow receiving beam with high gain and receive it in real time in the echo direction to achieve high receiving gain, so that a smaller transmitting antenna can be used to obtain a larger mapping swath, and the signal-to-noise ratio, optimize range ambiguity.

On the other hand, as the spaceborne millimeter wave SAR technology has developed from the systematic exploration and demonstration stage of simple technical verification to engineering development plans, many technical verifications is a necessary condition for the engineering development of spaceborne systems. The design and development and signal processing research of airborne mm-wave DBF SAR can be used as a necessary technical verification method for the deployment of advanced space-based millimeter wave SAR in the future, and Chap. 6 of this book will also introduce the imaging and InSAR processing methods of pulsed spaceborne millimeter wave DBF SAR.

This chapter first introduces the basic principles of DBF-SCORE technology, DBF processing combined with channel equalization and adaptive DBF processing methods, on this basis, the imaging processing method and measured data processing results of airborne millimeter wave DBF SAR are given, and finally the processing

H. Wang, *The Millimeter Wave Synthetic Aperture Radar Technology*,
https://doi.org/10.1007/978-981-97-1044-7_3

method and process of airborne azimuth multi-channel DBF SAR moving target detection are introduced and the processing and verification results of measured data are given.

3.2 DBF-SCORE Fundamentals

The fixed transmission power of the radar will be modulated to the entire width by the antenna pattern, and the larger the width, the less incident energy per unit area, and then the signal signal-to-noise ratio will decrease. According to the radar equation, the thermal noise of the radar system is first described (3.1)

$$P_n = kT_n B_n \tag{3.1}$$

thereinto, k is a Boltzmann constant $1.38 \times 10^{-23} J/K$, B_n for receiver bandwidth, T_n For temperature, the signal-to-noise ratio descriptive (3.2) is obtained

$$SNR^{raw} = \frac{P^{rec}}{P_n} = P_e \frac{\lambda^2}{(4\pi)^3} \frac{G^2}{R^4} \frac{1}{Loss} \frac{1}{kT_n B_n} \sigma \tag{3.2}$$

thereinto, P_e is the peak transmit power, G for antenna gain, R To observe the range, $Loss$ is system loss; σ is the target radar cross-sectional area, consider $\sigma^0 S$, that is, the product of the normalized radar cross-sectional area and the effective surface area of the stationary target. According to the observation geometry and radar parameters, the effective observation area (3.3) two-dimensional dimensions are respectively $R\lambda/D$ and $T^{pulse}c/2\sin\theta$:

$$S = \frac{R\lambda T^{pulse} c}{2D \sin\theta} \tag{3.3}$$

For the resolving unit of the SAR original echo, the signal-to-noise ratio is further expressed as Eq. (3.4)

$$SNR^{raw} = P_e \frac{\lambda^2}{(4\pi R)^3} G^2 \frac{1}{Loss} \frac{1}{kT_n B_n} \frac{\sigma^0 c T^{pulse}}{2D \sin\theta} \tag{3.4}$$

After the process of synthesizing the aperture, the signal-to-noise ratio will be greatly improved, because the range compression and the coherent processing of azimuth modulation can effectively suppress the independent noise. The number of independent samples accumulated in azimuth direction and the number of independent samples accumulated in range direction are (3.5) and (3.6), respectively

$$N_{az} = 2\frac{R_0}{D^2}\lambda \tag{3.5}$$

$$N_{rg} = B_d T^{pulse} \tag{3.6}$$

Then the signal-to-noise ratio of the complex image after SAR focusing is expressed as Eq. (3.7)

$$SNR^{slc} = P_e \frac{\lambda^2}{(4\pi R)^3} G^2 \frac{1}{Loss} \frac{1}{kT_n} T^{pulse} \frac{2\sigma^0}{D^2} \tag{3.7}$$

Under the sampling conditions $v/f_a = D/2$, the following Eq. (3.8).

$$SNR^{slc} = P_{av} \frac{\lambda^2}{(4\pi R)^3} G^2 \frac{1}{Loss} \frac{1}{kT_n} \frac{\sigma^0}{2v^2} \tag{3.8}$$

P_{av} In order to average transmit power, it is worth noting that the signal-to-noise ratio is modulated by negative 4 times of the range term before SAR processing, and after the azimuth is accumulated, due to the positive correlation between the synthetic aperture and slant range, it is only modulated by the negative 3 times of the range after synthesis. Definition for scatter target (3.9)

$$SNR_{ext}^{raw} = P_{av} \frac{\lambda^3}{(4\pi R)^3} \frac{G^2 \sigma^0 r_d}{Loss 2vkT_n} \tag{3.9}$$

Accordingly, the equivalent backscatter coefficient is defined $NE\sigma^0$ as Eq. (3.10) σ^0 with a signal-to noise ratio of 1

$$NE\sigma^0 = \frac{(4\pi R)^3}{\lambda^2} \frac{LosskT_n 2v}{P_{av} G^2 r_d} \tag{3.10}$$

Then the actual signal-to-noise ratio of the scene is Eq. (3.11)

$$SNR = \sigma - NE\sigma^0 \tag{3.11}$$

The scene swath is approximated $W_g = R\lambda/H_r$, H_r which is the antenna height. Through the above analysis, it is easy to know that there are three main ways to obtain a large width: first, a wider antenna main lobe width; Second, longer radar-scene range; Third, flexible beam pointing mode. Fundamentally, either approach reduces the energy gain of the target and thus the signal-to-noise ratio.

DBF-SCORE technology is a wide-swath imaging solution based on multi-channel, which is first equipped with a small-aperture transmitting antenna size according to the correspondence between beam width and antenna size, and has multiple small-aperture receiving antennas at the receiving end to meet the transmission and reception coverage of wide-format scenes. The signal quality is completed by the scanning reception of the receiving end, that is, SCORE, and the theoretical gain improvement is obtained through channel-to-channel phase modulation synthesis. Figure 3.1 is a schematic diagram of DBF-SCORE technology and echo ramp window as range profile. According to the principle of synthetic aperture radar, the detection width under the fixed beam direction is determined by the beam main lobe width, the lower viewing angle and the platform height, such as Eq. (3.12). Limited by factors such as platform power supply capability and component efficiency, the radiation power output of radar is limited. In order to improve the spatial coverage capability, whether it is increasing the beam width, increasing the center range, and increasing the viewing angle, it will greatly affect the reduction of echo power and affect the signal quality.

$$\theta_r \approx \frac{\lambda}{H_r} \tag{3.12}$$

where θ_r is the beam width in the range direction; λ is the wavelength; H_r is the aperture of the transmitting antenna range. θ_r The antenna beam width is proportional to the wavelength λ and H_r inversely proportional to the antenna size. so in order to achieve the same beam width, the shorter the wavelength, the smaller the antenna size required [2].

In the 4-channel DBF system shown in Fig. 3.1, d the height for one antenna channel, the ideal echo direction for fast time change $\theta(t)$, the actual direction of the front, and the difference between the normal angle of the β echo and the antenna, that is, the scanning angle, which $\alpha(t) = \beta - \theta(t)$ $R(t)$ represents the range; In Fig. 3.1b, set the width of the mapping belt along the range W_r; proximal echo range R_n; distal range R_f; R_e is the radius of the earth. Due to the spatial position difference of the phase center between channels, there is naturally a phase difference between echoes, and when the ideal compensation is made, the signals between channels can be accumulated in phase, further improving the gain and improving the signal-to-noise ratio. The figure below shows the equivalent antenna gain effect obtained by the simulation of multichannel synthesis, and DBF-SCORE produces a swept narrow receive beam.

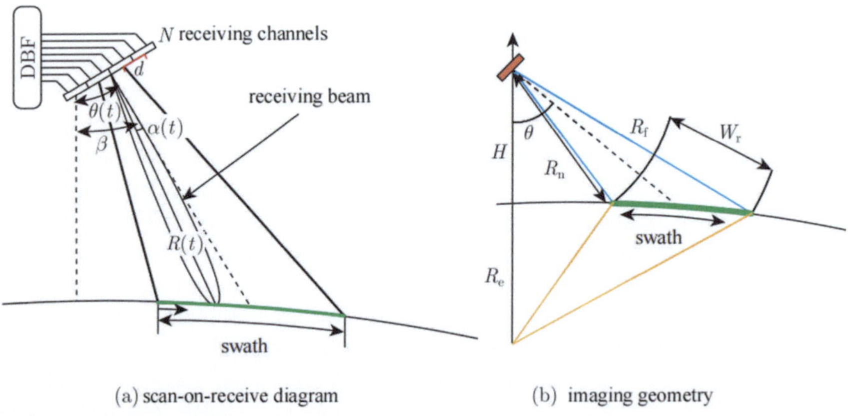

Fig. 3.1 Schematic diagram of SCORE

Ignoring the slow time, only the range time is analyzed t, the echo signal of the first n aperture is $s_n(t)$, the SCORE weight coefficient is, $\omega_n(t)$ and the synthesized signal is the vector sum of each signal (3.13).

$$s(t) = \sum_{n=1}^{N} \omega_n(t) \cdot s_n(t) \tag{3.13}$$

At t the moment, the beam direction is the angle between the center of the receiving beam and the antenna normal, where the angle between the antenna front direction and the bottom view direction is $\alpha(t) = \theta(t) - \beta$, and the receiving beam direction corresponding to the fast time is the theoretical wave arrival angle direction (3.14). β

$$\theta(t) = \arccos\left(\frac{4(H + R_e)^2 - 4R_e^2 + (ct)^2}{4(H + R_e)ct}\right) \tag{3.14}$$

It expresses the assumption that the echo reach angle varies with time and is only related to platform elevation when the ideal earth is flat (no elevation undulation on the surface). The schematic diagram of the time-varying wave arrival angle is Fig. 3.2, that is, the wave arrival angle has an approximately linear relationship with time.

The weighting coefficient of phase difference caused by the propagation process of different channels is $\omega_n(t)$ Eq. (3.15).

$$\omega_n(t) = \exp\{-j2\pi d_n \sin(\alpha(t))/\lambda\} \tag{3.15}$$

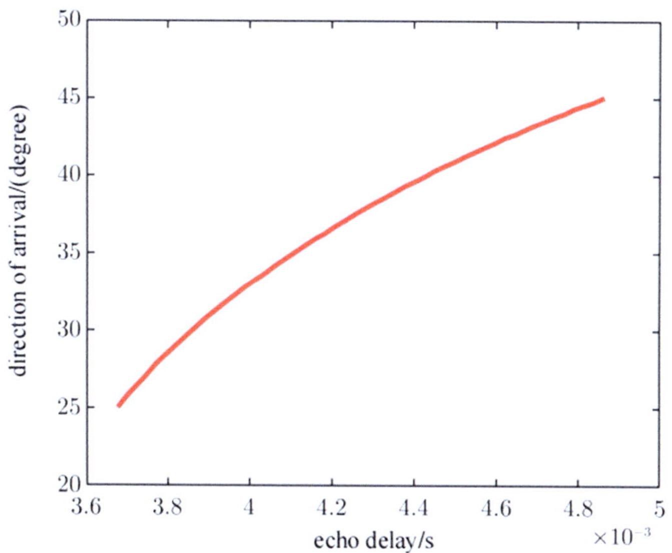

Fig. 3.2 Time-varying wave arrival angle

The antenna range direction height d is, which is the range of the first d_n receiving channel relative to the center of the antenna, expressed as Eq. (3.16). n

$$d_n = \left(n - \frac{N+1}{2}\right) \cdot d, \quad n = 1, \ldots, N \tag{3.16}$$

The weighted sum is equivalent to forming a large front synthesized by four small fronts on the pattern map, and has the characteristics of fast time variation, scanning the entire mapping belt within the repeat cycle, tracking and receiving ground echoes. In the DBF process, the echo signals of each independently sampled channel are coherently accumulated, while the random noise expectation remains unchanged, and the theoretical signal-to-noise ratio can be improved $10\log_{10}(N)$. The equivalent beamwidth improvement effect is shown in Fig. 3.3.

At the same time, under the wide mapping band, when the cross-cycle power enters the preset echo, the system performance will also be affected, and the range ambiguity (RASR) is serious. According to the range-time correspondence to analyze the echo timing relationship, the echo window width T_r needs to satisfy the Eq. (3.17).

$$T_r \approx T_p + \frac{2W_r}{c} = T_p + \frac{2R_f}{c} - \frac{2R_n}{c} \leq \frac{1}{2 \cdot PRF} \tag{3.17}$$

Only when the echo is sampled is the window large enough to fully receive the echo from the far end of the mapping band. According to the previous analysis of synthetic aperture resolution, the azimuth resolution of stripe mode in Eq. (3.18) ρ_a is greater than half of the antenna aperture

$$\rho_a \geq \frac{L_a}{2} \tag{3.18}$$

where L_a is the azimuth length of the antenna. The azimuth sampling criterion requires that the full effective resolution be obtained in order to achieve no aliasing of azimuth frequencies [3], and Eq. (3.19) constrains the pulse repetition rate

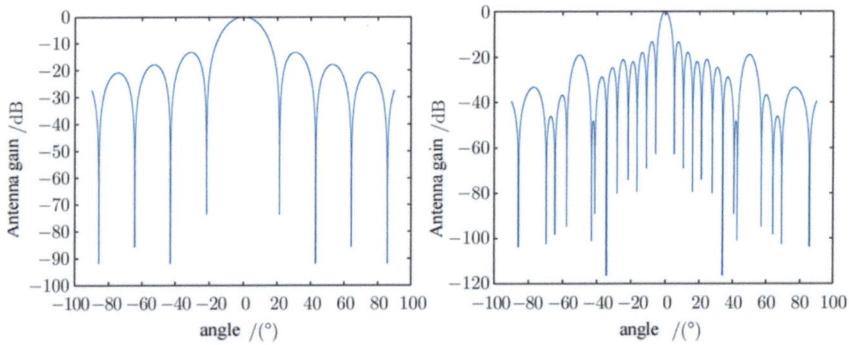

(a) Single-channel receive beams (b) Multi-channel equivalent receive beams

Fig. 3.3 Comparison of equivalent beamwidths

$$PRF \geq \frac{v}{\rho_a} \geq \frac{2v}{L_a} \tag{3.19}$$

v For radar platform speed, PRF exists in the lower bound. The maximum oblique width of the mapping band W_r is the formula (3.20).

$$W_r < c\frac{\rho_a}{2v} - c\frac{T_p}{2} \tag{3.20}$$

The contradiction between azimuth resolution, mapping bandwidth and launch width is obtained: under the same orbital height and launch width, the platform has a fixed speed, and the ideal azimuth resolution requires limited range mapping bandwidth. The radar equation also shows that sacrificing transmit width reduces echo power. Therefore, the single-transmit and single-receive system has inherent contradictions between mapping bandwidth, signal-to-noise ratio and resolution, while the DBF-SCORE system can avoid high transmit power and reduce system cost through gain improvement. As shown in Fig. 3.4, multi-channel reception can significantly reduce range ambiguity.

The specific DBF-SCORE implementation process can be shown in Fig. 3.5.

That is, range compression is carried out first, then fast time coefficient compensation, and finally time domain superposition is carried out.

Under the single-channel original echo signal-to-noise ratio of 20 dB, the signal-to-noise ratio improvement of pulse compression signal before and after multi-channel synthesis is shown in Fig. 3.6, and the signal-to-noise ratio improvement value is close to 6 dB, which is in line with the theoretical calculation.

DBF-SCORE has a serious gain drop in practical application, and the specific factors include channel amplitude and phase error and wave arrival angle estimation error affected by radar platform movement and ground elevation fluctuation [4]. In fact, when the ground has high undulation, the wave arrival angle estimation obtained by the ideal flat earth geometry has a non-negligible error, and the direction pointed

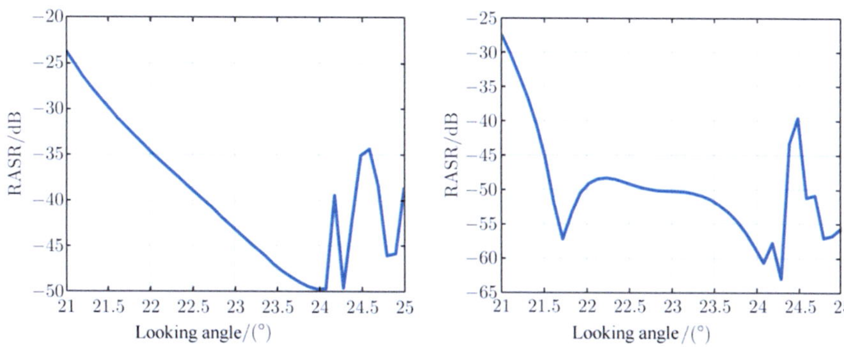

(a) single channel RASR (b) DBF equivalent RASR

Fig. 3.4 Ambiguity comparison of single-channel and multi-channel range

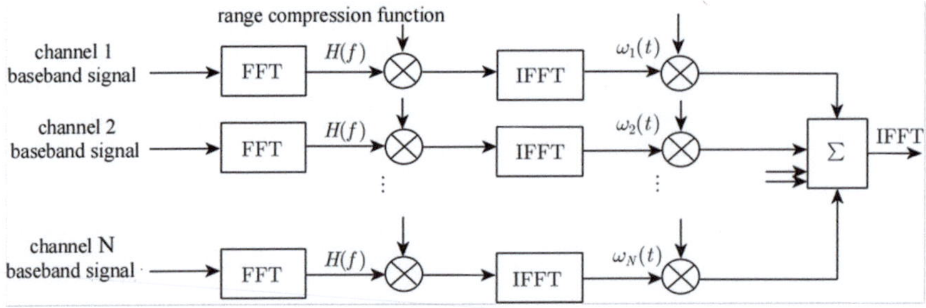

Fig. 3.5 DBF-SCORE implementation flow

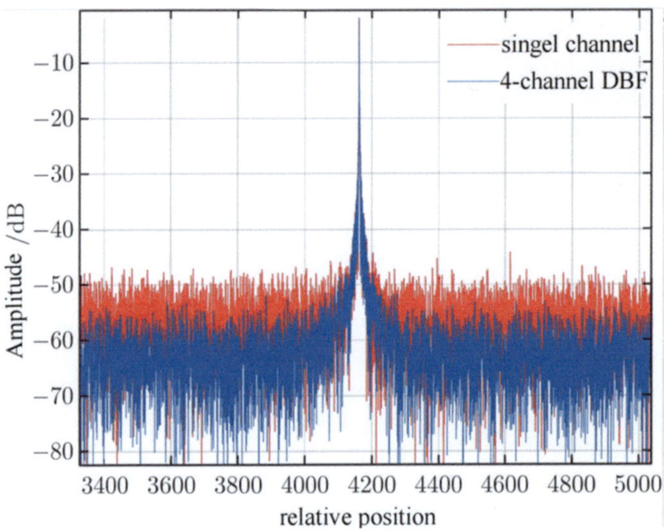

Fig. 3.6 Signal-to-noise ratio improvement

by the definition SCORE method is defined as, $\theta_s(t)$ and the target echo direction h with the true elevation is θ_0, then the angle estimation error is

$$\Delta\theta = \theta_0 - \theta_s(t_0) \tag{3.21}$$

thereinto t_0 Due to the true two-way delay of the echo, the echo is not received by the maximum gain synthesized by the multichannel according to the ideal state, but because the gain loss is caused by the terrain. G_0 For the equivalent pattern of the receiving antenna, define the loss amount as

$$G_1 = \frac{G_0(\theta_s + \Delta\theta)}{G_0(\theta_s)} \tag{3.22}$$

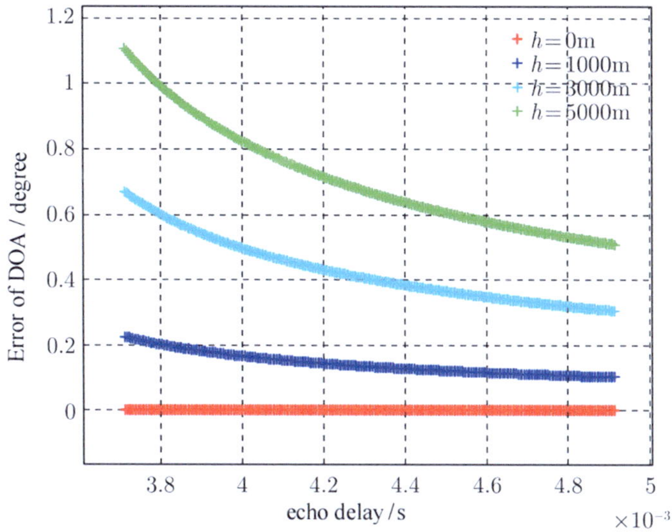

Fig. 3.7 Influence of terrain on beamforming

The terrain error for the wave arrival angle estimation is shown in Fig. 3.7.

According to the SCORE principle, the receiving pattern should be consistent with the direction of the wave arrival angle, and this value is only a fixed value set a priori in the traditional method, but the complex factors of the actual observation scene and radar platform cannot guarantee the stability of the wave arrival angle, so it is necessary to apply the adaptive method to improve the wave arrival angle parameters. One potential solution is to estimate the wave reach angle using actual echoes. The following applies a spectral estimation method under a narrow band, assuming that the echo slows down in azimuth upward wave reach angle (at the synthetic aperture of 1/16 and below scale, the wave arrival angle is considered constant at the same range gate), and the range direction has the property of changing with fast time. Set the SAR range echo model to

$$s(t) = rect\left(\frac{t}{T}\right)\cos\left[j2\pi\left(f_c t + \frac{k_r}{2}t^2\right)\right] \tag{3.23}$$

Assume that the two-way echo delay of the reference channel is $\tau_0 = 2R_0/c$, and the corresponding echo delay of each channel is

$$\tau_k = \tau_0 + \frac{dk\sin(\beta - \theta)}{c} \tag{3.24}$$

The echo of each channel is

$$r_k(t) = \alpha rect\left(\frac{t-\tau_k}{T}\right)\exp\left\{j\pi(t-\tau_k)^2\right\}\exp\{-j\pi f_c \tau_k\} \tag{3.25}$$

It is obtained after pulse compression

$$r_k(t) = \alpha \sin c\{kT(t - \tau_0)\} \exp\{-j2\pi f_c \tau_k\} \quad (3.26)$$

After simplification, the response envelope and fixed phase are ignored and obtained

$$\begin{aligned} \boldsymbol{r} &= \alpha \exp\left\{-j2\pi \frac{d}{\lambda} \sin(\beta - \theta)k\right\} \\ &= \boldsymbol{r} \cdot \boldsymbol{\alpha}(\theta) + \boldsymbol{v} \end{aligned} \quad (3.27)$$

thereinto $\boldsymbol{r}, \boldsymbol{\alpha}, \boldsymbol{v}$ is an N-dimensional complex vector. In the spaceborne flight system, the actual elapsed time of 500 PRTs is less than half a second, and it can be considered that the satellite platform remains in a relatively stable state. Assuming that the valid number of actual snapshots is NL, the matrix is expanded to a two-dimensional matrix of N*NL

$$\boldsymbol{r}(n) = \alpha \cdot \boldsymbol{\alpha}(\theta, n) + \boldsymbol{v}(n) \quad (3.28)$$

The target echo consists of a non-single scattering backscatter source

$$\boldsymbol{r}(n) = \sum_{i=1}^{N_s} \alpha_i \cdot \boldsymbol{\alpha}(\theta_i, n) + \boldsymbol{v}(n) \quad (3.29)$$

Therefore, the spatial spectral analysis based on the narrow band was carried out to complete the wave arrival angle estimation. Refactoring the above equation is

$$\boldsymbol{r}(n) = \sum_{i=1}^{N_s} \sqrt{\tau_i} \cdot \boldsymbol{\alpha}(\theta_i) \otimes \boldsymbol{x}_i(n) + \boldsymbol{v}(n), n = 1, \cdots, NL \quad (3.30)$$

thereinto, NL is independent vision, $\boldsymbol{v}$ It is thermal noise and has zero mean power σ_V^2, Ns is the number of stacked scattering sources, $\mathbf{a}(\varphi_i)$ for space beam pointing vectors, φ_i For interferometry phase—unknown, $\boldsymbol{x}_i(n)$ is a complex envelope stochastic Gaussian vector. Obtain the covariance matrix of multi-channel data

$$R_r = E\left[\boldsymbol{r}\boldsymbol{r}^H\right] \quad (3.31)$$

Bring in to get

$$\begin{aligned} R_x &= E\left[\left(\sum_{i=1}^{N_s} \sqrt{\tau_i} \cdot \boldsymbol{\alpha}(\theta_i) \otimes \mathbf{x}_i(n)\right)\left(\sum_{i=1}^{N_s} \sqrt{\tau_i} \cdot \boldsymbol{\alpha}(\theta_i) \otimes \boldsymbol{x}_i(n)\right)^H\right] \\ &= AE\left[SS^H\right]A^H + E\left[NwNw^H\right] \\ &= AR_sA^H + R_{Nw} \end{aligned} \quad (3.32)$$

thereinto, The items are stated as follows

$$
\begin{aligned}
A &= [\boldsymbol{\alpha}(\theta_1), \boldsymbol{\alpha}(\theta_2), \ldots, \boldsymbol{\alpha}(\theta_{N_s})]^T \\
S &= [\sqrt{\tau_1} \cdot \boldsymbol{x}_1(n), \sqrt{\tau_2} \cdot \boldsymbol{x}_2(n), \ldots, \sqrt{\tau_{N_s}} \cdot \boldsymbol{x}_{N_s}(n)]^T \\
X &= [\boldsymbol{r}_1(1), \boldsymbol{r}_2(2), \ldots, \boldsymbol{r}_{\mathrm{N}}(n)]^T \\
Nw &= [n_1(1), n_2(2), \ldots, n_{\mathrm{N}}(n)]^T
\end{aligned}
\tag{3.33}
$$

And the signal correlation $R_s = E\left[SS^H\right]$ matrix, the noise matrix $R_{Nw} = E\left[NwNw^H\right]$. The feature decomposition of the covariance matrix is carried out, and then the eigenvalues are arranged from small to large, and the result is

$$\lambda_1 \geq \lambda_2 \geq \ldots \geq \lambda_N \geq 0 \tag{3.34}$$

Among them, the signal and noise correspond to the large and small eigenvalues, respectively λ_i, and the corresponding eigenvectors v_i

$$R_x v_i = \lambda_i v_i \tag{3.35}$$

Can get

$$
\begin{aligned}
AR_sA^H v_i &= 0 \\
R_s^{-1}\left(A^H A\right)^{-1} A^H A R_s A^H v_i &- 0
\end{aligned}
\tag{3.36}
$$

Define the noise matrix $E_n = \left[v_{N_s+1}, v_{N_s+2}, \ldots, v_N\right]$ with a spatial spectrum of

$$P(\theta) = \frac{1}{\mathbf{a}^H(\theta) E_n E_n^H E_n} \tag{3.37}$$

The maximum value is obtained when the noise matrix space is orthogonal to the wave reach angle function, but since the noise is always present, there is a peak in the spectrum in the search space. Table 3.1 shows the parameters of the simulation experiment.

The test simulation is performed in the setting wave direction of 7° and the actual wave arrival angle direction is 10.155°. The spectral search space is 4.5° to 11.5°. The search result is 10.1553°, the search error is 0.0003°, and Fig. 3.8 is an angle estimation plot.

Analyzing the synthetic gain under the angle estimation error, as shown in Fig. 3.9, it can be found that the intensity of the synthesized signal pulse after compression can be greatly improved by angle correction under the existing error, and the gain improvement reaches 4.5 dB.

Table 3.1 Parameter settings for simulation experiments

Project	Parameter	Remark
Track height	550 km	The geocentric range is constant
Peak transmit power	13000 W	
Duty cycle	7.9%	
Minimum look angle angle	4°	
Maximum look angle	12°	
FM pulse width	20 us	
Carrier frequency bandwidth	200 MHz	
Antenna range direction size	0.3 m	

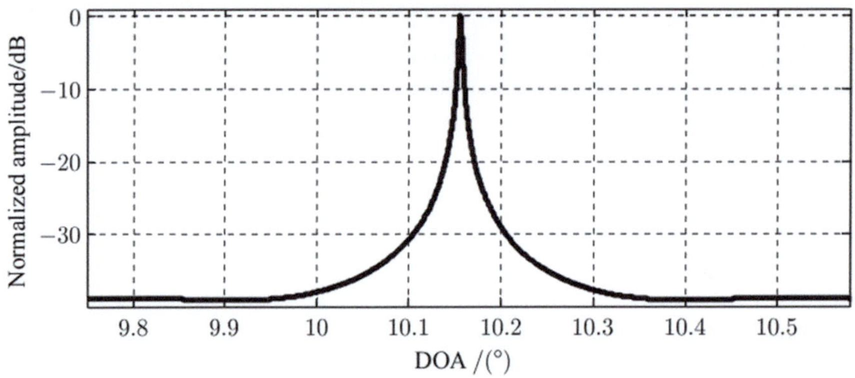

Fig. 3.8 Wave angle estimation results

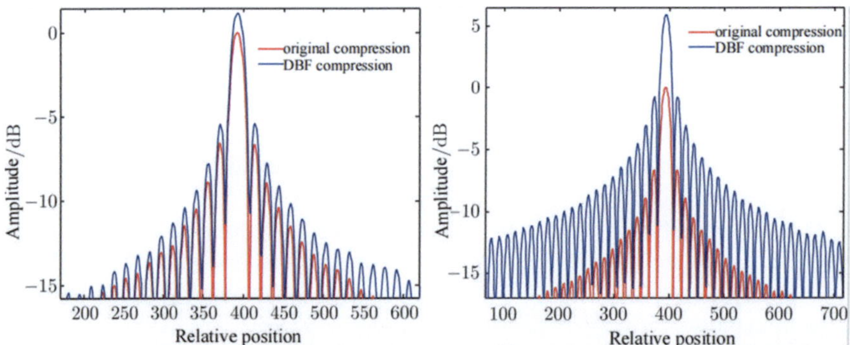

(a) Gain improvement with DOA error (b) Gain improvement with DOA estimation

Fig. 3.9 Angle estimation effect on synthetic gain

3.3 Channel Balanced DBF Processing Method

The relationship between the echo time variable t and $\theta(t)$, $\alpha(t)$ is one-to-one. The weighted summation is equivalent to forming a time-varying high-gain narrow beam that sweeps the entire mapping band in a pulse repetition period, tracking the received ground echo. Let's take two channels as examples to verify the effect of weighting coefficient compensation. As shown in Fig. 3.10, take one channel as the reference channel, observe the interferometry phase of the signal received by the other channel and the data received by the reference channel in the range frequency domain (as shown in Fig. 3.10a), compensate the weighting coefficient, and observe the interferometry phase again (as shown in Fig. 3.10b). In order to clearly reflect the compensation effect, a profile view is made along the range (as shown in Fig. 3.10c, d).

As can be seen from the figure, the original phase difference between the selected channel and the reference channel has been roughly compensated by the weighting coefficient compensation, and there are other influencing factors besides this error. Therefore, the following non-reference channel also needs to be channel equalization.

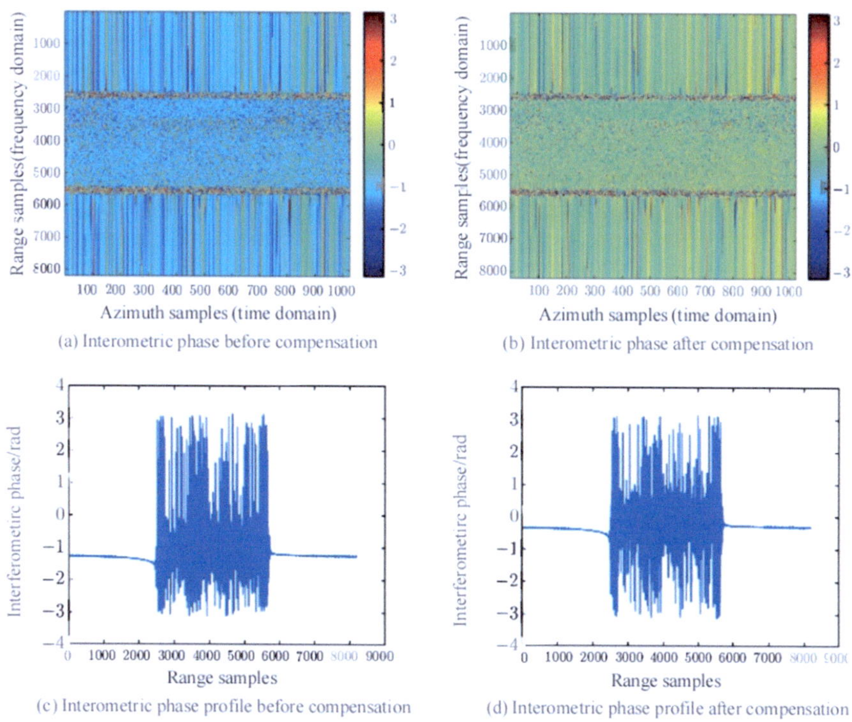

Fig. 3.10 Weighting coefficient compensation effect

1. **SAR image description with channel imbalance**

(1) **Difference in beam pattern**

In the actual SAR system, the beam pattern sometimes changes with the position of the radar (that is, the slow time domain), and sometimes the beam pattern changes between different transmitted pulses due to the internal hardware circuit affected by temperature and some natural factors. For single-channel SAR systems, the distortion of the beam pattern will lead to the point spread function or image point response spatial, while for the multi-channel SAR-GMTI, we are more concerned about the difference in the beam pattern of each channel, because this difference will cause registration errors and reduce the coherence between complex images [5]. The antenna error is a (ψ, φ) multiplicative error in the plane, where ψ the azimuth cone angle is denoted and φ the high and low angles are debited. When given the speed of the carrier aircraft along the track, the azimuth Doppler corresponds to the azimuth cone angle of the antenna system one-to-one, and similarly, when the carrier aircraft is given the flight altitude, the range gate corresponding to the range fast time corresponds to the high and low angles. Therefore, when there is an array error, the first m antenna receive pattern can be modeled as:

$$\widetilde{H}_m^A(\tau, f_a) \approx H_m^A(\tau, f_a)(1 + \delta_m^A(\tau, f_a)) \exp(j\phi_m^A(\tau, f_a)) \tag{3.38}$$

where $\delta_m^A(\tau, f_a)$ and $\phi_m^A(\tau, f_a)$ are expressed as (τ, f_a) amplitude and phase errors in the plane, respectively.

(2) **Frequency response error**

Another source of channel error is inconsistencies in the frequency response of the individual receive channels. The receive channel is usually composed of link links such as RF, IF, baseband, sample-and-hold and A/D devices, and if any of the loops in the link links of each channel are inconsistent, it will cause channel mismatch and reduce coherence. Considering the time-varying nature of the channel frequency response, we use echo data (or SAR complex images) as the correction source instead of the known test signal. At the same time, since each link of the receiving channel only plays the role of frequency transmission of the baseband echo signal, directly using the echo signal as the correction source can compensate for the frequency response inconsistency in all links of each receiving channel. Remember m that the time-varying frequency transfer function of the first channel is, and the disturbance time-varying frequency transfer function based on the first-order zero-pole error model can be expressed as: $H_m^T(f_r, t)$

$$\widetilde{H}_m^T(f_r, t) = H_m^T(f_r)(1 + \Delta H_m^T(f_r, t)) \tag{3.39}$$

thereinto

$$H_m^T(f_r) = \frac{P(f_r)}{Q(f_r)} = \frac{(j2\pi f_r - s_1^p)\cdots(j2\pi f_r - s_l^p)}{(j2\pi f_r - s_1^q)\cdots(j2\pi f_r - s_n^q)} \tag{3.40}$$

$$\Delta H_m^T(f_r, t) = \sum_{k=1}^{n} \frac{\Delta s_k^q(t)}{j2\pi f_r - s_k^q} - \sum_{k=1}^{l} \frac{\Delta s_k^p(t)}{j2\pi f_r - s_k^p} \tag{3.41}$$

where, the $s_1^q, s_2^q \cdots s_n^q$ Russ transform $s_1^p, s_2^p \cdots s_l^p$ root of the function $Q(f_r)$ sum, and $P(f_r)$ the perturbation term of $\Delta s_k^q(t)$ the root $\Delta s_k^p(t)$ sum $s_k^q s_k^p$, respectively. The error perturbation term $\Delta H_m^T(f_r, t)$ can be modeled as $\delta_m^T(f_r, t) \exp(j\phi_m^T(f_r, t))$ the $\delta_m^T(f_r, t)$ $\phi_m^T(f_r, t)$ perturbation terms here and amplitude and phase, respectively. We get:

$$\begin{aligned} \widetilde{H}_m^T(f_r, t) &\approx H_m^T(f_r)(1 + \delta_m^T(f_r, t) + j\delta_m^T(f_r, t)\phi_m^T(f_r, t) + j\phi_m^T(f_r, t)) \\ &\approx H_m^T(f_r)(1 + \delta_m^T(f_r, t)) \exp(j\phi_m^T(f_r, t)) \end{aligned} \tag{3.42}$$

As can be seen from the above equation, the channel frequency response error is in two dimensions (f_r, t) The plane is the multiplicative error.

2. **Two-dimensional amplitude phase filtering**

SAR complex images are perfectly coherent or have high coherence to meet the following three conditions: (1) for two images obtained from two different channels (e.g., channels 1 and 2), the ground scattering distribution must be basically the same, that is, satisfied $a_1(x, y) = a_2(x, y)$; (2) The registration accuracy of the two images must be quite high; (3) Two complex images must have the same impulse response function, that is $\tilde{h}_1(x, y) = \tilde{h}_2(x, y)$. The first condition requires that the SAR image pair have the same radar parameters and observation geometry, which is easily met for a single-base system. The second condition requires a good registration method or prior knowledge of geometric configuration, and through motion error analysis and compensation, the two SAR images have high registration accuracy, which basically meets the second condition. The third condition requires identical antenna patterns, channel frequency response, and SAR focusing algorithms [6]. Since the imaging algorithm of multi-channel SAR data is often completed on the same algorithm processor, here we ignore the difference in the response of the algorithm processor. To meet the third condition, amplitude and phase correction must be done in two different SAR image domains, namely the range time domain, Doppler in the time domain, and the time domain in the range frequency azimuth time.

The phase error is a multiplicative error, and we can use the interferometry phase planar filtering technique to correct the impulse response difference. The most commonly used phase filters are complex mean filtering and rotary mean or median filtering. These filters are not adaptive to local noise level changes. For places with high noise power, it is necessary to ensure that enough noise is filtered out to achieve the smoothing effect of phase correction, and for places with low noise power, only a small amount of noise can be filtered out to save the moving target information. The amplitude noise of SAR images can be modeled as additive models rather than multiplicative models, and intensity and noise can be regarded as two correlated and coherent interferometry processes in interferometric SAR images. Therefore, adaptive noise filtering should consider coherence. This chapter uses adaptive weighted rotary mean filtering to complete noise reduction before phase correction, which uses a fixed window and fully considers the periodicity of phase, which is an improvement on the rotary mean filter. The weight of the filter changes according to the coherent information in the interferogram, and the specific design of the filter is as follows:

$$\hat{\varphi}(k,l) = \frac{1}{M \times N} \sum_{k=-(M-1)/2}^{(M-1)/2} \sum_{l=-(N-1)/2}^{(N-1)/2} w(k,l) \cdot \arg(\frac{S'(k,l)}{S_{sum}}) + \arg(S_{sum}) \quad (3.43)$$

thereinto

$$S_{sum} = \sum_{k=-(M-1)/2}^{(M-1)/2} \sum_{l=-(N-1)/2}^{(N-1)/2} w(k,l) \cdot S'(k,l) \quad (3.44)$$

$$S'(k,l) = \frac{S(k,l)}{|S(k,l)|} \quad (3.45)$$

$$w(k,l) = \frac{\frac{1}{\sigma_{\varphi}^2(k,l)}}{\frac{1}{M \times N} \sum_{k=-(M-1)/2}^{(M-1)/2} \sum_{l=-(N-1)/2}^{(N-1)/2} \frac{1}{\sigma_{\varphi}^2(k,l)}} \quad (3.46)$$

where $S(k,l)$ the (τ, f_a) interferometric SAR complex image representing the or domain is the (f_r, t) $\sigma_{\varphi}(k,l)$ standard deviation of noise and is a function of the correlation coefficient. The weight corresponding to the pixel can be calculated from the local mean variance and variance $S(k,l)$, that is $w(k,l)$.

In the two-dimensional frequency domain and the range time domain azimuth frequency domain, the m-channel and the reference channel are filtered out of noise, and the interferometry phase is obtained by coherent processing, and then the original data of the m-channel is compensated. In this way, the coherence of the signal is improved, so that the DBF accumulation effect is better, and the noise is not processed.

For amplitude correction, averaging filtering on the real plane can be used, and the size of the filter window is the same as the size of the phase correction, which can be expressed as

$$AMP(k,l) = \frac{1}{M \times N} \sum_{k=-(M-1)/2}^{(M-1)/2} \sum_{l=-(N-1)/2}^{(N-1)/2} w(k,l) \cdot |S(k,l)| \quad (3.47)$$

The weight factor is the $w(k,l)$ same as the phase filtering.

As in the previous section, a channel is also selected as the reference channel to verify the two-dimensional amplitude phase filtering with the data from another channel. As shown in Figs. 3.11 and 3.12, after compensating for the weighting coefficients, there are still some detailed differences in amplitude and phase between

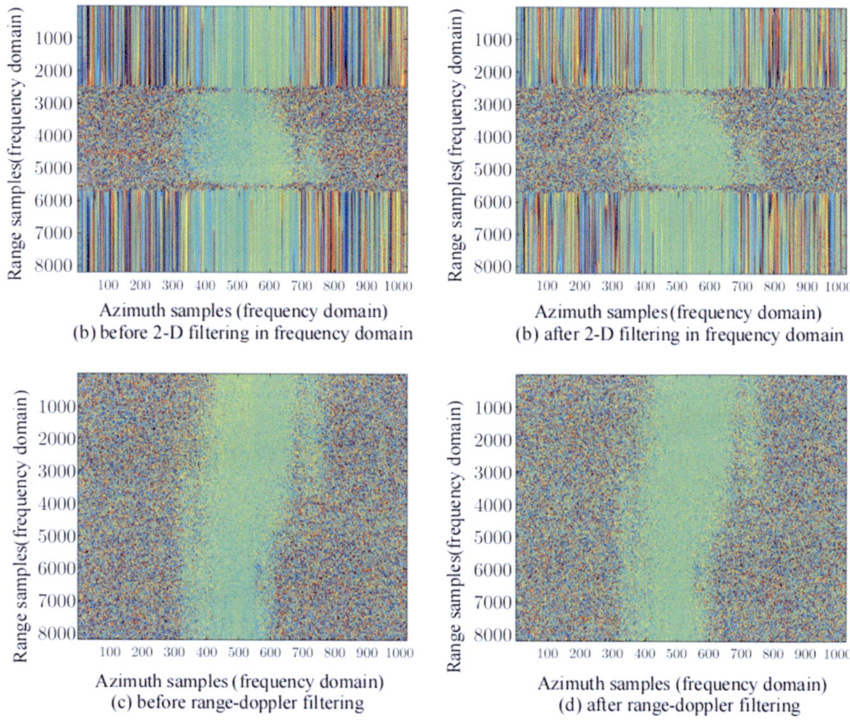

Fig. 3.11 Phase comparison before and after filtering

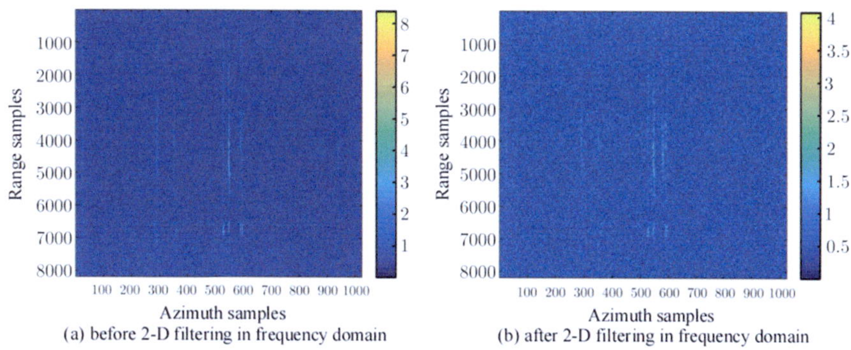

Fig. 3.12 Amplitude comparison before and after filtering

the selected channel and the reference channel, and most of these differences are compensated for after two-dimensional amplitude phase filtering.

Combined with the channel equalization step, the improved DBF processing model is shown in Fig. 3.13.

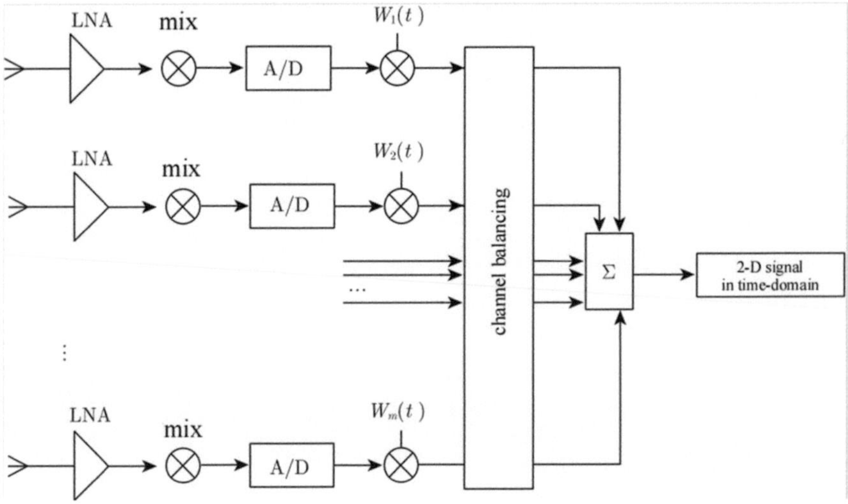

Fig. 3.13 Range to DBF processing model combined with channel equalization

If you want to obtain the direction diagram of the normal direction of the array antenna, add the signals of each array element directly, that is, the signal envelope vector and the unit vector $\mathbf{1}(=[1,\cdots,1]^T)$ are normalized as the dot product to obtain:

$$\begin{aligned} G(\alpha) &= \frac{1}{N}|(\mathbf{s}_r(t)\cdot\mathbf{1})| \\ &= \frac{1}{N}\left[1+e^{j\frac{2\pi}{\lambda}d\sin\alpha(t)}+\cdots+e^{j\frac{2\pi}{\lambda}(N-1)d\sin\alpha(t)}\right] = \frac{\sin(\frac{\pi Nd}{\lambda}\sin\alpha(t))}{N\sin(\frac{\pi d}{\lambda}\sin\alpha(t))} \end{aligned} \tag{3.48}$$

If you want the direction in which the beam is pointed θ_d, you should change the vector as a weight **1** vector to:

$$\mathbf{a}(\theta_d)=\left[1,e^{-j\frac{2\pi}{\lambda}d\sin\theta_d},\cdots,e^{-j\frac{2\pi}{\lambda}(N-1)d\sin\theta_d}\right]^T \tag{3.49}$$

At this point, the entire DBF processing work is completed, and the traditional SAR imaging processing is carried out to obtain SAR images with improved signal-to-noise ratio. Figure 3.14 shows a complete process flowchart, in which the post-processing part is not a necessary operation step.

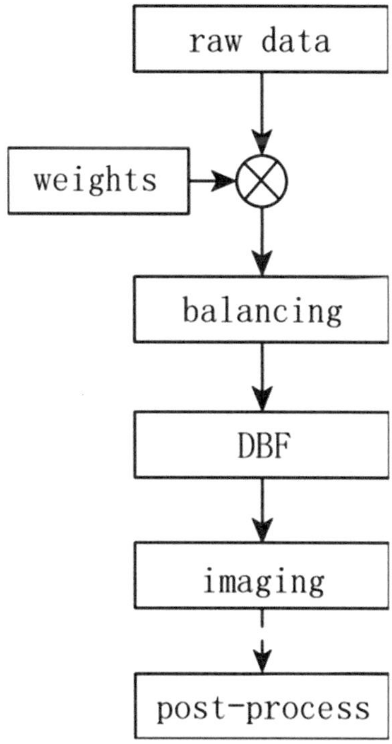

Fig. 3.14 Process flow chart

3.4 Airborne Millimeter Wave DBF SAR Imaging

Under the non-ideal condition of multi-channel imaging, the motion state, ground scene, signal gain, and time–frequency characteristics of the radar platform are not fully known, and it is necessary to carry out imaging processing methods with robustness and high gain.

The proposed multi-channel synthetic imaging process based on echo data and inertial guide data is shown in Fig. 3.15, starting from the original data after unpacking, the digital processing links include three parts: preprocessing, DBF synthesis, and single-channel processing: 1. Quadrature demodulation, half-band filtering, decimation and pulse compression after preprocessing AD sampling; Second, single-channel processing takes the reference channel as the data input, and according to the compensation requirements of airborne data processing, the motion error fitting including Doppler center estimation and frequency modulation estimation is completed, and finally serves the envelope and phase compensation of multi-channel synthetic data focusing; 3. DBF synthesis uses the inertial guided data input in the timing to update the DBF-SCORE right, and compensates for the phase gradient between the residual channels for less than three times. Finally, the fixed amplitude correction between channels, channel synthesis and focusing processing are carried out in turn.

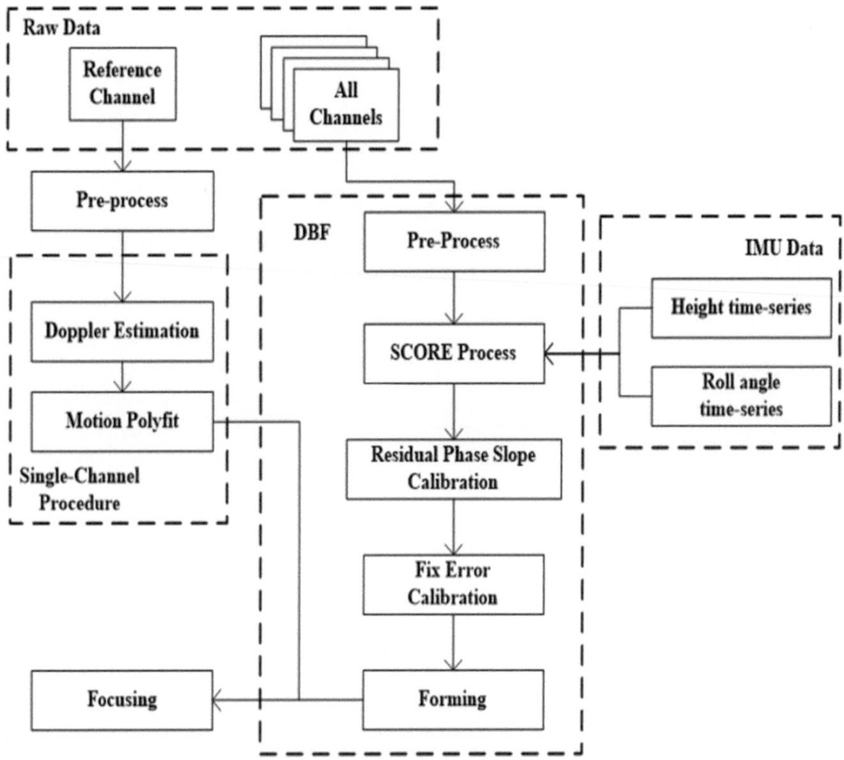

Fig. 3.15 DBF-SCORE measured data processing method

3.4.1 Single Channel Processing

The ideal SAR vehicle flight track is a straight line with a constant altitude and a uniform motion along the linear line in the antenna phase center. In practice, the position change of the antenna phase center will cause the range change, and the change of beam pointing in the roll direction will mainly cause the actual direction change of the antenna pattern, and then reduce the effective width under the fixed receiving gate, if there is a large pitch angle and three-dimensional position fluctuation at the same time, it may cause data "tearing": the same point target echo window jumps between the range gates, resulting in the azimuth dimension can not obtain effective accumulation, difficult to focus [7]. As shown in Fig. 3.16, the two-dimensional compressed image in the time domain can analyze the beam direction situation through the image defocusing situation, and the main characteristics are: (1) the intensity is uneven along the longitudinal (azimuth); (2) The point target is torn in multiple directions along the range gate defocus. Depending on the time–frequency correspondence, similar characteristics will be present in the range Doppler domain.

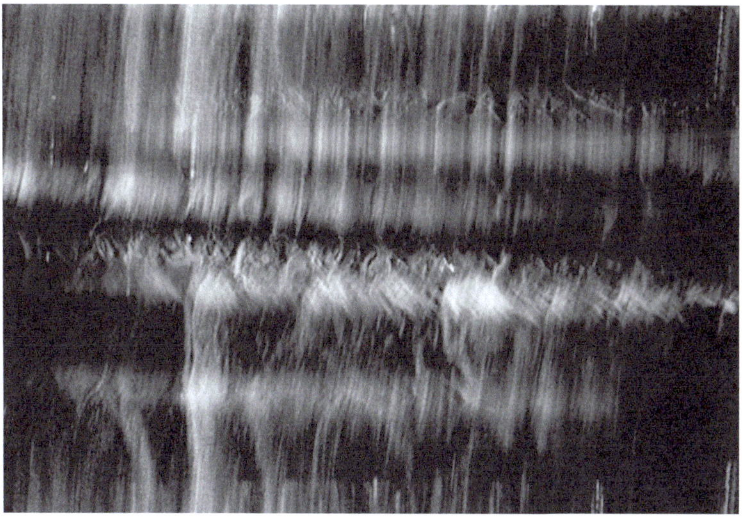

Fig. 3.16 Time domain focusing results of strong beam pointing error

The phenomenon of "tearing" in the azimuth frequency domain is reflected in the Doppler center not having stable focus, but spreading out in the sampling band or showing multiple peaks. As shown in Fig. 3.17, the azimuth frequency domain characteristics of the overall 65,536 frame echo can be clearly found that a strong energy of about 800 Hz wide appears in the effective frequency band of 1600 Hz, and if the imaging process is carried out according to a single Doppler frequency, the overall effective correction of the wide beam cannot be achieved, and it can only be processed by the sub-aperture method.

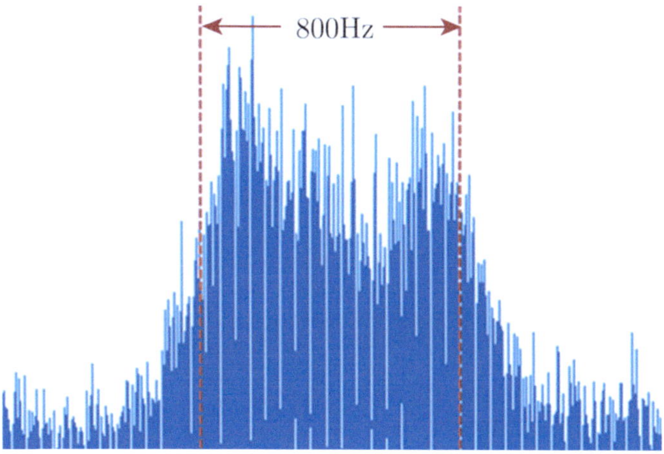

Fig. 3.17 Doppler center diffusion phenomenon

Doppler tuning frequency offset will cause phase error, which directly affects the broadening of the main lobe and the loss of gain after matching filtering. The phase of the matching filter can be approximated as Eq. (3.50)

$$\phi(t_{\mathrm{m}}) = \pi K_{\mathrm{a}}(t_{\mathrm{m}} - t_{\mathrm{c}})^2 \tag{3.50}$$

where is the t_{c} beam center crossing time, and the azimuth frequency is expressed as the Eq. (3.51)

$$K_{\mathrm{a}} = \frac{2v^2 \cos(\theta)}{\lambda R(t_{\mathrm{c}})} \tag{3.51}$$

The azimuth adjustment frequency error caused by speed mismatch is mainly composed of quadratic term (3.52)

$$\Delta\phi(t_{\mathrm{m}}) = \frac{4\pi v^2 \cos(\theta)}{\lambda R(t_{\mathrm{c}})}(t_{\mathrm{m}} - t_{\mathrm{c}})^2 \Delta v \tag{3.52}$$

Then the amount of secondary phase error at the edge of the synthetic aperture has the formula (3.53)

$$\begin{aligned} \Delta\phi\left(t_c + \frac{T_{\mathrm{a}}}{2}\right) &= \frac{4\pi v^2 \cos(\theta)}{\lambda R(t_{\mathrm{c}})} \Delta v \left(\frac{T_{\mathrm{a}}}{2}\right)^2 \\ &= \pi \Delta K_{\mathrm{a}} \left(\frac{T_{\mathrm{a}}}{2}\right)^2 \end{aligned} \tag{3.53}$$

By simulating the typical values of four atmospheric windows in the millimeter wave band [8], it can be found in Fig. 3.18 that the higher the frequency band with the same antenna length and the same speed error, the smaller the secondary phase error at the edge of the synthetic aperture.

In general, the speed and acceleration along the heading will cause the slope of frequency modulation and the high phase change, respectively. Radial velocity and acceleration cause changes in the Doppler center and FM slope, respectively. The processing flow is Fig. 3.19.

The data-based motion extraction method is used for measured data processing, and the single-channel focusing results are obtained as shown in Fig. 3.20, which estimates the central oblique viewing angle of 1.35° based on the parameters of echoes, and the fluctuation of the viewing angle during flight can be estimated to be 3.4° through the width of the Doppler main lobe, and the stability is high.

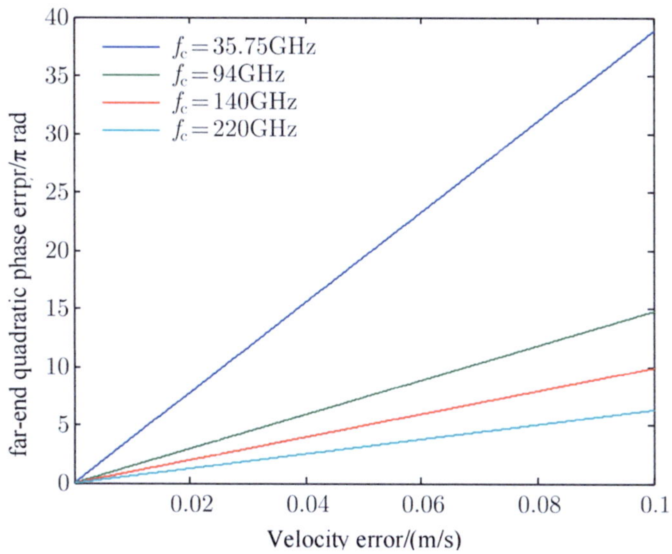

Fig. 3.18 Quadratic error versus frequency band

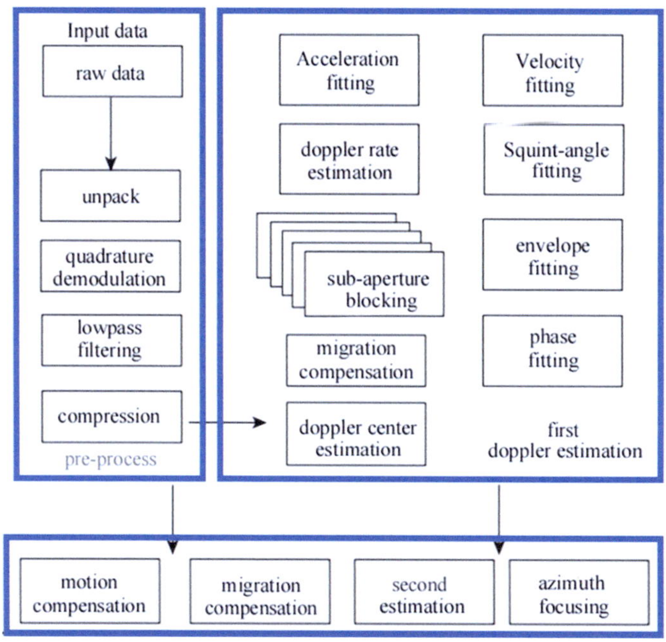

Fig. 3.19 Measured processing process of airborne data

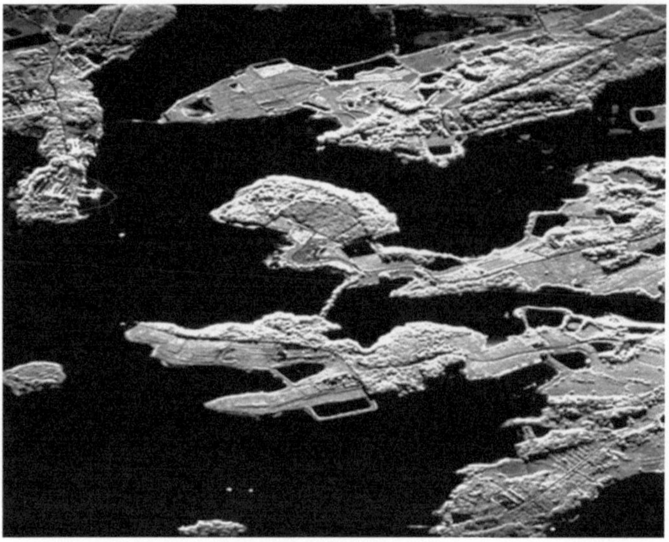

Fig. 3.20 Imaging results after motion compensation

3.4.2 Two-Dimensional Multi-stage Phase Gradient DBF-SCORE Synthesis

This section uses the airborne 4-channel measured data to analyze multi-channel calibration data, echo data, and inertial guidance data. The parameters of the airborne system are shown in Table 3.2.

In the calibration data analysis, the amplitude-frequency characteristics of multiple channels, the difference of peak intensity and peak point stability after

Table 3.2 Airborne system parameters

Parameter items	Content	Remark
Carrier	35.75 GHz	
Bandwidth	200 MHz	Chirp
Flight altitude	3 km	
Sampling frequency	550 MHz	
PRF	1600 Hz	
Emission pulse width	6 us	
Number of transmit channels	1	
The number of receive channels	4	
Mapping bandwidth	2 km	
Carrier	35.75 GHz	
Bandwidth	200 MHz	

pulse compression were observed. In the performance of the calibration data shown in Fig. 3.21, it can be found that the amplitude consistency between channels is good, and the amplitude difference relative to the reference channel is within 0.1dB; the phase stability is relatively stable, and the phase difference between channels is stable.

Observing the compression profile of the calibration data shown in Fig. 3.22, the peak sidelobe ratio between channels is consistent, that is, the secondary and high error terms are approximate, and the channels are more balanced.

Affected by airflow and driving, the airborne SAR platform cannot maintain a uniform linear motion, which in turn affects the process of synthetic aperture. The inertial guidance subsystem equipped with the airborne SAR system can provide platform attitude $\alpha_{\text{imu}}(t_{\text{a}})$, speed, $v_{\text{imu}}(t_{\text{a}})$ and altitude values in time series $H_{\text{imu}}(t_{\text{a}})$

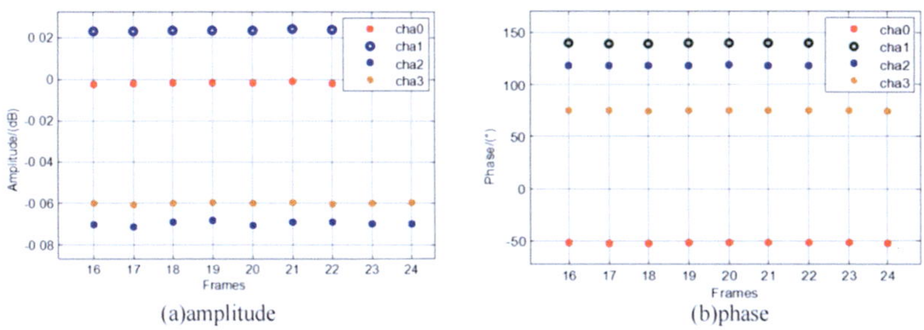

(a) Inter-channel amplitude characteristics (b) Inter-channel phase characteristics

Fig. 3.21 Calibration multi-frame data phase characteristics

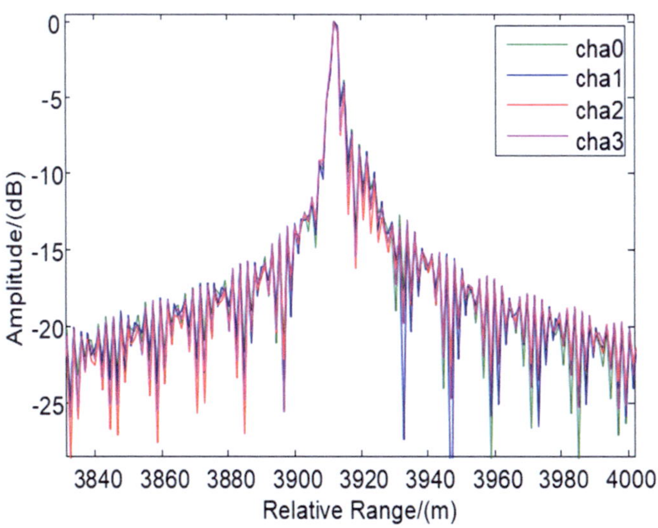

Fig. 3.22 Calibration data compression profile

to serve the imaging focus of SAR. In multi-channel synthesis, because the phase difference between channels is also modulated by the above typical parameters, the SCORE weight supplemented by inertial guided data $\omega_{\text{imu}}(t, t_a)$ is extended to a two-dimensional matrix of azimuth and range.

In the measured data analysis, the one-dimensional compressed scene special point is first used to analyze the target. First, due to the short mmWave multichannel baseline, the same point target remains within the same range gate, and because there is no down-orbit baseline, registration processing is not required; This analysis can be verified by the actual one-dimensional range image [9].

Analyzing the peak intensity and phase values of the specific display points, it is easy to find that there is a large amplitude gap between the various channels. Figure 3.23 shows that there is a gap of 1 order of magnitude in the case of unbalanced.

As shown in Fig. 3.24, there is a phase difference between the phases, which can be considered as a fixed phase difference.

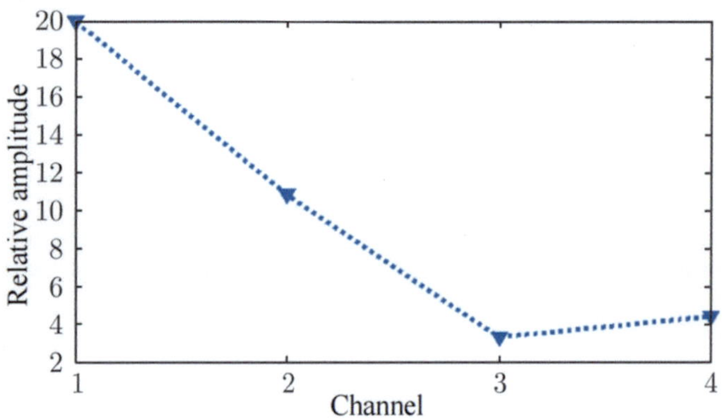

Fig. 3.23 Channel relative intensity

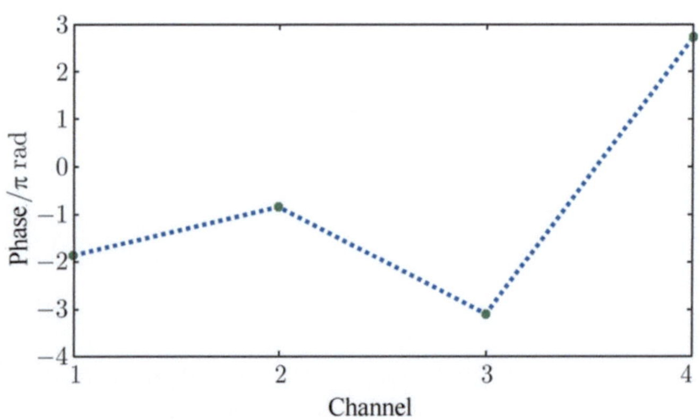

Fig. 3.24 Channel phase difference

In practice, for multi-channel echo data that accumulates hundreds of thousands of frames, the complexity of the signal and the richness of information will be more unique. The interferometry phase state of the two-dimensional data after range compression should be analyzed to obtain the signal difference linked to the baseline of the channel.

Set the transmitted antenna radiation signal to a wideband FM signal, as in Eq. (3.54)

$$s_t(t) = W\left(\frac{t}{T_{\mathrm{p}}}\right)\exp\left(j2\pi f_{\mathrm{c}}t + j\pi k_{\mathrm{r}}t^2\right) \tag{3.54}$$

The formula is the W transmit window, T_{p} which is the emission width, the f_{c} carrier frequency, and the k_{r} modulation frequency. According to the setting of one transmission and multiple reception, the baseband echo with the echo delay difference between channels after frequency conversion, filtering and decimation $s_n(t)$ is such as the formula (3.55)

$$s_n(t) = A \cdot W\left(\frac{t - t_0 - \Delta t_{\mathrm{n}}}{T}\right)\exp\left[-j2\pi f_{\mathrm{c}}(t_0 + \Delta t_{\mathrm{n}}) + j\pi k_{\mathrm{r}}(t - t_0 - \Delta t_{\mathrm{n}})^2\right] \tag{3.55}$$

where t_0 is the reference channel delay. Set up the antenna normal viewing angle β, and the delay difference is Δt_n obtained as Eq. (3.56)

$$\Delta t_{\mathrm{n}} = \frac{d_{\mathrm{n}}}{c}\sin\left\{\left[\beta - arc\cos\left(\frac{4(H + R_{\mathrm{e}})^2 - 4R_{\mathrm{e}}^2 + (ct)^2}{4(H + R_{\mathrm{e}})ct}\right)\right]\right\} \tag{3.56}$$

The angle between the echo and antenna normal is $\beta - arc\cos\left(4(H + R_{\mathrm{e}})^2 - 4R_{\mathrm{e}}^2 + (ct)^2/4(H + R_{\mathrm{e}})ct\right)$ unfolded by the reference channel to the central delay of the scene t_{c}, and the angle variable is obtained as in Eq. (3.57)

$$\theta(t) \approx \beta - \theta(t_{\mathrm{c}}) - \left.\frac{d\theta(t)}{dt}\right|_{t=t0}(t - t_{\mathrm{c}}) = -\left.\frac{d\theta(t)}{dt}\right|_{t=t0}(t - t_{\mathrm{c}}) = k_\theta(t - t_{\mathrm{c}}) \tag{3.57}$$

The angle of the heart tone is as in Eq. (3.58)

$$k_\theta = -\frac{c\sqrt{(H + R_{\mathrm{e}})^2\left(1 - \cos^2\beta\right) + R_{\mathrm{e}}^2}}{2(H + R_{\mathrm{e}})\sin\beta\left[(H + R_{\mathrm{e}})\cos\beta - \sqrt{(H + R_{\mathrm{e}})^2\left(1 - \cos^2\beta\right) + R_{\mathrm{e}}^2}\right]} \tag{3.58}$$

Easy-to-know frequency range angular tuning frequency $K_\theta \approx k_\theta / k_r$. The first treatment is range compression in the frequency domain, performing the Fourier transform of the echo data, and obtaining the Eq. (3.59).

$$S_n(f_r) = AW_{f_r}\left(\frac{f_r}{B_r}\right)\exp\left(-j\pi\frac{f_r^2}{k_r}\right)\exp\left(-j2\pi f_r\frac{d_n \sin\theta}{c}\right)\exp\left(j2\pi\frac{d_n K_\theta}{\lambda k_r}f_r\right) \tag{3.59}$$

In the range frequency domain, use equations such as (3.60) The expressed frequency domain pulse compression filter $H_r(f_r)$ is multiplied to remove the secondary frequency modulation term

$$H_r(f_r) = \exp\left(j\pi\frac{f_r^2}{k_r}\right) \tag{3.60}$$

After completing the range compression, the inverse Fourier transform obtains the time domain signal $c_n(t)$. Considering the multi-frame sampling based on prior flight parameters and inertial guidance $\omega_{imu}(t, t_a)$ data, the $c_n(t, t_a)$ signal between multi-channel complex data after SCORE compensation still has residual phase due to vehicle motion error, channel characteristics, and topographic and geographical factors [10]., after direct synthesis, the data complex vector is canceled out, and the gain is lost. For real-time synthesis systems, a complete calibration and verification module is necessary, but it is easier to estimate the echo parameters based on the data block. Since the main error is slowed in the azimuth direction, the azimuth sub-aperture can be used to reduce the estimation and compensation times. According to the SAR principle, Eq. (3.61) The number of pulse accumulations for the point target depends on the synthetic aperture, platform velocity, and pulse repetition rate PRF

$$N_{sample} = PRF\frac{\lambda R_0}{L_a v} \tag{3.61}$$

where R_0 is the range between the center of the scene; L_a is the azimuth antenna aperture. The spectral analysis of the triaxial angle of inertial conduction, the peak frequency of the flight altitude fluctuation, f_{max} and the number of blocks in the synthetic aperture is at least the formula (3.62) Expressions

$$N_{sub} \geq N_{sample}\frac{f_{max}}{PRF} = \frac{\lambda R_0}{L_a v}f_{max} \tag{3.62}$$

It is calculated that the $s'_{n,i}$ interferometry phase between sampling corresponding to the data block $s'_{1,i}$ and the reference channel in the $\phi_{n,i}$ whole scene is calculated as (3.63)

$$\phi_{n,i} = \arg\left(s'_{n,i} \cdot s'_{1,i}*\right), n = 1, \ldots, N \tag{3.63}$$

$i \in [1, 2, \ldots I]$ is the number of data blocks, I which is the total number of blocks; Constant and $\phi_{\mathrm{n,i}}$ primary, quadratic, and cubic polynomial fittings along the range direction are synthesized to obtain the formula (3.64) Residual phase gradient, fixed phase error correction amount

$$\Phi_{\mathrm{n,i}} = \varphi_{\mathrm{n,i}} + b_{\mathrm{n,i}}t + c_{\mathrm{n,i}}t^2 + d_{\mathrm{n,i}}t^3 \tag{3.64}$$

where is the fixed phase error, $\varphi_{\mathrm{n,i}}$ which is the primary phase contrast $b_{\mathrm{n,i}}$ coefficient, the quadratic $c_{\mathrm{n,i}}$ phase difference coefficient, $d_{\mathrm{n,i}}$ and the third phase difference coefficient. The global estimation matrix is obtained by merging the block estimators along the azimuth direction, and the Φ_{n} SCORE weight compensation and multi-channel synthesis based on the inertial guided data and the measured data are completed, and the synthesized signal is as shown in Eq. (3.65)

$$S(t) = \sum_{n=1}^{N} \omega_{\mathrm{n}}(t, t_{\mathrm{a}}) \cdot c_{\mathrm{n}}(t, t_{\mathrm{a}}) \cdot \exp(-\Phi_{\mathrm{n}}), n = 1, \ldots, N \tag{3.65}$$

The core purpose of the data verification test for DBF-SCORE imaging method is to verify multi-channel imaging methods and achieve high-quality image acquisition. The focus is on how to balance transmit power, angle of incidence, flight altitude, flight speed, pulse repetition rate, transmit width, transmit bandwidth, data rate to achieve satisfactory resolution, width, reception window, range ambiguity, azimuth ambiguity, and ground echo intensity.

In general, there is a significant degradation in image quality at the effective width boundary of SAR: the main reasons include the low two-way gain caused by the range towards the edge of the main lobe and the receive window is not fully acquired, resulting in a loss of signal-to-noise ratio and effective resolution, respectively. The system-level verification is carried out by millimeter wave multi-channel SAR measured data, and the imaging observation area is the urban area of Jingmen City, Hubei Province, and the remote scene area of the intercepted mapping belt is about 0.5 km^2. The scene shown in Fig. 3.25 is a focused image of a single-channel imaging mapping strip at the far end.

In multichannel direct synthesis, even degradation in image quality can be observed in Fig. 3.26, mainly due to phase modulation along the width due to the interchannel spatial baseline and the existence of fixed phase errors between channels. When the dynamic value of signal phase angle difference under comprehensive factors is too large, the multi-channel data is canceled out and the effective signal is weakened. The presence of more than one high order term for phase difference is reflected in the fluctuation of the gray value along the width width. Compared with the single-channel result, the composite image has a gain improvement of 4dB ~ 7dB.

The two-dimensional multi-stage phase gradient processing method of the above measured data was applied to synthesize the 4-channel data. By analyzing the signal-to-noise ratio improvement of the building shadow area as the ground noise and the

Fig. 3.25 Single-channel imaging result

Fig. 3.26 SCORE imaging result

unshadowed area, it can be obtained that the signal-to-noise ratio of the image after 4-channel synthesis is improved by 10.7 dB compared with the single channel, of which the intensity of the shadow part of the building is only increased by 1.5 dB. The synthesis effect is shown in Figs. 3.27 and 3.28.

Validation analysis was performed using another piece of data, which corresponds to a total area of about 4 km^2. As shown in Fig. 3.28, it can be found that a good

Fig. 3.27 Result with proposed method

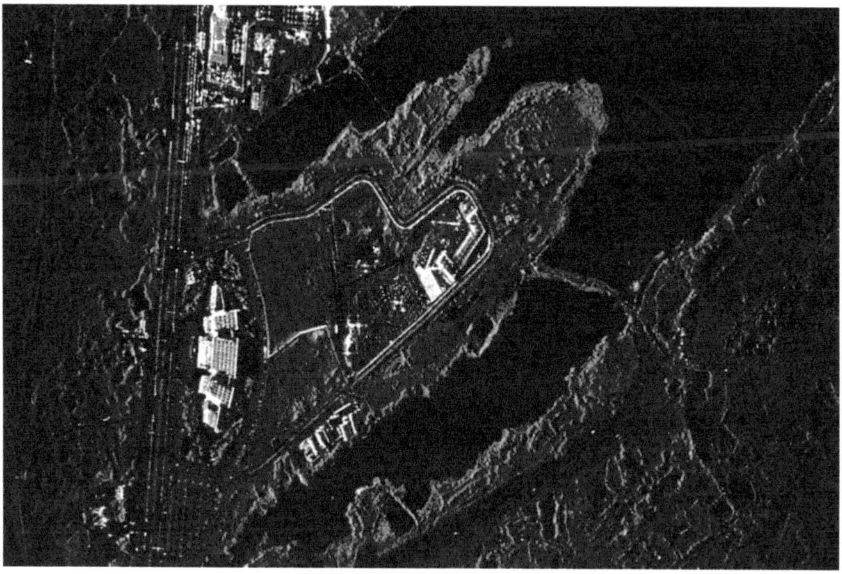

Fig. 3.28 4-channel full-width imaging results of this chapter method

signal-to-noise ratio is obtained after imaging, and the water surface noise is low. During the verification process, it is worth noting that the water surface echo, which usually has very weak backscatter, has also obtained a certain gain accumulation at a large angle of incidence, showing the potential of DBF-SCORE technology in ocean wide observation and earth water resources observation.

3.5 Airborne Millimeter Wave DBF SAR Moving Target Detection Method

3.5.1 Multi-channel Moving Target Detection Method

The first few sections of this chapter focused on range multichannel technology, which enables clutter suppression by increasing the number of channels and distinguishing between stationary and moving targets. Multi-channel detection methods based on azimuth multi-channel can be divided into two categories: the first type is to use interference processing methods to achieve the detection of moving targets, the most typical is the interferometric along the trajectory (ATI) method [11, 12]; The second type is to use the space–time characteristics of the target to complete the suppression of clutter, and then detect the moving target, and the more commonly used methods include the phase center bias antenna (DPCA) method and the space–time adaptive processing (STAP) method.

1. **Displaced Phase Center Antenna**

The classical DPCA method uses the principle of pulse cancellation to observe the same area twice at a certain interval of time at the same location. Since the center position of the antenna phase is unchanged in the two observations, the stationary target information can be eliminated. In the interval between the two observations, the information of the moving target changes due to the movement of the target, so that the moving target information is retained, so the detection of the slow-moving target in the main clutter range can be completed.

In order for different antennas to obtain the same clutter information at different times, the data received by the first antenna must have the same phase center as the data received by the other antenna after a certain number of pulses. Therefore, the antenna spacing *d*, carrier speed, V_a and pulse repetition rate *PRF* must meet the conditions of the following formula:

$$d = 2 \cdot m \cdot V_a / PRF, \text{ (} m \text{ is a integer number)} \tag{3.66}$$

The DPCA schematic diagram is shown in Fig. 3.29, two antennas are set along the flight direction of the radar platform, antenna 2 transmits signals, and both antennas receive echo signals at the same time. When receiving the first echo signal O_2, the receiving phase center of antenna 2 is at the point, and the receiving phase center of antenna 1 is at the point O_1.

m After transmitting a pulse, the receiving phase center O_4 of antenna 2 is moved to the place, the receiving phase center of O_3 antenna 1 is in, O_2 and is located in the same O_3 azimuth position, so that the first pulse echo received by antenna 2 is the same as the first pulse received by antenna $1m+1$ Each pulse echo contains the same stationary target information, but the moving target produces additional information due to its

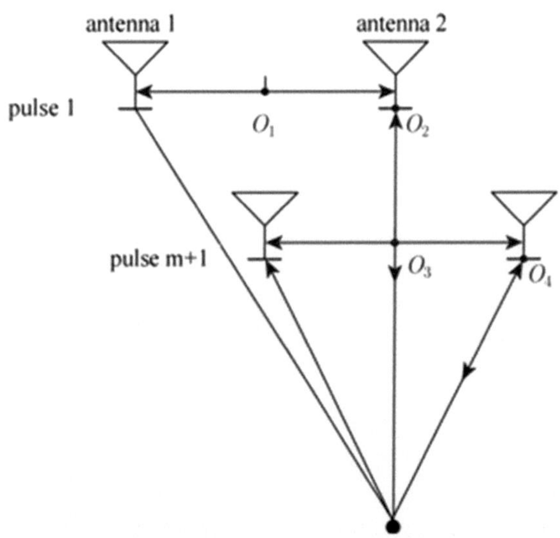

Fig. 3.29 Schematic diagram of the working principle of DPCA

own motion, so the two contain different moving target information. Subtracting the two echo signals eliminates the stationary target information and retains the moving target information.

Satisfying the constraints between antenna spacing d, carrier speed, and V_a pulse repetition rate PRF is the key to achieving the ideal clutter suppression effect of the DPCA method, and this strict spatiotemporal relationship makes the DPCA method lack adaptability and robustness. In the actual radar system, due to the influence of many factors, the constraints are difficult to be met, so it is difficult to obtain good clutter cancellation performance. If the echo signal is first imaged, and the two complex images obtained are interpolated and registered so that the clutter information of the two images is the same, and then the two images after registration are used for cancellation, the limitation of this condition can be avoided, the clutter can be effectively suppressed, and the moving target information can be retained. Figure 3.30 shows a flowchart for implementing DPCA in the complex image domain.

2. **Along Track Interferometry**

Similar to the DPCA method, the ATI method also sets two phase centers along the direction of the radar platform and receives echo signals at the same time. However, unlike the DPCA method, ATI does not detect moving targets by subtracting the echo signals of the two channels to eliminate the clutter and then detect moving targets, but by calculating the interference phase of two images of the same scene, so the ATI method does not require strict constraints between the antenna baseline, aircraft

Fig. 3.30 DPCA flowchart of the complex image domain

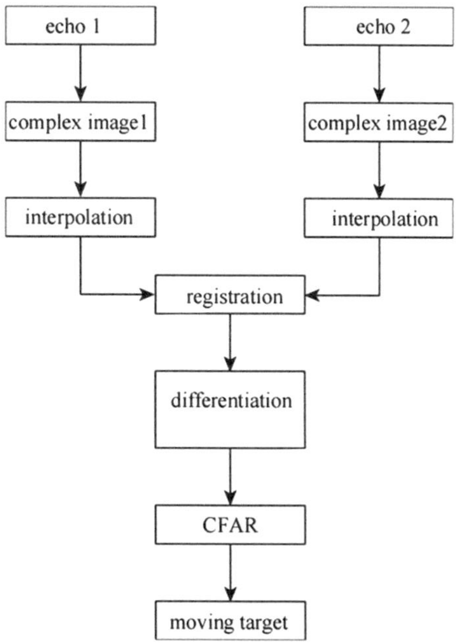

velocity and pulse repetition rate [13, 14]. Application examples of ATI are covered in Chaps. 6 and 7.

In the ATI method, we refer to the range between the phase centers of the two antennas as the interference baseline. When the radar is operating, the phase centers of the two antennas illuminate the target area with approximately the same pointing and receive echo signals at the same time. There is a time difference when the two antenna phase centers irradiate the same target, and in the echo signal obtained by the two channels, the phase of the same moving target is different due to the movement of the target. By using the interference method to detect this phase change, the detection of the moving target can be completed.

Based on the geometric relationship between the two images, the interference phase of the two images can be calculated as:

$$\phi_{\mathrm{ATI}} = \phi_2 - \phi_1 = \frac{4\pi}{\lambda} R_1 - \frac{4\pi}{\lambda} R_2 \tag{3.67}$$

It can be found that since the difference between the range of the stationary target and the radar platform when imaging each channel is fixed, its interference phase should also be fixed; The interference phase due to target motion in the two images can be expressed as:

$$\phi_{ATI\_m} = \frac{4\pi v_r B}{\lambda V_a} \tag{3.68}$$

where B is the baseline of the interference phase along the track. From the above equation, it can be found that the interference phase of the moving target is proportional to its radial velocity, so by setting a phase threshold, the interference phase of the two images is detected: if the interference phase of a certain point is greater than the phase threshold, it is considered that there is a moving target at the point; Conversely, it is considered that there is no movement target at that point. At the same time, according to the obtained interference phase, the radial velocity of the moving target can also be estimated

$$v_{r\_est} = \frac{\phi_{ATI\_m} \cdot \lambda \cdot V_a}{4\pi \cdot B} \tag{3.69}$$

3. **Space–time adaptive signal processing**

Space–time adaptive signal processing (STAP) is an important method for multi-channel moving target detection. By analyzing the space–time two-dimensional signal, it can be found that the space–time two-dimensional spectrum of clutter has strong coupling, and the two-dimensional plane in space time is manifested in the form of clutter ridges, and the projection of clutter ridges in the spatial domain and Doppler domain makes the clutter main lobe widened, so it is difficult to obtain good moving target detection performance (especially slow targets) by one-dimensional processing alone (spatial filtering or Doppler filtering). The STAP method uses the

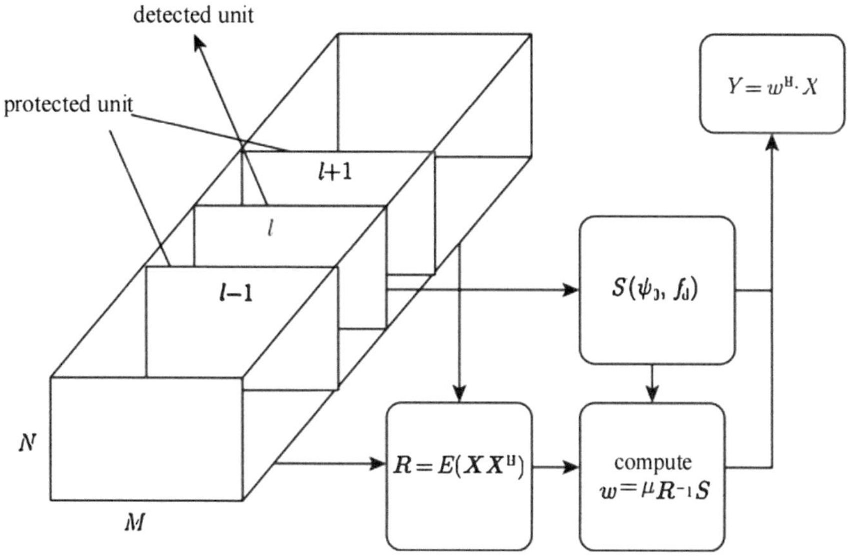

Fig. 3.31 Adaptive processing flowchart when all empty

two-dimensional information of the spatial domain and the Doppler domain at the same time, which can form grooves in the clutter ridge and interference direction, so as to effectively suppress the clutter and improve the performance of moving target detection. Figure 3.31 shows the flowchart of space–time adaptive processing.

Suppose a radar antenna has channels, each N channel receives pulses during the beam dwell time, M and the range of each pulse is the number of sampling points L. We arrange the space–time sampled data to get a $MN \times 1$ vector form:

$$\mathbf{X} = \left[x_{1,1}, \cdots, x_{N,1}, x_{1,2}, \cdots, x_{N,2}, \cdots, x_{1,M}, \cdots, x_{N,M}\right]^{\mathrm{T}} \tag{3.70}$$

Among them, the two subscripts represent the channel sequence and then the pulse train. Under the $\mathbf{H_0}$ binary assumptions of (no target signal, only clutter and noise) and (both target signal, clutter and noise $\mathbf{H_1}$), it $\mathbf{X}$ can be expressed as:

$$\mathbf{X} = \begin{cases} c + n, & \mathbf{H_0} \\ b \cdot \mathbf{S} + c + n, & \mathbf{H_1} \end{cases} \tag{3.71}$$

where, b represents the complex amplitude of the target echo c, clutter n and noise, respectively, $\mathbf{S} = \mathbf{S}_l / \sqrt{\mathbf{S}_l^{\mathrm{H}} \mathbf{S}_l}$ for normalized space–time guidance vectors, which $\mathbf{S}_l$ can be expressed as:

$$\begin{cases} \mathbf{S}_l = \mathbf{S}_s(\psi_s) \otimes \mathbf{S}_t(f_d) \\ \mathbf{S}_t(f_d) = \left[1, e^{j\varphi_t(f_d)}, \cdots, e^{j(M-1)\varphi_t(f_d)}\right]^{\mathrm{T}} \\ \mathbf{S}_s(\psi_s) = \left[1, e^{j\varphi_s(\psi_s)}, \cdots, e^{j(N-1)\varphi_s(\psi_s)}\right]^{\mathrm{T}} \end{cases} \tag{3.72}$$

where $\otimes$ represents the Krodecker product $\varphi_t(f_d)$, denoting the angular phase shift $\varphi_s(\psi_s)$ between pulses and elements at the corresponding f_d sum, respectively ψ_s. Therefore, the ratio of the power of the input signal to the clutter plus noise (i.e., the signal–noise ratio,) can be obtained $SINR_i$:

$$SINR_i = \frac{|b|^2}{\sigma_n^2 + \sigma_c^2} \tag{3.73}$$

where σ_n^2 is the input noise power and σ_c^2 the input clutter power.

The all-empty adaptive filter pair $\mathbf{X}$ is filtered, and if its weight vector $\mathbf{w}$ is, the filtering result is:

$$y = \mathbf{w}^H \cdot \mathbf{X} \tag{3.74}$$

where $(\cdot)^H$ is the complex conjugate transpose. Therefore, the signal-to-noise ratio of the filter output can be expressed as:

$$SINR_o = \frac{\left|\mathbf{w}^{\mathrm{H}}\mathbf{S}\right|^2}{\mathbf{w}^{\mathrm{H}}\mathbf{R}\mathbf{w}} = |b|^2 \cdot \mathbf{S}^{\mathrm{H}}\mathbf{R}^{-1}\mathbf{S} \tag{3.75}$$

The optimal weight vector can $\mathbf{w}_{opt}$ be obtained by the optimization problem of linear constraints as follows:

$$\begin{cases} \min\limits_{\mathbf{w}} \ \mathbf{w}^H \mathbf{R}\mathbf{w} \\ s.t. \ \mathbf{w}^H \mathbf{S} = 1 \end{cases} \tag{3.76}$$

Then the optimal adaptive weight vector processed by STAP can be written as:

$$\mathbf{w}_{opt} = \mu \mathbf{R}^{-1}\mathbf{S} \tag{3.77}$$

thereinto $\mathbf{R} = E\left(\mathbf{X}\mathbf{X}^{\mathrm{H}}\right)$ covariance matrix for clutter and noise, μ is a complex constant.

Then the system improvement factor, that is, the ratio of the output SINR to the input SINR, can be expressed as:

$$\begin{aligned} IF = \frac{SINR_o}{SINR_i} &= \frac{|b|^2 \cdot \mathbf{S}^{\mathrm{H}}\mathbf{R}^{-1}\mathbf{S}}{\frac{|b|^2}{\sigma_{ni}^2 + \sigma_{ci}^2}} = \mathbf{S}^{\mathrm{H}}\mathbf{R}^{-1}\mathbf{S} \cdot \left(\sigma_{ni}^2 + \sigma_{ci}^2\right) \\ &= \mathbf{S}^{\mathrm{H}}\mathbf{R}^{-1}\mathbf{S} \cdot (CNR_i + 1) \cdot \sigma_{ni}^2 \end{aligned} \tag{3.78}$$

thereinto σ_{ni}^2 and σ_{ci}^2 Indicates the power of input noise and clutter, respectively, $CNR_i = \sigma_{ci}^2/\sigma_{ni}^2$ Indicates the input noise ratio CNR (Clutter to Noise Ratio).

In theory, if the covariance matrix **R** is known, adaptive filtering in all-space can obtain good clutter suppression. However, in practice, the characteristics of clutter are unknown, and we can only estimate the covariance matrix from the adjacent range unit sample data with independent iso-distribution (IID) characteristics. In the context of gaussian clutter plus noise, the maximum likelihood estimation yields:

$$\hat{\mathbf{R}} = \frac{1}{L}\sum_{l=1}^{L}\mathbf{x}(l)\mathbf{x}^H(l) = \frac{1}{L}\mathbf{X}\mathbf{X}^H \tag{3.79}$$

thereinto $\hat{\mathbf{R}}$ represents the covariance matrix of the sampled data, L indicates the number of range gates, **X** Represents a matrix of sampled data.

Although under ideal circumstances, the adaptive processor can obtain good clutter suppression performance in all-empty time, the serious shortcomings of excessive computation and limited sampling data vector limit its practical application. In order to be applied to practical projects, STAP must adopt dimensionality reduction treatment scheme.

The dimensionality reduction processing of the STAP method can be carried out directly in the array element domain, pulse domain, and also in its Fourier transform domain (beam domain, Doppler domain). Regardless of how the dimensionality reduction processing is performed, it can be equivalent to **X** the process of linear transformation of the sampled data vector through a $MN \times K$ dimensionality reduction matrix B, where MN and represent the dimensionality K of the data before and after the dimensionality reduction, respectively. The data vector before and after dimensionality reduction and the signal guide vector have the following relationship:

$$\begin{cases} \mathbf{X}_r = \mathbf{B}_r^H\mathbf{X} \\ \mathbf{S}_r = \mathbf{B}_r^H\mathbf{S} \end{cases} \tag{3.80}$$

According to the linear constraint minimum variance criterion, dimensionality reduction STAP processing can be equivalent to the following optimization problem:

$$\begin{cases} \min\limits_{\mathbf{w}} \ \mathbf{w}_r^H\mathbf{R}_r\mathbf{w}_r \\ s.t.\ \mathbf{w}_r^H\mathbf{S}_r = 1 \end{cases} \tag{3.81}$$

Different dimensionality reduction structures can be used to obtain different filtering performance. Even if the same dimensionality reduction structure is used, it will exhibit different performance under different system parameters and operating conditions. Therefore, no dimensionality reduction structure is optimal in any case, and the selection of dimensionality reduction structure is analyzed according to the actual situation.

3.5.2 DBF SAR Moving Target Detection Processing

Considering the complexity and effect of processing, this section uses the CSI (clutter suppression interference) method combining DPCA and ATI to process and analyze the DBF SAR data obtained from the flight test of the Key Laboratory of Millimeter Wave Imaging Technology of Shanghai Academy of Aerospace Technology [15, 16]. The processing flow is shown in Fig. 3.32.

1. **Range single channel processing results**

The range single-channel data is used for processing, one channel is selected as the reference channel, and the other channel is used to perform differential phase estimation with the reference channel. Figure 3.33 is a schematic diagram of range single-channel difference phase estimation, wherein Fig. 3.33a is the interference phase of the channel and the reference channel in the range frequency domain, and Fig. 3.33b is the curve after the interference phase is intercepted, smoothed and fitted. Figure 3.33c is a schematic diagram of the interference phase of the two-channel data in the two-dimensional frequency domain after compensating (b) fitting the phase curve, and Fig. 3.33d is the smoothing result of (c). Using the results of Fig. 3.33d, a two-dimensional search rough estimation is performed, as shown in Fig. 3.34, the

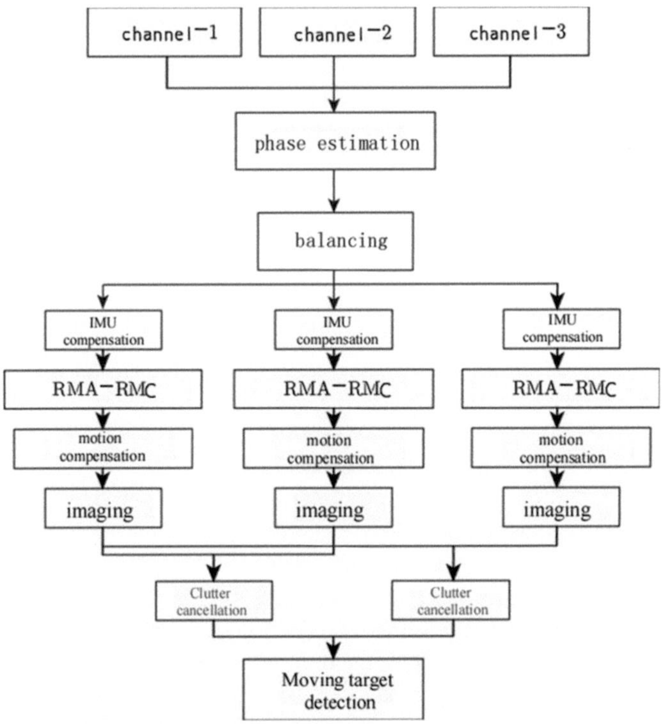

Fig. 3.32 Moving target processing flow chart

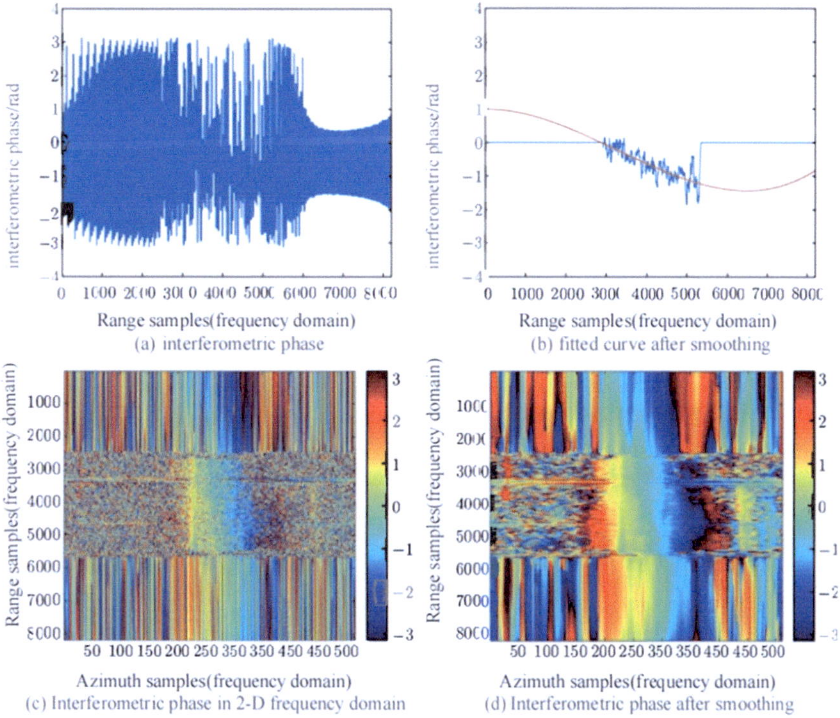

Fig. 3.33 Schematic diagram of phase estimation of difference from single channel

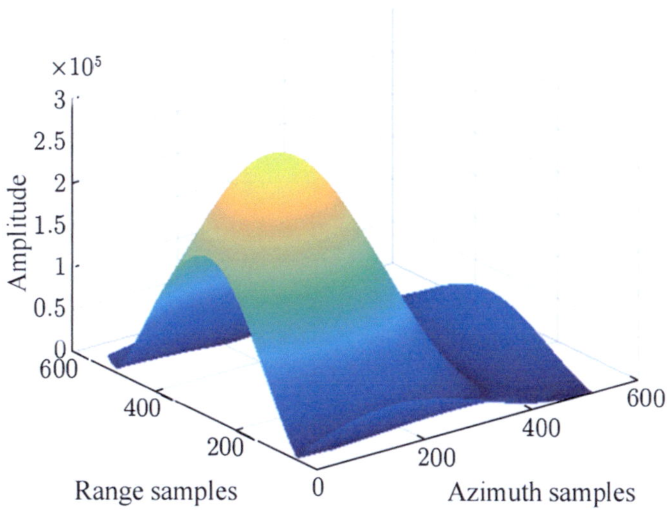

Fig. 3.34 Rough estimation based on two-dimensional search (range single channel)

Table 3.3 Estimates the resulting channel differences

Difference in range direction (m)	0.0508
Azimuth difference (m)	0.0479

maximum value of the search shows the difference in range and azimuth of the two channels.

The search results are shown in Table 3.3.

2. **DBF post-processing results**

Similarly, the data processed after DBF is used here, and the processing result is shown in Fig. 3.35. The channel differences obtained by the two-dimensional search rough estimation are shown in Table 3.4.

The coarse estimation (after DBF) simulation based on two-dimensional search is shown in Fig. 3.36.

The estimated channel differences are shown in Table 3.4.

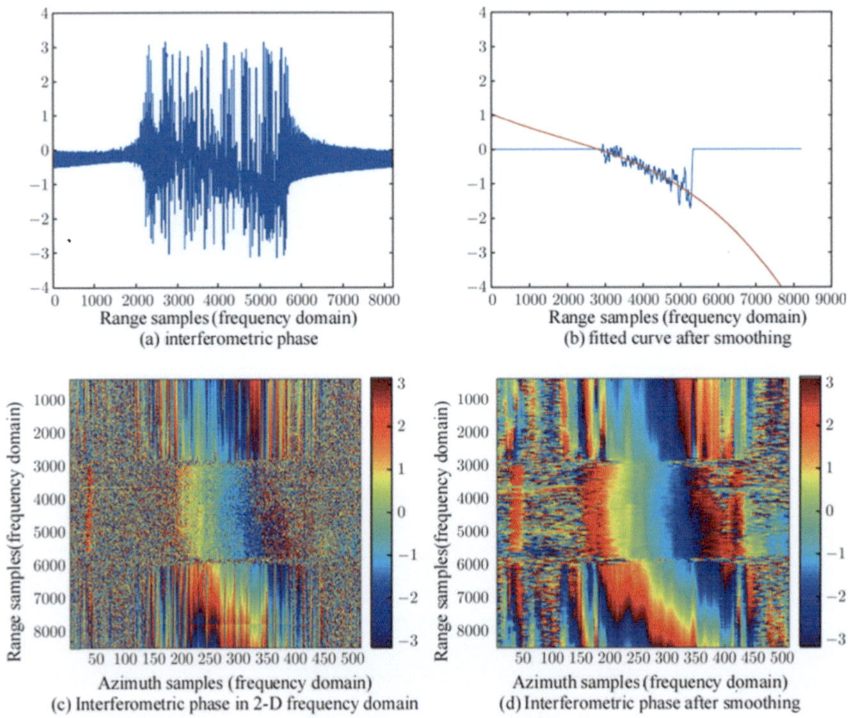

Fig. 3.35 Schematic diagram of difference phase estimation after DBF

Table 3.4 Estimates the resulting channel differences

Difference in range direction (m)	0.0392
Azimuth difference (m)	0.0482

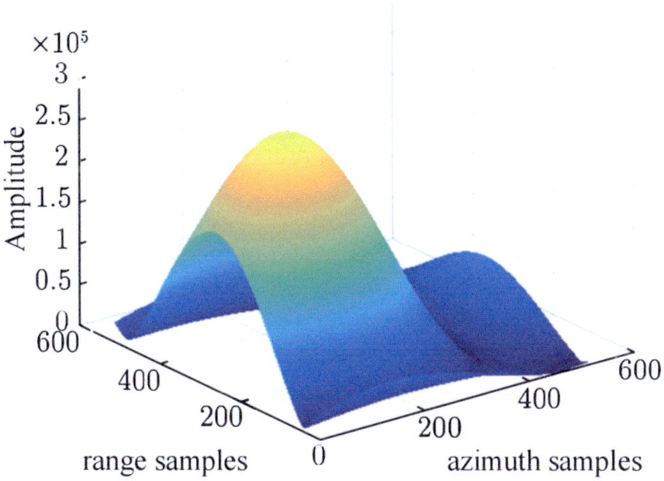

Fig. 3.36 Rough estimation based on two-dimensional search (after DBF)

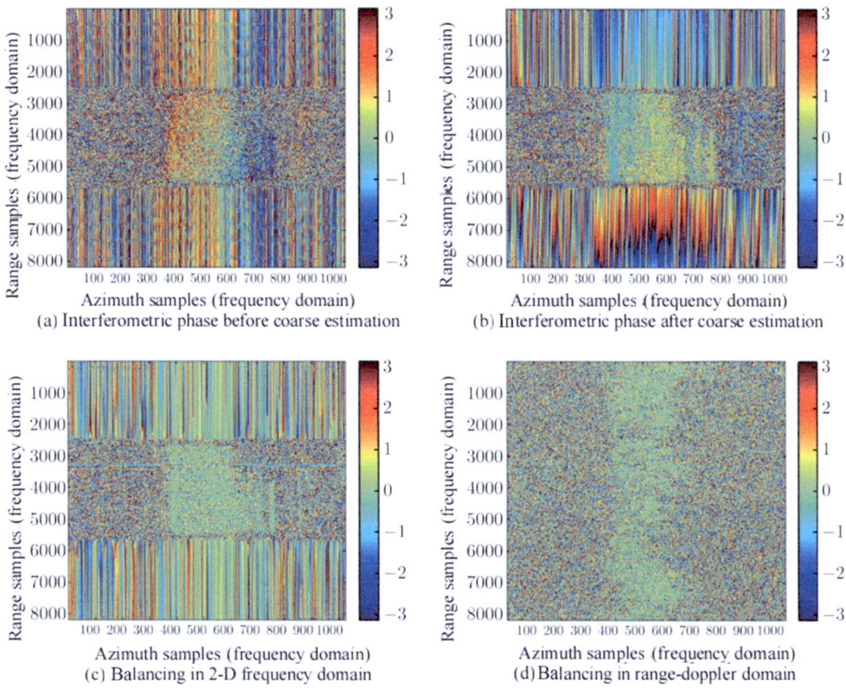

Fig. 3.37 Range single-channel azimuth channel equalization interference phase

3. **Range single-channel azimuth channel equalization**

Using a single channel of range, the data of the two channels in azimuth is used for channel equalization, one channel is selected as the reference channel, and the other channel is compensated for the channel difference, the interference phase before compensation is shown in Fig. 3.37a, and the effect after compensation is shown in Fig. 3.37b. Then interfere with the reference channel in the two-dimensional frequency domain and the azimuth frequency domain of the range time domain, and then perform channel equalization after smoothing [17, 18]. The effect after two-dimensional frequency domain channel equalization is shown in Fig. 3.37c, and the interference phase after frequency domain channel equalization in the range time domain is shown in Fig. 3.37d. Visible channel differences are significantly eliminated.

4. **DBF rear azimuth channel equalization**

The results obtained by the same processing of DBF data are shown in Fig. 3.38, and it can be seen that the channel balance between the azimuth channel data after DBF processing can also achieve good results.

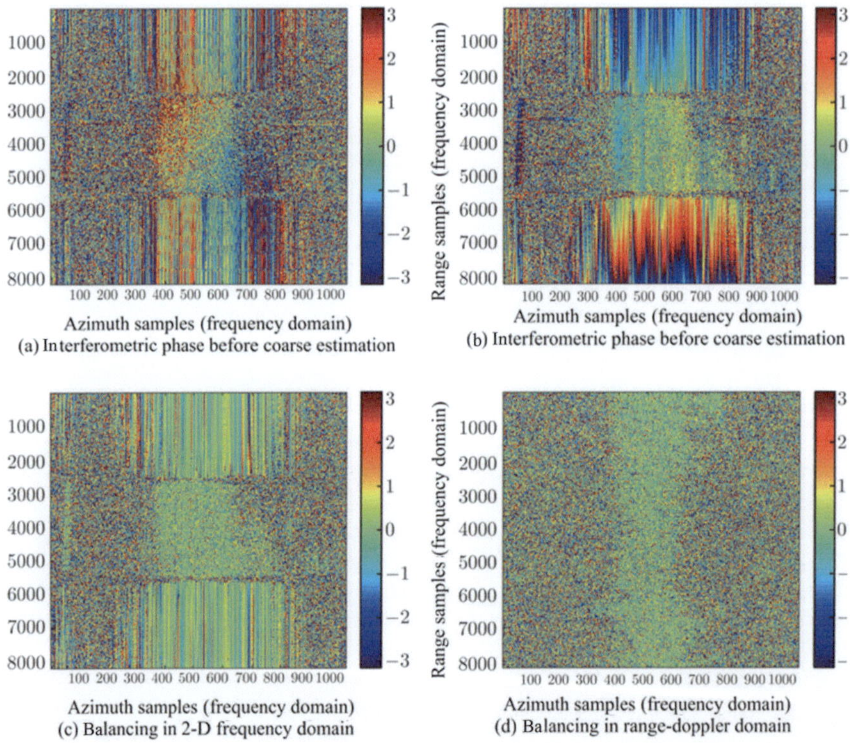

Fig. 3.38 Equalization interference phase of the single-channel azimuth channel from the range

The data of each channel through channel equalization is imaged, and clutter cancellation is performed in the image domain. Taking the two-channel imaging results and the cancellation results as an example, the effect is shown in Fig. 3.39.

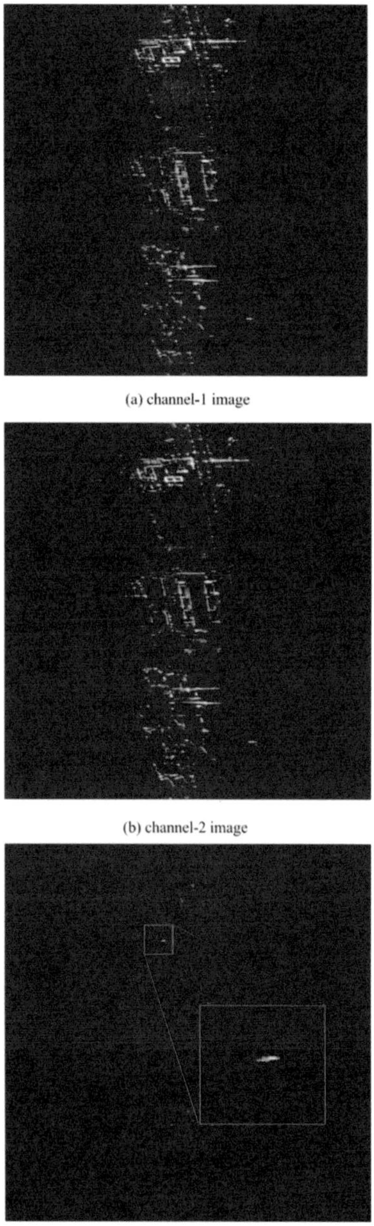

(a) channel-1 image

(b) channel-2 image

(c) clutter cancellation

Fig. 3.39 Clutter cancellation results

From the elimination results, the target can be clearly identified, which verifies the effectiveness of the method.

References

1. C. W. Sherwin, J. P. Ruina, R. D. Rawcliffe. Some early developments in synthetic aperture radar system [J]. IRE Trans. on MIL, 1962, 6(2): 111-115.
2. I. G. Cumming, F. H. Wong, Hong Wen, Hu Donghui et al. Synthetic aperture radar imaging——Algorithm and implementation[M]. Beijing: Publishing House of Electronics Industry, 2007:36–68
3. C. A. Wiley. Synthetic Aperture Radar [J]. IEEE Trans. on AES, 1985, 21(3): 440-443.
4. L. J. Cutrona, W. E. Vivian, E. N. Leith, G. O. Hall. A high-resolution radar combat-surveillance systems [J]. IRE Trans. on MIL, 1961, 5(2): 127-131.
5. L. J. Cutrona, G.O. Hall. A comparison of techniques for achieving fine azimuth resolution [J]. IRE Trans. on MIL, 1962, 6(2): 119—121.
6. D. A. Ausherman, A. Kozma, J. L. Walker, H. M. Jones, E. C. Poggio. Developments in radar imaging [J], IEEE Trans on AES, 1984, 20(4): 363—400.
7. Bao Zheng, Xing Mengdao, Wang Tong Radar imaging technology[M]. Beijing: Publishing House of Electronics Industry, 2008:25-28
8. W. M. Brown, L. J. Porcello. An introduction to synthetic-aperture radar [J]. IEEE Spectrum, 1969, 6(9): 53-63. 。
9. Goldstein RM, Zebker H. Interferometric radar measurement of ocean surface currents [J]. Nature. 1987;328(6132):707-709.
10. Zebker HA, Villasenor J. Decorrelation in interferometric radar echoes[J]. IEEE Transactions on Geo-science Remote sensing 1992;30(5):950-959.
11. Ferretti A, Prati C, Rocca F. Multibaseline InSAR DEM reconstruction: The wavelet approach[J]. IEEE Transactions on Geoscience Remote Sensing of Environment. 1999;37(2):705-715.
12. Xu W, Cumming I. A region-growing algorithm for InSAR phase unwrapping [J]. IEEE Transactions on Geoscience Remote Sensing. 1999;37(1):124-34.
13. Süß M, Grafmüller B, Zahn R, editors. A novel high resolution, wide swath SAR system[C]. IGARSS 2001 Scanning the Present and Resolving the Future Proceedings IEEE 2001 International Geoscience and Remote Sensing Symposium; 2001: 133–139
14. Breit H, Eineder M, Holzner J, Runge H, Bamler R, editors. Traffic monitoring using SRTM along-track interferometry[C]. IGARSS 2003 2003 IEEE International Geoscience and Remote Sensing Symposium Proceedings; 2003.
15. Younis M, Fischer C, Wiesbeck W. Digital beamforming in SAR systems[J]. IEEE Transactions on Geoscience Remote Sensing. 2003;41(7):1735-1739.
16. Ferraiuolo G, Pascazio V, Schirinzi G. Maximum a posteriori estimation of height profiles in InSAR imaging[J]. IEEE Geoscience Remote Sensing Letters. 2004;1(2):66-70.
17. Wright TJ, Parsons BE, Lu Z. Toward mapping surface deformation in three dimensions using InSAR[J]. Geophysical Research Letters. 2004;31(1).
18. Lohman RB, Simons M. Some thoughts on the use of InSAR data to constrain models of surface deformation: Noise structure and data downsampling[J]. Geochemistry, Geophysics, Geosystems. 2005;6(1).

Chapter 4
Ground-Based Millimeter-Wave Inverse Synthetic Aperture Radar Technology

4.1 Overview

Inverse synthetic aperture radar (ISAR) imaging technology is an important development direction of today's space measurement radar, which can be used to achieve detection, capture, tracking, orbiting, identification, and provide information on space target activities and various target characteristics for various space targets (satellites, spacecraft, space shuttles, launch vehicles, space debris, etc.). Although the existing ISAR can provide high distance resolution, the characteristics of space-important targets are usually smaller than the distance resolution that can be obtained by conventional processing. Moreover, with the continuous development of satellite manufacturing technology in recent years, the miniaturization of satellites has become a trend in order to save launch costs and improve stealth. Therefore, higher resolution requirements are imposed on space target surveillance radar.

The high resolution of the images obtained by millimeter wave W-band and the rich detail information of electromagnetic scattering characteristics make it very suitable for ISAR spatial surveillance, and the accuracy of target classification and identification can be greatly improved by using high-resolution ISAR images obtained by W-band compared with low-band radar.

However, W-band ISAR imaging has the following difficulties compared with conventional band ISAR imaging, which need to be addressed. First, the target envelope changes a lot during the imaging process. In the ISAR imaging of non-cooperative targets, the translational motion of the target needs to be compensated first, so that the imaging of the target is equivalent to the imaging of the turntable target. The motion compensation processing is mainly to estimate and compensate the target translation by using the similarity of ISAR echo signals. However, in W-band ISAR imaging, the high resolution leads to more significant changes in the envelope during the imaging process, which makes the envelope alignment difficult. Second, the migration caused by target rotation is serious. Due to the high resolution of W-band radar imaging systems, the migration of trans-resolution cells caused

H. Wang, *The Millimeter Wave Synthetic Aperture Radar Technology*,
https://doi.org/10.1007/978-981-97-1044-7_4

by target rotation is much more serious during the imaging process. Many targets that can be imaged without compensation in conventional bands may cross tens of resolution cells in W-band, which can cause severe defocussing of imaging results if not compensated accurately. Furthermore, the amount of W-band ISAR echo data is large. As the resolution rises, the higher data volume creates a significant processing burden for the W-band imaging process, and the imaging processing time will be significantly longer if the computational complexity of the image formation algorithm is high.

This chapter introduces the solutions to the above-mentioned problems in W-band ISAR imaging technology. Based on the analysis of the echo model of ISAR imaging with the system parameters, the motion compensation technique and the imaging migration compensation method of W-band ISAR are introduced, and the imaging method with two-dimensional compensation for speed-stable targets and the improved sub-aperture imaging method for speed-varying targets are presented. Finally, the situation of simulation data processing was introduced.

4.2 W-Band ISAR Echo Modeling

In order to investigate the image formation algorithm of W-band ISAR, a mathematical model of the imaging has to be developed first. In this section, the preliminary imaging system parameters for W-band ISAR are given, the echo model and geometric model are derived, the approximation formulas for W-band are given, and the approximation errors are estimated.

4.2.1 W-Band ISAR Modeling

1. W-band ISAR Radar System Parameters

In order to facilitate the approximation and error estimation calculations, this section gives a brief analysis of the preliminary W-band ISAR imaging system and the parameters of the imaging target based on the existing W-band ISAR imaging radar.

The band range of the W-band is determined by the atmospheric absorption bands. The two nearest atmospheric absorption bands near the W-band range are the two oxygen absorption bands at 71.05 and 118.75 GHz, so the W-band system can be used as long as the band range is within this atmospheric window. The center frequency can be selected according to the hardware conditions. The center frequency of devices such as W-band gyrotrons is generally 94 GHz, so the center frequency of the system is selected as 94 GHz. The pulse duration of the system can be selected according to the hardware system requirements, referring to the American WARLOC radar system, the pulse duration here is 100 μs.

The bandwidth of the system and the imaging accumulation angle are determined by the required resolution, which is set at 3 cm for both azimuth and range.

Then the system bandwidth can be obtained according to the distance resolution formula for ISAR imaging:

$$B = \frac{c}{2\rho_\mathrm{r}} = 5\mathrm{GHZ} \tag{4.1}$$

For the complex signal system, the system sampling rate only needs to be larger than the system bandwidth, and the sampling rate is selected as 6 GHz.

The imaging accumulation angle is obtained based on the imaging azimuth resolution of ISAR:

$$\Delta\theta = \frac{\lambda}{2\rho_\mathrm{a}} = 0.05\mathrm{rad} \tag{4.2}$$

Set for aircraft target imaging, the rotation speed of the target is 0.1 rad/s, and the imaging accumulation time is 0.5 s. Set the maximum imaging size of the target is 50 m, and the distance between the radar and the target is 5 km.

The parameters of the collated system are given in Table 4.1.

2. Echo signal model

Let the signal transmitted by the radar be $s(t)$ and the signal received after τ time, the received signal is

$$s_\mathrm{r}(t) = ks(t - \tau) \tag{4.3}$$

where k is the attenuation coefficient of the echo. When the radar is a single-station radar and the target is stationary, the echo delay is $\tau = 2R/C$.

If the target is moving away from the radar at a speed of v and the delay of the echo is τ, then the echo received at time t is emitted at time $t - \tau$ and the time of irradiation on the target is $t' = t - \tau/2$. If the position of the target at time 0 is R_0,

Table 4.1 The parameters of W-band ISAR system and target

Simulation parameters	Value
Center frequency	94 GHz
Signal bandwidth	5 GHz
Sampling rate	6 GHz
Pulse repetition frequency	4000
Pulse duration	100 us
Measurement duration	0.5 s
Distance to target	5000 m
Maximum scale of the target	50 m
Angular velocity of the target	0.1 rad/s

then the distance between the target and the radar at the time of irradiation on the target is

$$R(t') = R_0 + vt' \tag{4.4}$$

Combined with the time of the target reflex, which is:

$$\tau = \frac{2R(t')}{c} \tag{4.5}$$

We can obtain:

$$\tau = 2\frac{R_0 + vt}{c + v} \tag{4.6}$$

Substituting τ into $s_r(t)$, we get

$$s_r(t) = ks(t - 2\frac{R_0 + vt}{c + v}) = ks(\frac{c - v}{c + v}t - \frac{2R_0}{c + v}) = ks\frac{c - v}{c + v}(t - \frac{2R_0}{c - v}) \tag{4.7}$$

From formula (4.7), it can be seen that when the target has movement along the direction of the line with the radar, the received echoes will stretch accordingly, and when the target moves away from the radar, the echoes stretch in the time axis, which in turn leads to the compression of the spectrum. When the target approaches the radar, the echoes are compressed in the time axis, while the spectrum will stretch.

Considering the case of coherent accumulation of M pulses, the moment of emission of the mth pulse of the radar $t = mT_r$, T_r is the pulse repetition period and the radar time can be expressed as a combination of the fast time (intra- pulse time) t_k and the slow time (inter- pulse time) t_m:

$$t = t_k + t_m \tag{4.8}$$

As a result, the distance between the center point of the target and the radar at the time of (t_k, t_m) is:

$$R_O(t_k, t_m) = R_O(t_m) + v(t_m)t_k \tag{4.9}$$

where $R(t_m)$ and $v(t_m)$ are the initial distance and average velocity of the target within the mth pulse. The exact model of the echo of the chirp can be obtained as follows:

$$\begin{aligned} s(t_k, t_m) = {} & A\mathrm{rect}[\frac{\alpha^2(t_m)(t_k - \tau(t_m))}{T_p}]\exp[j\pi\mu\alpha^2(t_m)(t_k - \tau(t_m)] \\ & \times \exp[j2\pi f_0\alpha(t_m)t_k]\exp[-j2\pi f_0\alpha(t_m)\tau(t_m)] \end{aligned} \tag{4.10}$$

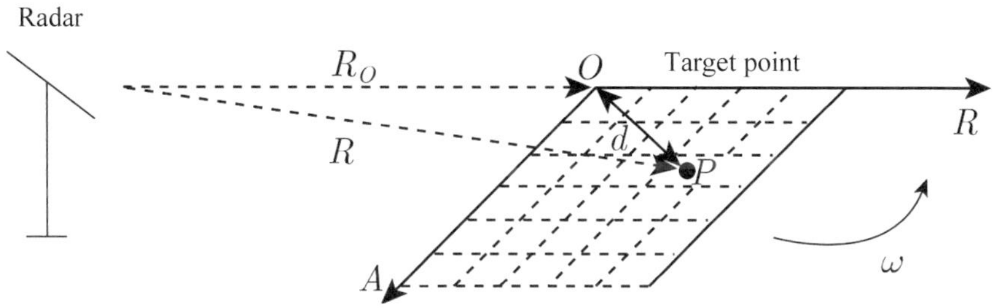

Fig. 4.1 Schematic diagram of the imaging geometry relationship of ISAR

where $\alpha(t_{\mathrm{m}}) = \frac{c-v(t_{\mathrm{m}})}{c+v(t_{\mathrm{m}})}$ is the expansion coefficient caused by the moving speed of the target; T_{p} is the pulse width; μ is the rate, which is $B = \mu T_{\mathrm{p}}$, $\tau'(t_{\mathrm{m}}) = \frac{2R_O(t_{\mathrm{m}})}{c-v(t_{\mathrm{m}})}$.

3. Calculation and approximation of the geometric relationship of scattering points

For a scattering point P on the target, let the distance between this point and the target center point be d_{OP}, and let the angle between OP and the positive direction of R-axis be $\theta(t_{\mathrm{m}}, t_{\mathrm{k}})$. Figure 4.1 shows the position of the scattering point in relation to the radar.

Firstly, without considering the movement speed of the target, we can get

$$\begin{aligned} R_P(t_{\mathrm{m}}, t_{\mathrm{k}}) &= \sqrt{R_O^2(t_{\mathrm{m}}, t_{\mathrm{k}}) + 2d_{OP}R_O\cos\theta(t_{\mathrm{m}}, t_{\mathrm{k}}) + d_{OP}^2} \\ &= R_O(t_{\mathrm{m}}, t_{\mathrm{k}})\sqrt{1 + 2\frac{d_{OP}}{R_O(t_{\mathrm{m}}, t_{\mathrm{k}})}\cos\theta(t_{\mathrm{m}}, t_{\mathrm{k}}) + \frac{d_{OP}^2}{R_O^2(t_{\mathrm{m}}, t_{\mathrm{k}})}} \\ &\approx R_O(t_{\mathrm{m}}, t_{\mathrm{k}}) + \frac{d_{OP}^2}{2R_O(t_{\mathrm{m}}, t_{\mathrm{k}})} + d_{OP}\cos\theta(t_{\mathrm{m}}, t_{\mathrm{k}}) \end{aligned} \tag{4.11}$$

To ignore the effect of the second term, it is necessary that the displacement due to the second term is less than the resolution of the imaging during the time of imaging processing,the following equation is satisfied:

$$\Delta\left(\frac{d_{OP}^2}{2R_O(t_{\mathrm{m}}, t_{\mathrm{k}})}\right) = \frac{d_{OP}^2}{2R_O^2(t_{\mathrm{m}}, t_{\mathrm{k}})}\Delta R_O(t_{\mathrm{m}}, t_{\mathrm{k}}) \le \rho \tag{4.12}$$

According to the previously set system parameters:

$$\Delta R_O(t_{\mathrm{m}}, t_{\mathrm{k}}) \le \rho\frac{2R_O^2(t_{\mathrm{m}}, t_{\mathrm{k}})}{\left(\frac{D_a}{2}\right)^2 + \left(\frac{D_r}{2}\right)^2} \approx \rho\frac{8R_O^2}{D^2} = 2400\mathrm{m} \tag{4.13}$$

That is, the maximum distance displacement of the target during the whole imaging process does not exceed 2400 m, from which the velocity of the target is obtained:

$$V \leq \frac{\Delta R}{T_{\text{mesure}}} = 4800\,\text{m/s} \tag{4.14}$$

It can be seen that this approximation is generally satisfied for aircraft targets imaged at a distance of 5 km. In the case of satellite target imaging, assuming the orbit height of the satellite is 500 km, it can be calculated that

$$V \leq \frac{\Delta R}{T_{\text{mesure}}} \leq \rho \frac{8R_0^2}{D^2 T_{\text{mesure}}} = 4.8 \times 10^7\,\text{m/s} \tag{4.15}$$

The satellite target cannot reach this speed, so the effect caused by this item in the imaging process can be neglected.

For the third term of formulaequation (4.11), further calculation and expansion of sin and cos using Taylor series can get:

$$\begin{aligned} d_{OP}\cos\theta(t_{\text{m}}, t_{\text{k}}) &= d_{OP}\cos[\theta(t_0) + \theta(t_{\text{m}} + t_{\text{k}})] \\ &= d_{OP}\cos[\theta(t_0)]\cos[\theta(t_{\text{m}} + t_{\text{k}})] - d_{OP}\sin[\theta(t_0)]\sin\theta[(t_{\text{m}} + t_{\text{k}})] \\ &= a\cos\theta[(t_{\text{m}} + t_{\text{k}})] - r\sin[\theta(t_{\text{m}} + t_{\text{k}})] \\ &= r[1 - \frac{\theta^2(t_{\text{m}} + t_{\text{k}})}{2!} + \ldots] - a\left[\theta(t_{\text{m}} + t_{\text{k}}) - \frac{\theta^3(t_{\text{m}} + t_{\text{k}})}{3!} + \ldots\right] \end{aligned} \tag{4.16}$$

where a, r are the coordinates where the initial moment P point is located respectively, considering the maximum values of a and r as $D_a/2$ and $D_r/2$ espectively, and the maximum value of the rotation angle is the required angle for imaging, from which the error order of each order approximation can be calculated as follows

1st order: $D\theta = 2.5\text{m} > \rho$
2nd order: $\frac{D\theta^2}{2} = 0.0625\text{m} > \rho$
3rd order: $\frac{D\theta^3}{6} = 1.04 \times 10^{-3}\text{m} < \rho$

It can be seen that the resolution of the W-band imaging can be satisfied after the 2nd order error compensation, so the second order approximation is used in the W-band ISAR imaging, from which the distance between the points on the target and the radar can be obtained when the target as a whole has no translational velocity, which can be approximated as follows

$$R_P(t_{\text{m}}, t_{\text{k}}) = R_0(t_{\text{m}}, t_{\text{k}}) + r - a \times \theta(t_{\text{m}} + t_{\text{k}}) - r \times \frac{\theta^2(t_{\text{m}} + t_{\text{k}})}{2} \tag{4.17}$$

Here $\theta(t_{\text{m}})$ is used to approximate $\theta(t_{\text{m}} + t_{\text{k}})$, and the error caused by the approximation is estimated as

$$T_P\omega\frac{D}{2} = 5 \times 10^{-4} m < \rho \tag{4.18}$$

The error can be ignored, so we can get

$$R_P(t_\mathrm{m}, t_\mathrm{k}) = R_0(t_\mathrm{m}) + V(t_\mathrm{m})t_\mathrm{k} + r - a \times \theta(t_\mathrm{m}) - r \times \frac{\theta^2(t_\mathrm{m})}{2} \tag{4.19}$$

Substituting the above equation into the echo signal, we get the delay of the echo

$$\tau_P(t_\mathrm{m}) = \frac{2R_P(t_\mathrm{m})}{c - v(t_\mathrm{m})} = \frac{2R_O(t_\mathrm{m}) + 2r - 2a\theta(t_\mathrm{m}) - r\theta^2(t_\mathrm{m})}{c} \times \left(1 + \frac{v}{c - v}\right) \tag{4.20}$$

If approximation is performed and the second term in the above equation is removed, the effect on the imaging process is that all imaging points will move, If the error caused by the approximation is required not to cross the resolution cell, then it is required that

$$\frac{v}{c - v} D < \rho \tag{4.21}$$

which is

$$v < \frac{\rho c}{D + \rho} \approx \frac{\rho c}{D} = 90000\,\mathrm{m/s} \tag{4.22}$$

It can be seen that the general target is difficult to reach this velocity, so it can be approximated to obtain

$$\tau_P(t_\mathrm{m}) = \frac{2R_O(t_\mathrm{m}) + 2r - 2a\theta(t_\mathrm{m}) - r\theta^2(t_\mathrm{m})}{c} \tag{4.23}$$

4.2.2 Echo Signal Processing

ISAR imaging is generally divided into several parts: range compression, motion compensation, and imaging processing. Figure 4.2 gives the flow of the imaging process.

Section 4.3 will focus on motion compensation processing, and Sect. 4.4 will focus on the migration compensation part of imaging processing. Therefore, the commonly used range compression in ISAR imaging processing is introduced here,

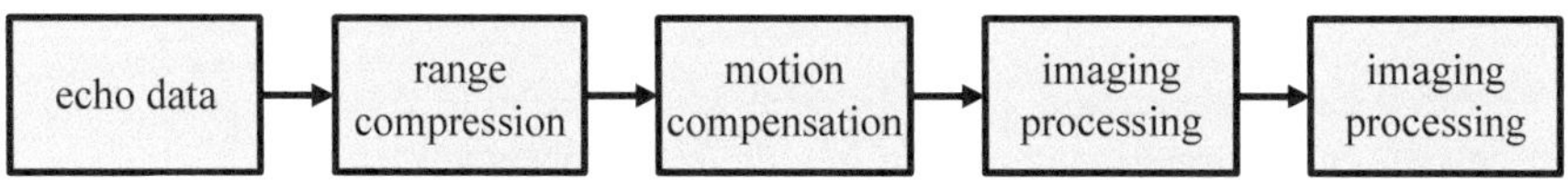

Fig. 4.2 ISAR imaging processing flowchart

and the role of the subsequent signal processing process after range compression is analyzed from the echo model.

1. Range compression processing

After obtaining the echo signal, the first step in imaging processing is to perform distance-oriented compression. The range compression can be achieved by matching the filtering in the distance direction. According to Eq. (4.10), the echo at point P, without considering the effect caused by the target velocity, can be approximated as

$$s(t_k, t_m) = A\mathrm{rect}\left[\frac{t_k - \tau_P(t_m)}{T_P}\right]\exp[\mathrm{j}\pi\mu(t_k - \tau_P(t_m))^2]\exp[\mathrm{j}2\pi f_0(t_k - \tau_P(t_m))] \tag{4.24}$$

In the imaging process of the ISAR imaging system, the target is tracked using a narrowband beam to obtain the distance of the target center point relative to the radar. The reception window of the echo is then adjusted so that the reflected echo of the target is located in the reception window. Assuming that the measured distance to the target at t_m is $R_c(m)$, the received echo signal can be expressed as

$$\begin{aligned} s(t_k, t_m) = {} & A\mathrm{rect}\left[\frac{t_k - \tau_P(t_m)}{T_P}\right]\exp\left\{\mathrm{j}\pi\mu\left[t_k - \left(\tau_P(t_m) - \frac{2R_c(m)}{c}\right)\right]^2\right\} \\ & \times \exp\left\{\mathrm{j}2\pi f_0\left[t_k - \left(\tau_P(t_m) - \frac{2R_c(m)}{c}\right)\right]\right\} \end{aligned} \tag{4.25}$$

Use

$$\tau'_P(t_m) = \tau_P(t_m) - \frac{2R_c(m)}{c} \tag{4.26}$$

to simplify (4.25) to get

$$s(t_k, t_m) = A\mathrm{rect}\left[\frac{t_k - \tau_P(t_m)}{T_P}\right]\exp[\mathrm{j}\pi\mu(t_k - \tau'_P(t_m))^2]\exp[\mathrm{j}2\pi f_0(t_k - \tau'_P(t_m))] \tag{4.27}$$

After pulse compression of the echo, the waveform is obtained as

$$s(t_k, t_m) = A'\mathrm{rect}\left[\frac{t_k - \tau_P(t_m)}{2T_P}\right]\sin c\left[\mu T_P\left(t_k - \tau'_P(t_m)\right)\right]\exp\left[-\mathrm{j}2\pi f_0\tau'_P(t_m)\right] \tag{4.28}$$

From the above equation, it can be seen that the echo of a certain pulse for the scattering point P, after matched filtering, the peak appears at

$$t_k = \tau'_P(t_m) = \frac{2[R_0(t_m) - R_c(t_m)] + 2r - 2a\theta(t_m) - r\theta^2(t_m)}{c}$$
$$= \frac{2[R_0(t_m) - R_c(t_m)]}{c} + \frac{2r}{c} - \frac{2a\theta(t_m) + r\theta^2(t_m)}{c} \tag{4.29}$$

And in different echoes, the phase of the echo at point P changes at the peak as follows:

$$\varphi(t_m) = 2\pi f_0 \tau_P(t_m) = 2\pi f_0 \frac{2[R_0(t_m) - R_c(t_m)] + 2r - 2a\theta(t_m) - r\theta^2(t_m)}{c}$$
$$= 2\pi f_0 \left(\frac{2[R_0(t_m) - R_c(t_m)]}{c} + \frac{2r}{c} - \frac{2a\theta(t_m)}{c} - \frac{r\theta^2(t_m)}{c} \right) \tag{4.30}$$

The azimuthal phase variation obtained from the above equation can be used to obtain the azimuthal Doppler variation of the target, and then calculate the PRF required in the W-band ISAR imaging system. In an ISAR system, the PRF of the system is the sampling rate in the azimuth direction. When the PRF is low, it can cause azimuthal aliasing. Considering the target located at the azimuthal coordinate a, the phase change due to rotation is calculated according to the above equation, and we can get

$$\Delta\varphi(t_m) = \frac{4\pi f_0 a\omega \Delta t_m}{c} = \frac{2\pi a\omega \Delta t_m}{\lambda} \tag{4.31}$$

Calculate the Doppler frequency in the azimuth direction as

$$f = \frac{1}{2\pi}\frac{\Delta\varphi(t_m)}{\Delta t_m} = \frac{a\omega}{\lambda} \tag{4.32}$$

$$B_a = f_{max} - f_{min} = \frac{(a_{max} - a_{min})\omega}{\lambda} = \frac{D_a \omega}{\lambda} \tag{4.33}$$

It can be seen that the azimuthal Doppler bandwidth is mainly related to the width of the azimuthal scene, the rotational angular velocity of the target, and the chosen wavelength. According to the sampling theorem, from PRF $\geq 2B$ we can obtain

$$\text{PRF} > \frac{2D_a\omega}{\lambda} = 3333\,\text{Hz} \tag{4.34}$$

As can be seen from the above equation, for W-band ISAR imaging, a large PRF is generally required to avoid azimuthal aliasing of the target due to the short wavelength. The larger the angular velocity of the target and the larger the scene area to be imaged, the higher the required PRF.

2. High-speed target range compression

According to Eq. (4.10), when the target has a certain moving speed, comparing the transmitted signal shows that the speed of the target causes the frequency of the echo to change from f_0 to $\alpha(t_{\mathrm{m}})f_0$, the rate to change from μ to $\alpha^2(t_{\mathrm{m}})\mu$ and the corresponding pulse duration to change from T_P to $T_P/\alpha^2(t_{\mathrm{m}})$, thus affecting the imaging of the target.

When using the matched filtering algorithm for range compression, the frequency of the echo signal and the change of the rate will cause the mismatch of the matched filter. According to the stop–go model, the rate of the matched filter should be μ, while the actual rate is $\alpha^2(t_{\mathrm{m}})\mu$. Using the principle of stationary phase, the spectrum of $u_c(t)$ can be calculated as

$$u_c(f) = \frac{1}{\sqrt{|\mu|}}\exp\left[\mathrm{j}\frac{\pi}{4}\operatorname{sgn}(\mu)\right]\exp\left[-\mathrm{j}\pi\frac{f^2}{\mu}\right],\quad |f| < \frac{B}{2} \tag{4.35}$$

According to the matched filter of the first-order approximate echo, the output signal spectrum is

$$U_c'\left(\frac{f}{\alpha(t_{\mathrm{m}})}\right) \approx \frac{1}{|\mu|}\exp\left[\mathrm{j}\pi\frac{f^2}{\mu}\left(\frac{1}{\alpha^2(t_{\mathrm{m}})} - 1\right)\right] \tag{4.36}$$

The phase mismatch error is

$$\Delta\varphi = \left|\pi\frac{f^2}{\mu}\left(\frac{1}{\alpha^2(t_{\mathrm{m}})} - 1\right)\right| \approx \left|\pi\frac{f^2}{\mu}\frac{4v}{c}\right| \le \frac{\pi v B T_P}{c} \tag{4.37}$$

When the phase mismatch error is less than π/2, the mismatch-induced pulse compressed output envelope is considered negligible, whereby the velocity of the target needs to satisfy the following inequality:

$$v \le \frac{c}{2BT_P} \tag{4.38}$$

For the W-band, based on the previously set system parameters with a bandwidth of 5 GHz and a T_P of 100 μs, it can be calculated that the required target velocity is less than 300 m/s if the matched filter for imaging is not mismatched. Because of the higher frequency and larger bandwidth of the W-band, the target velocity required for the stop–go model in the W-band is also relatively stringent.

When the target is moving faster than the speed required by the stop–go model, a first-order approximation model is used to construct the matched filter. In the first-order approximation model, what is more important is the radial motion speed of the target within the current echo. This velocity can be measured by using the narrow-wave tracking mode of the radar. After obtaining this velocity, the matched filter is constructed as follows:

$$s(t) = \mathrm{rect}\left[\frac{\alpha^2 t}{T_P}\right]\exp[\mathrm{j}\pi(2f_0\alpha t + \mu\alpha^2 t^2)] \tag{4.39}$$

where $\alpha = \frac{c-v}{c+v}$. By using this matched filter to match the echoes, the filter mismatch caused by the radial velocity of the target can be eliminated and better processing results can be obtained.

3. Subsequent signal processing

Analyzing Eq. (4.29), it can be seen that in the fast time dimension, the influence on the peak position of the echoes can be divided into three main factors: the first one is the movement caused by the target translation, the second one is the time shift caused by the position of the P point with respect to the distance direction of the target coordinate system, and the third one is the influence of the migration caused by the target rotation. The second term is required to be retained for the final imaging, and the remaining two terms need to be compensated.

The analysis of equationequation (4.30) shows that in the slow time dimension, the factors affecting the echo phase can be divided into four terms. The first one is also the phase change caused by the target translation. The second term is the initial phase change due to the target distance to the coordinates, which has no real impact on the imaging process and can be ignored. The third term is the phase change of the target azimuthal coordinates with the rotation of the target itself. In this term, if the angular velocity of the target is constant, a stable phase change is produced, and after the Fourier transform of the azimuthal direction, the energy of the P-point in the azimuthal direction can be concentrated to the corresponding frequency point for imaging. If the angular velocity of the target is not constant, the change in the angular velocity will cause ambiguities of the image and needs to be compensated during the imaging process. The fourth item is the azimuthal migration, which also needs to be compensated during the imaging process.

$$\begin{aligned}\varphi(t_\mathrm{m}) &= 2\pi f_0\tau_p(t_\mathrm{m})\\ &= 2\pi f_0\frac{2[R_0(t_\mathrm{m}) - R_c(t_\mathrm{m})] + 2r - 2a\theta(t_\mathrm{m}) - r\theta^2(t_\mathrm{m})}{c}\\ &= 2\pi f_0\left(\frac{2[R_0(t_\mathrm{m}) - R_c(t_\mathrm{m})]}{c} + \frac{2r}{c} - \frac{2a\theta(t_\mathrm{m})}{c} - \frac{r\theta^2(t_\mathrm{m})}{c}\right)\end{aligned} \tag{4.40}$$

In the envelope alignment and phase correction process of motion compensation, the first term of Eq. (4.29) and the first term of Eq. (4.30) will be compensated. In the migration compensation process of imaging processing, the range migration caused by the third term of Eq. (4.29), the angular velocity variation part in the third term of Eq. (4.30), and the azimuth migration caused by the fourth term will be compensated.

After motion compensation and imaging migration compensation, the ideal peak point of the target echo signal in the fast time direction is

$$t_\mathrm{k} = \frac{2r}{c} \tag{4.41}$$

In the slow time direction, the phase change of the target is

$$\varphi(t_{\mathrm{m}}) = 2\pi f_0\left(\frac{2r}{c}\right) - 2\pi f_0\frac{2a\omega t_{\mathrm{m}}}{c} = \varphi_0 - \frac{4\pi f_0}{c}a\omega t_{\mathrm{m}} \tag{4.42}$$

A high-resolution 2D image of the target can be obtained by simply performing Fourier transform in the azimuthal direction.

4.3 W-Band ISAR Motion Compensation Technology

In the ISAR imaging process, as long as the target tends to rotate with respect to the radar, the Doppler frequency change caused by the target rotation with respect to the radar can be used for azimuthal imaging. In practical situations, besides the rotation of the target, the impact on the echoes of the target also includes the radial displacement of the target Fig. 4.3 shows the geometric relationship between the target and the radar. Therefore, the effect of this velocity needs to be compensated f before the target can be imaged by ISAR.

According to the range compression processing results obtained in Sect. 4.2, the peak and the phase of the compressed single echo of a point on the target are as follows

$$t_{\mathrm{k}} = \frac{2[R_0(t_{\mathrm{m}}) - R_c(t_{\mathrm{m}})]}{c} + \frac{2r}{c} - \frac{2a\theta(t_{\mathrm{m}}) + r\theta^2(t_{\mathrm{m}})}{c} \tag{4.43}$$

$$\varphi = 2\pi f_0\left(\frac{2[R_0(t_{\mathrm{m}}) - R_c(t_{\mathrm{m}})]}{c} + \frac{2r}{c} - \frac{2a\theta(t_{\mathrm{m}})}{c} - \frac{r\theta^2(t_{\mathrm{m}})}{c}\right) \tag{4.44}$$

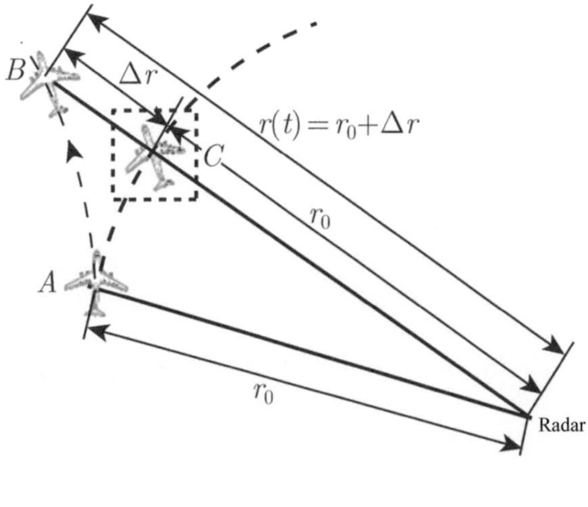

Fig. 4.3 Schematic diagram of motion compensation principle

It can be seen that the effect caused by the narrowband tracking error of the ISAR imaging system, that is, the $R_0(t_\mathrm{m}) - R_c(t_\mathrm{m})$ term causes a shift in the peak point of the target point after matched filtering in the t_k direction. Therefore, if imaging is performed, it is first necessary to compensate the same target point in the range direction to within a distance gate. This process is mainly achieved by the envelope alignment method.

After the matched filtered peaks of the same target are compensated within the same rrange cell, it can be seen that the phase of the polarization point is also affected by the influence brought by the tracking error, which causes the azimuthal ambiguities. Therefore, after the envelope alignment, the signal needs to be processed for phase correction to eliminate the effect of tracking error on the azimuthal phase change.

In this section, we analyze the envelope alignment and phase correction algorithms for motion compensation processing of ISAR images, and discuss the problems arising from the W-band motion compensation process. For the problem of large variations in the W-band echo envelope during the envelope alignment process, the quality of the image is improved by first performing envelope regularization.

4.3.1 Envelope Alignment

Envelope alignment is a processing method that eliminates the misalignment between ISAR echo signals after range compression processing by a translation operation in the range direction. Figure 4.4 shows the schematic diagram of envelope alignment. As shown in the figure, the left side is the echoes without processing. Due to the tracking error of the radar, the echoes of the same scattering point will be misaligned in the range direction. After the envelope alignment process, the echoes of the same scattering point in each echo are aligned to the same range cell, which makes it possible to further use the phase information in the azimuthal direction of the echoes for azimuthal imaging [1–4]. In this subsection, the envelope alignment method is analyzed and discussed with respect to the characteristics of the W-band.

Correlation Method The correlation method is the earliest proposed envelope alignment method, which mainly uses the strong correlation between adjacent echoes to calculate the distance to error between echoes and compensate for it. The time interval between two adjacent echoes in the ISAR imaging process is very short, which is 2.5×10^{-4} s for the system parameters used in this chapter. In such a short time, the translation and rotation of inert targets such as aircraft are very small. is very small. As a result, there is a large change in phase between two adjacent echoes, but the change in envelope is very small. When adjacent envelopes align in the distance direction, the cross-correlation coefficient will be very close to 1. The correlation method uses this principle to align the echoes in the range direction.

The envelope of the signal is divided into real envelope and complex envelope, and since the phase between adjacent echoes has a large variation, the complex envelope

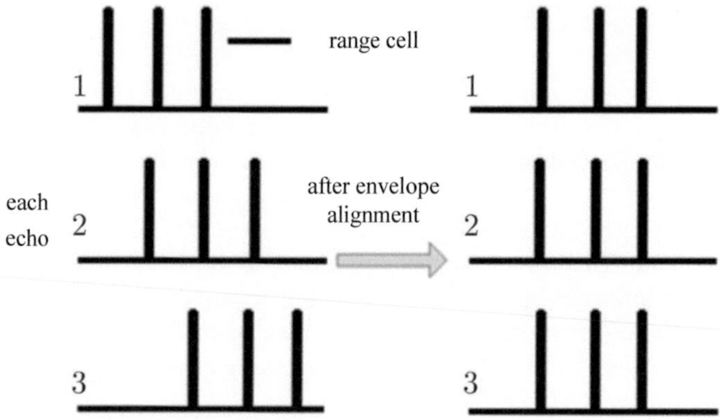

Fig. 4.4 Schematic diagram of envelope alignment principle

will also have a certain variation, so the correlation method mainly deals with the real envelope of the echo signal.

For the echo signal, the azimuthal coordinates are transformed to the range domain, and let $e_{\mathrm{m}}(t_{\mathrm{k}})$ and $e_{\mathrm{m}+1}(t_{\mathrm{k}})$ be the distance images of two adjacent echoes, while $E_{\mathrm{m}}(t_{\mathrm{k}})$ and $E_{\mathrm{m}+1}(t_{\mathrm{k}})$ are their Fourier transforms, respectively. According to Eq. (4.34), we can obtain

$$e_{\mathrm{m}}(t_{\mathrm{k}}) = \sum_{P} A'\mathrm{sinc}\left[\mu T_P\left(t_{\mathrm{k}} - \tau_P'(t_{\mathrm{m}})\right)\right]\exp\left[-\mathrm{j}2\pi f_0 \tau_P'(t_{\mathrm{m}})\right] \tag{4.45}$$

$$e_{\mathrm{m}+1}(t_{\mathrm{k}}) = \sum_{P} A'\mathrm{sinc}\left[\mu T_P\left(t_{\mathrm{k}} - \tau_P'(t_{\mathrm{m}+1})\right)\right]\exp\left[-\mathrm{j}2\pi f_0 \tau_P'(t_{\mathrm{m}+1})\right] \tag{4.46}$$

Combine

$$\tau_P'(t_{\mathrm{m}}) = \frac{2[R_0(t_{\mathrm{m}}) - R_c(t_{\mathrm{m}})]}{c} + \frac{2r}{c} - \frac{2a\theta(t_{\mathrm{m}}) + r\theta^2(t_{\mathrm{m}})}{c} \tag{4.47}$$

It can be seen that the variation of $e_{\mathrm{m}+1}(t_{\mathrm{k}})$ with respect to $e_{\mathrm{m}}(t_{\mathrm{k}})$ is mainly composed of the error in range and the variation in phase, which are labeled by Δt and φ_i, respectively, then we can obtain

$$e_{\mathrm{m}+1}(t_{\mathrm{k}}) \approx e_{\mathrm{m}}(t_{\mathrm{k}} - \Delta t_{\mathrm{km}})\exp(j\varphi_{\mathrm{m}}) \tag{4.48}$$

$$E_{\mathrm{m}+1}(t_{\mathrm{k}}) \approx E_{\mathrm{m}}(f)\exp(-\mathrm{j}2\pi f \Delta t_{\mathrm{km}})\exp(\mathrm{j}\varphi_{\mathrm{m}}) \tag{4.49}$$

Among them.

$$\Delta t_{\mathrm{km}} = \tau_P'(t_{\mathrm{m}+1}) - \tau_P'(t_{\mathrm{m}})$$

$$= \frac{2\Delta[R_0(t_m) - R_c(t_m)]}{c} - \frac{2a\Delta\theta(t_m) + r\Delta\theta^2(t_m)}{c}$$
$$\approx \frac{2\Delta[R_0(t_m) - R_c(t_m)]}{c} \tag{4.50}$$

Since the time is short, the range shift brought by the angle change in the adjacent envelope is negligible. Their cross-correlation function is thus obtained as

$$R(s) = \int |e_m(t_k)| \cdot |e_{m+1}(t_k + s)| dt_k \approx \int |e_m(t_k)| \cdot |e_m(t_k + s - \Delta t_{km})| dt_k \tag{4.51}$$

According to Schwartz inequality, $R(s)$ can be obtained as a maximum at $s = \Delta t_{km}$. Therefore, the error of the adjacent envelope Δt_{km} can be obtained from the location of the maximum value of the cross-correlation function, and this error is then used to translate the envelope to compensate for the translation of the target in the range direction caused by the term $R_0(t_m) - R_c(t_m)$.

The envelope shifts calculated by the above method is an integer multiple of the range sampling, and interpolation can be used to further improve the accuracy of the compensation.

1. 1st Order distance minimization method

The distance minimization method is another envelope alignment algorithm in ISAR imaging motion compensation, which uses a different alignment determination criterion than the correlation method.

Consider two envelope-similar signals $e_1(t)$ and $e_2(t)$ for which the translation error compensation is performed, if the maximum likelihood criterion is used, it should be calculated as follows:

$$\tau = \arg\min_{\tau} |e_1(t) - e_2(t - \tau)|^2 \tag{4.52}$$

The computed τ is the error amount, and the compensation result can be obtained after the translation. Further expansion of the above equation can be obtained

$$\begin{aligned} \min_{\tau} |e_1(t) - e_2(t - \tau)|^2 &= \min_{\tau} [e_1(t) - e_2(t - \tau)]\left[e_1^*(t) - e_2^*(t - \tau)\right] \\ &= \min_{\tau} 2\{1 - \mathrm{Re}[\rho(\tau)]\} \end{aligned} \tag{4.53}$$

where $\rho(\tau) = e_1(t)e_2(t - \tau)$ is the correlation function of the two input signals. It can be seen that using the maximum likelihood criterion to find the method that minimizes the 2nd order parametrization is equivalent to the correlation method to find the maximum of the cross-correlation function. The maximum likelihood criterion for the 2nd order parametrization can be further extended to obtain the expression for the P-order parametrization.

$$\tau = \arg\min_{\tau} |e_1(t) - e_2(t-\tau)|^P \tag{4.54}$$

Considering this criterion as the standard for envelope alignment, it can be seen that when P increases, the point with the larger value in the two data sets contributes more to the calculation of the alignment error. When P is infinite, the above equation can be considered as a direct alignment of the maximum values of the two sets of signals. Considering that in radar echoes, there are often strong scintillation points, which have a greater impact on the envelope alignment, and the increase of P will strengthen the impact of such scintillation points on the alignment.

As P decreases, the contribution of larger values to the calculation of the alignment error decreases if $P > 1$. When $P = 1$, all values have the same contribution. And when $P < 1$, smaller values have a larger contribution, which is clearly not what we need. Therefore, $P = 1$ is the ideal choice. In contrast to the correlation method, the distance minimization method not only improves the processing of scintillation points, but also uses the direct addition and subtraction of neighboring cells in the calculation of the evaluation function, which is less complex.

2. Minimum entropy algorithm

Both the correlation method and the 1st-order distance minimization method are used to estimate the delay error by making a difference between the adjacent echoes and using the corresponding evaluation method for the obtained difference. As can be imagined, when the delay error of the adjacent echoes is small, the value obtained from the difference is mainly composed of the different parts of the two waveforms, that means a large part is made up of random errors. Thus, when compensated to a high accuracy, the main influence on the error estimation will not be the main part of the envelope but the random error, which affects the accuracy of further estimation. As a result, the envelope minimum entropy algorithm was proposed in the 1990s. For signals $e_1(t)$ and $e_2(t)$, the envelope minimum entropy algorithm uses the minimum value of the sharpening degree $H(\tau)$ of the envelope as the criterion for determining the envelope alignment of the.

$$H(\tau) = -\sum_{t \in T_P} [e_1(t) + e_2(t-\tau)] \ln[e_1(t) + e_2(t-\tau)] \tag{4.55}$$

The schematic diagram of the principle of the minimum entropy algorithm is shown in Fig. 4.5. Considering two adjacent envelopes that are not aligned, as shown on the left side of the figure, it can be seen that after doing the sum, the obtained envelope is not sharp enough due to staggering, and the entropy obtained from this calculation is larger. When these two envelopes are aligned as on the right, the resulting envelope and is sharper, which means that the entropy is smaller. This shows the effectiveness of the minimum entropy algorithm. The minimum entropy algorithm makes full use of all the information of the target through the summation process, and has better robustness than the echo similarity-based envelope alignment algorithm.

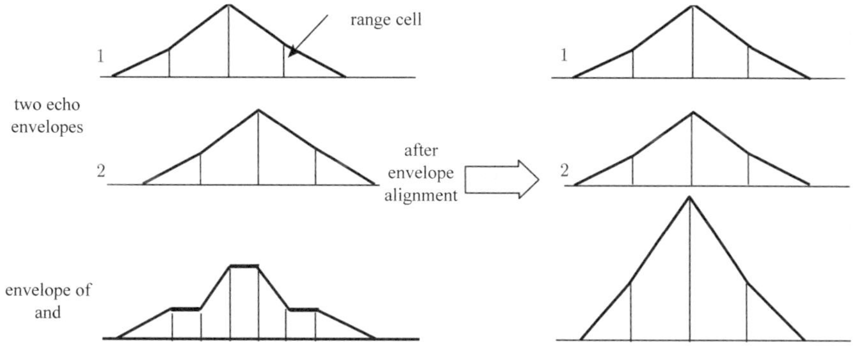

Fig. 4.5 Schematic diagram of the principle of minimum entropy algorithm

4.3.2 Envelope Regularization Processing

1. W-band envelope variation problem

In the process of aligning the envelopes of ISAR echoes, due to the movement and rotation of the target, scintillation points often appear in one of the echoes of the image, causing a large gap between adjacent envelopes and affecting the alignment results of the echoes. At the same time, if only the adjacent echoes are processed, if a small error is generated, the error will accumulate because there is no error correction mechanism, causing the echoes to shift continuously with the change of slow time, so the weighted average of the previous multiple echoes can be used as the alignment standard. The longer the window length, the better the suppression of scintillation points and accumulated errors, so a rectangular window with half sides is generally chosen.

In conventional band ISAR imaging, such processing does not pose any problems because the target point does not generally cross the resolution cell during the imaging time, causing significant changes in the envelope. However, in the W-band ISAR imaging process, due to the increased resolution and shorter range cells, the rotation of the target will cause the reflecting points on the target to cross multiple range cells and causing significant migration. If a longer rectangular window is used directly to accumulate compensation, it will result in more serious errors, or even completely wrong alignment results.

The echo delay obtained from the echo model analysis in Sect. 4.2 is shown below

$$t_{\mathrm{k}} = \frac{2R_0(t_{\mathrm{m}})}{c} + \frac{2r}{c} - \frac{2a\theta(t_{\mathrm{m}}) + r\theta^2(t_{\mathrm{m}})}{c} \tag{4.56}$$

Assuming there are P and Q points on the target, and assuming that the target is rotating at a uniform angular velocity, the delay between the two points in the imaging process can be obtained as follows:

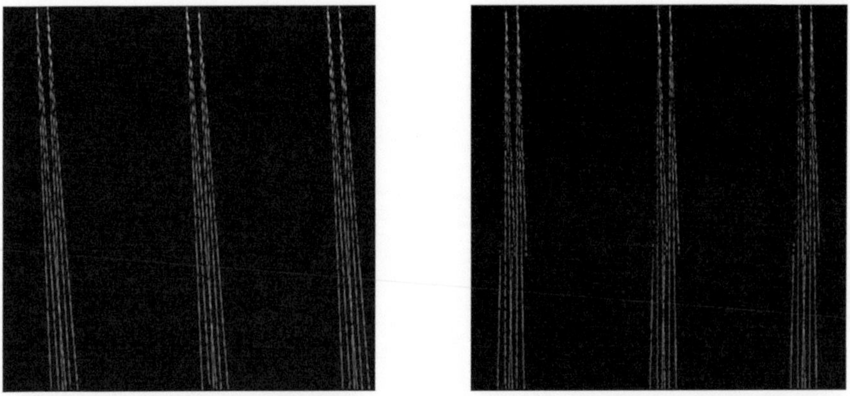

(a) result of distance compression (b) result of envelope alignment

Fig. 4.6 Schematic diagram of alignment errors caused by envelope changes

$$\Delta t_{\mathrm{k}} = \left|t_{\mathrm{kp}} - t_{\mathrm{kq}}\right| = \left|\frac{(2+\omega^2 t_{\mathrm{m}}^2)}{c}\Delta r - \frac{2\omega t_{\mathrm{m}}}{c}\Delta a\right| \tag{4.57}$$

Assume that point P and point Q are at the same echo position, tand then they have the same echo delay, and are on both sides of the target in azimuthal direction. Then, the range shift of the target during the imaging process is estimated based on the system parameters given in Sect. 4.2 as follows:

$$\Delta r_{\mathrm{image}} = \frac{2\omega t_{\mathrm{m}}}{c} D \times c = 2\omega D T_P = 5\mathrm{m} \tag{4.58}$$

It can be seen that this shift is well beyond the resolution of the W-band ISAR imaging and results in a significant change in the echo envelope after range compression. Figure 4.6 shows the results of range compression and direct envelope alignment for the 9-point target.

From Fig. 4.6a, it can be seen that the envelope of the target at the beginning moment changes very significantly from the envelope of the target at the end moment. The results of the direct minimum entropy envelope alignment are shown in Fig. 4.6b. It can be seen that although the target displacement is aligned to a certain extent, the alignment is clearly wrong in the case of a large envelope change, aligning the echo signals that are completely out of phase into the same range cell. The final imaging results are shown in Fig. 4.7, and such an alignment is completely unsuccessful in obtaining imaging results. Therefore, it is necessary to perform an envelope rectification process before the envelope alignment to compensate for the absence of significant changes in the echo envelope in the slow time dimension, making the next step of envelope alignment possible.

Fig. 4.7 Schematic diagram of imaging defocus caused by envelope alignment error

2. Signal Coherence Processing

In response to this phenomenon, the method of first compensating for distance migration by Keystone transform on the echo signal, and then aligning the envelope of the half-edge rectangular window is adopted to address the problem of large envelope transformation in the W-band. The Keystone transform can compensate the range migration of the target echoes more accurately, so that the echo envelopes of different slow times are similar again. Since the Keystone transform mainly compensates for the migration of the echoes, it is described in detail in Sect. 4.4.

Since Keystone processing requires the signal to be coherent, the echoes need to be coherent before the Keystone transform is performed. For inert targets such as aircrafts and satellites, the translational motion of the target relative to the radar can be considered to be continuously varying after removing the tracking error of the radar. And the exact echo time of the target is easier to be recorded. Therefore, if a continuously moving echo window is used in the process of tracking, the target motion in the window can be considered to be continuously varying. Therefore, the coherent processing of the echo signal is achieved by fitting the trajectory of the target's motion in the echo window and then using the fitted result to compensate for the phase change caused by its motion.

Since the resolution of W-band is high and the scattering characteristic of the target is good, the scattering points on the target are generally distributed more uniformly. It can be assumed that the weighted center of gravity of the target echo envelope in a certain echo is the center of gravity of the target in the echo window, and the trajectory of the target in the echo window can be obtained by tracking the center of gravity of the target.

However, during the actual aircraft flight, small-amplitude vibration occurs due to the engine and air turbulence, and this vibration will make the actual radar echo phase deviate from the fitted trajectory. Here, we can use a shorter window length alignment process to obtain the information of small-amplitude vibration of the target by searching in a small range in the vicinity according to the fitted motion trajectory. By combining the fitted trajectory of the target's center of gravity with the information of the small-range vibration of the target obtained from the search, the phase change between adjacent envelopes caused by the target translation can be calculated, and the coherence of the echo signal can be completed.

The envelope similarity of the W-band ISAR echo signal is greatly improved by coherent and Keystone processing of the echoes. After this, the envelope alignment algorithm with a half-edge rectangular window can be used without any significant variation in the echo envelope.

3. Envelope regularization processing flow

Based on the above analysis, the processing flow of the alignment method based on envelope regularization is given here.

Step 1: Estimate the target center of gravity position from the result of range compression and fit the center of gravity shift curve.

Step 2: Based on the movement curve of the center of gravity, an alignment process with shorter window length and smaller range is performed near the curve to obtain the amplitude of the target's vibration in a small range of range.

Step 3: Based on the results obtained in the first two steps, the phase to be compensated is calculated and superimposed into the range compression result to make it coherent.

Step 4: Keystone transform is performed on the coherent data to compensate for the migration in the range direction and complete the envelope regularization process.

Step 5: Using a longer window length, align the data after envelope regularization process to get the alignment result.

Figure 4.8 shows the detailed processing flow.

4.3.3 Phase Correction Technology

After performing the envelope alignment, the target echo has been aligned to the resolution cell for imaging, but such an alignment result does not allow for further imaging. From Eq. (4.36) in Sect. 4.2, it can be obtained that for a given scattering point, its phase change in the slow time direction after range compression is as follows

$$\phi(t_{\mathrm{m}}) = 2\pi f_0\left(\frac{2[R_0(t_{\mathrm{m}}) - R_c(t_{\mathrm{m}})]}{c} + \frac{2r}{c} - \frac{2a\theta(t_{\mathrm{m}})}{c} - \frac{r\theta^2(t_{\mathrm{m}})}{c}\right) \tag{4.59}$$

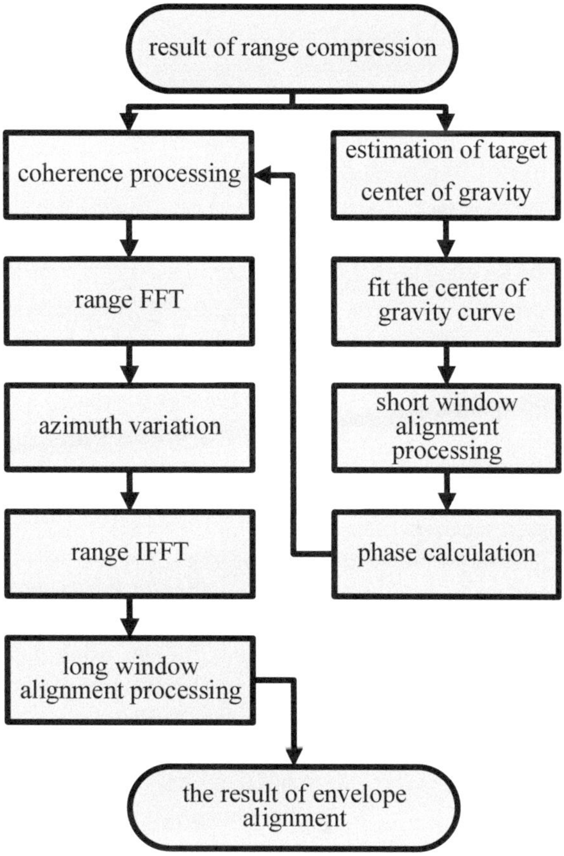

Fig. 4.8 Alignment processing flowchart based on envelope regularization

As can be seen, the tracking error $R_0(t_m) - R_c(t_m)$ affects the phase of the aligned echo signal, resulting in azimuthal defocussing and unavailable imaging results. Therefore, further phase correction is needed to compensate for the tracking error term.

1. Prominent Point method

The method of the prominent point is the first proposed phase correction method. For Eq. (4.59), if a point P is assumed to be located at the center of the target, that is, at the point $r = 0, a = 0$ Substituting it into Eq. (4.59), we can obtain

$$\phi_P(t_m) = 2\pi f_0\left(\frac{2[R_0(t_m) - R_c(t_m)]}{c}\right) \tag{4.60}$$

It can be seen that only the tracking error term will exist for the Doppler variation at this point. Thus, this point can be used to compensate for the phase tracking error. Taking

$$\Delta\phi(t_m) = -\phi_P(t_m) = -2\pi f_0\left(\frac{2[R_0(t_m) - R_c(t_m)]}{c}\right) \tag{4.61}$$

Superposition the signals after envelope alignment. Considering the echo signal of a scattering point, we can obtain

$$\begin{aligned}\varphi'(t_m) &= \varphi(t_m) + \Delta\varphi(t_m)\\ &= 2\pi f_0\left(\frac{2[R_0(t_m) - R_c(t_m)]}{c} + \frac{2r}{c} - \frac{2a\theta(t_m)}{c} - \frac{r\theta^2(t_m)}{c}\right)\\ &\quad - 2\pi f_0\left(\frac{2[R_0(t_m) - R_c(t_m)]}{c}\right)\\ &= 2\pi f_0\left(\frac{2r}{c} - \frac{2a\theta(t_m)}{c} - \frac{r\theta^2(t_m)}{c}\right)\end{aligned} \tag{4.62}$$

It can be seen that the tracking errors in the images are compensated by performing phase superposition.

However, in the actual echo signal, it is basically impossible to have a scattering point that is exactly at the center of the target. The prominent point method is to select a prominent point in the image, which is considered to be at the center of the target, and use it for phase correction. Consider using a prominent point $P(r_P, a_P)$ for phase correction, and the corrected echo signal of a scattering point can be obtained according to the above derivation as follows:

$$\begin{aligned}\phi'(t_m) &= \phi(t_m) + \Delta\phi(t_m)\\ &= 2\pi f_0\left(\frac{2[R_0(t_m) - R_c(t_m)]}{c} + \frac{2r}{c} - \frac{2a\theta(t_m)}{c} - \frac{r\theta^2(t_m)}{c}\right)\\ &\quad - 2\pi f_0\left(\frac{2[R_0(t_m) - R_c(t_m)]}{c} + \frac{2r_p}{c} - \frac{2a_p\theta(t_m)}{c} - \frac{r_p\theta^2(t_m)}{c}\right)\\ &= 2\pi f_0\left(\frac{2(r - r_P)}{c} - \frac{2(a - a_P)\theta(t_m)}{c} - \frac{(r - r_P)\theta^2(t_m)}{c}\right)\end{aligned} \tag{4.63}$$

The first term in equation is a constant value and has no effect on the imaging results. The second term in equation introduces an additional Doppler frequency shift, which is reflected in the imaging as an azimuthal shift of the image, moving the P point in the azimuthal direction to the center of the image. The third term in equation is a migration term that needs to be dealt with in further migration compensation. Therefore, if we want to avoid the effect on the migration compensation, we can avoid the effect of the third term by shifting the image in the range direction before the migration compensation process and then shifting it in the opposite direction after the compensation. The phase error caused by the image center shift can also be superimposed directly during the phase compensation process.

From the above derivation, it can be seen that the prominent point method has very high requirements on the variation of the azimuthal phase of the selected special display point, so that a high result can be obtained only if there is only one scattering point in that range cell. The method of calculating the variance of the normalized envelope amplitude within the range cell is generally used for screening. The calculation method is as follows:

$$\sigma^2(u_n) = \overline{(u_n - \overline{u}_n)^2}/\overline{u_n^2} = 1 - \overline{u}_n^2/\overline{u_n^2} \tag{4.64}$$

If a very good single prominent point cannot be found, try to find several good prominent points and use the weighted average of multiple prominent points to combine them to come up with a better reference point.

2. Doppler centroid tracking method

For W-band ISAR imaging, since the target has a better echo scattering effect, the prominent points in the echoes are more difficult to obtain than in conventional bands, thus requiring processing methods that do not depend on a prominent point. For a given target, if the center of rotation is in the target, some of the scattering points of the target must be close to the radar motion, and some of the scattering points are far from the radar motion. Then there must exist a certain point on the target, when the target rotates around the point, the overall echo Doppler sum is zero. The Doppler centroid tracking method is used to achieve the phase correction by forcing the total Doppler of each slow time cell of the envelope-aligned echo to zero.

For adjacent echoes, the average phase difference between them can be expressed as follows

$$e^{\mathrm{j}\varphi} = \frac{\int e_{\mathrm{m}}^*(t_{\mathrm{k}})e_{\mathrm{m+1}}(t_{\mathrm{k}})dt_{\mathrm{k}}}{\left|\int e_{\mathrm{m}}^*(t_{\mathrm{k}})e_{\mathrm{m+1}}(t_{\mathrm{k}})dt_{\mathrm{k}}\right|} \tag{4.65}$$

where $e_{\mathrm{m}}(t_{\mathrm{k}})$ and $e_{\mathrm{m+1}}(t_{\mathrm{k}})$ are the echo signals of two adjacent slow time cells after envelope alignment, respectively. After the phase difference is calculated using the above equation, the phase difference is superimposed on the echoes of each slow time cell to compensate for the phase difference caused by the compensation error of the envelope alignment.

However, in the above processing, the same weighting factor is used for each range cell in the echo signal, that is, a range cell with no scattering point at all has the same contribution to the final phase difference estimate as the echo signal of a scattering point with better phase information, which can introduce a large error. Therefore, the weights of each range cell can be further calculated so that the scattering point with better phase information contributes more to the result.

There are two main types of factors that affect the echo phase of a range cell, the interference effect caused by multiple scattering points within the same range cell, and the effect caused by jamming signals dominated by the sidelobes of range compression.

For the former effect, it can be weighted by the value of the normalized envelope amplitude variance, so that the range cells with more serious interference have less influence on the calculation results. Since the smaller the variance is taken, the better the phase information is retained, the following weights can be taken:

$$W_{1n} = \frac{1}{\left(\overline{\sigma}_n^2 \sum_{i=1}^{N_R} \frac{1}{\sigma_i^2}\right)} \tag{4.66}$$

where $\overline{\sigma_n^2}$ is the normalized envelope amplitude variance of the nth range cell; N_R is the total number of azimuthal points of the image.

For the latter effect, a threshold A can be chosen to remove the sidelobes of range compression and other jamming signals so that they do not have an impact on the estimation of the total phase difference. That is

$$\begin{cases} W_{2n} = 1, A_n > A \\ W_{2n} = 0, A_n \le A \end{cases} \tag{4.67}$$

After the above calculation of the weighting coefficients, substituting them into the phase calculation formula, we get

$$e^{j\varphi} = \frac{\int e_{\mathrm{m}}^*(t_{\mathrm{k}}) e_{\mathrm{m}+1}(t_{\mathrm{k}}) W_{1n} W_{2n} dt_{\mathrm{k}}}{\left|\int e_{\mathrm{m}}^*(t_k) e_{\mathrm{m}+1}(t_{\mathrm{k}}) W_{1n} W_{2n} dt_{\mathrm{k}}\right|} \tag{4.68}$$

By doing so, the accuracy of estimating the phase difference between adjacent slow time cells can be further improved.

3. Phase gradient autofocus method

When tracking the Doppler center of a target, the Doppler centroid tracking method only considers the phase variation of the target due to different distances, and it obtains the required phase difference between different range cells by summing the phases of the points on a certain echo cell. However, in the ISAR echo data, the position difference of the scattering points in the azimuthal direction also causes the phase variation of the echoes. Considering two scattering points P and Q symmetric in the range direction with respect to the target Doppler center, their coordinates are assumed to be (r, a_P) and $(-r, a_Q)$, respectively, then the phase sum of these two scattering points in a certain echo can be obtained as

$$\begin{aligned} \sum \varphi &= \varphi_P + \varphi_Q \\ &= 2\pi f_0 \left(\frac{2[R_0(t_{\mathrm{m}}) - R_c(t_{\mathrm{m}})]}{c} + \frac{2r}{c} - \frac{2a_P\theta(t_{\mathrm{m}})}{c} - \frac{r\theta^2(t_{\mathrm{m}})}{c} \right) \\ &\quad + 2\pi f_0 \left(\frac{2[R_0(t_{\mathrm{m}}) - R_c(t_{\mathrm{m}})]}{c} - \frac{2r}{c} - \frac{2a_Q\theta(t_{\mathrm{m}})}{c} + \frac{r\theta^2(t_{\mathrm{m}})}{c} \right) \\ &= 2\varphi_0 + 2\pi f_0 \left(\frac{2(a_P + a_Q)\theta(t_{\mathrm{m}})}{c} \right) \end{aligned} \tag{4.69}$$

According to the Doppler centroid estimation method, the phase sum of two points symmetric with respect to the Doppler center should be zero. From the above equation, it is clear that the azimuthal position of the scattering point will cause errors in the initial phase estimation, resulting in degradation of the imaging quality. The phase gradient autofocus (PGA) algorithm is proposed to address this shortcoming of the Doppler centroid tracking method [5–10]. The PGA method avoids the effect of azimuthal position by compensating the phase error generated by the azimuthal position of the prominent point of a range cell and then performing Doppler centroid tracking.

The PGA method is divided into the following steps:

Step 1: Perform a Doppler centroid tracking phase correction and Fourier transform imaging to obtain the initial image of the target.

Step 2: Find the strong scattering point within each range cell and compensate the phase caused by the azimuthal position of this strong scattering point, and perform windowing in the frequency domain to isolate the strong scattering point.

Step 3: Perform inverse Fourier transform on the compensated data in the azimuthal direction back to the signal domain, align it again for Doppler centroid tracking phase correction, and image to obtain the initial image for the next iteration of processing.

Step 4: Repeat steps 2 and 3, and keep reducing the window length used to isolate the strong scattering points until the desired number of iterations is reached.

The strong scattering point in step 2 of the above process can be found by using criteria such as maximum amplitude and maximum contrast. The phase compensation can be achieved by superimposing the phase in the frequency domain or by a direct cyclic shift in the time domain. Also, a weighting method similar to that in the Doppler centroid tracking method can be used to improve the accuracy of the phase correction process before the PGA processing.

4.4 W-Band ISAR Imaging and Migration Compensation Technology

Because of the high resolution and small resolution cells in W-band, the migration problem of trans-resolution cells is much more serious than in conventional bands. The migration compensation problem is not very serious in the conventional band, and in many cases, the target only migrates a few or even less than one range cell, and good imaging results can be obtained without compensation. However, in the W-band, the target rotation may bring about migration of dozens or even tens of range cells, therefore, in order to perform ISAR imaging in the W-band, the migration compensation needs to be studied thoroughly.

With the increase in imaging resolution, the amount of data in W-band images is also much larger than in conventional bands. If the computational complexity of the algorithm used for imaging is higher, it might bring a larger increase in computational time. Therefore, the computational complexity of the imaging algorithm needs to be considered.

4.4.1 Conventional Imaging Methods

In traditional band imaging, there are many mature methods. Here, we mainly introduce the Range Doppler (RD) algorithm, Keystone transform algorithm, polar format algorithm (PFA), and time–frequency algorithm. The RD algorithm is the most basic algorithm in ISAR imaging, which is based on the direct Fourier transform of the motion-compensated data to obtain the imaging results. The PFA requires the exact angular velocity of the target rotation as the motion parameter and is widely used in stable target imaging. The time–frequency algorithm can obtain the imaging results of targets with rotational speed variations by time–frequency transformation of the echo data.

In this section, each algorithm is described in turn, and its problems in W-band ISAR imaging are pointed out.

1. RD Imaging algorithm

According to the results obtained in Sect. 4.2, the phase of the echo at a scattering point after range compression at the point where the distance to the maximum is taken, and at the point of the maximum, is as follows

$$t_{\mathrm{k}} = \frac{2[R_0(t_{\mathrm{m}}) - R_c(t_{\mathrm{m}})]}{c} + \frac{2r}{c} - \frac{2a\theta(t_{\mathrm{m}}) + r\theta^2(t_{\mathrm{m}})}{c} \tag{4.70}$$

$$\varphi(t_m) = 2\pi f_0\left(\frac{2[R_0(t_m) - R_c(t_m)]}{c} + \frac{2r}{c} - \frac{2a\theta(t_m)}{c} - \frac{r\theta^2(t_m)}{c}\right) \tag{4.71}$$

After the motion compensation process in Sect. 4.3, the tracking error term in the phase of the echo is compensated,which is:

$$t_{\mathrm{k}} = \frac{2r}{c} - \frac{2a\theta(t_{\mathrm{m}}) + r\theta^2(t_{\mathrm{m}})}{c} \tag{4.72}$$

$$\varphi(t_{\mathrm{m}}) = 2\pi f_0\left(\frac{2r}{c} - \frac{2a\theta(t_{\mathrm{m}})}{c} - \frac{r\theta^2(t_{\mathrm{m}})}{c}\right) \tag{4.73}$$

The RD algorithm is a method to image them directly by azimuthal Fourier transform without further compensation.

From Eq. (4.72), we can see that since the azimuthal scale of the target in the second term will bring about displacement in the range direction with rotation, and the range scale will also bring about displacement in the range direction with second-order rotation, these displacements will make the target migrate in the range direction and cause ambiguities in the range direction of the image.

As can be seen in Eq. (4.73), its third term, the phase of the range scale of the target with rotation, has an impact on the imaging. This term affects the image in the W-band due to the high center frequency, which can cause severe ambiguities in the azimuthal direction.

The RD algorithm gives poor imaging results in W-band ISAR imaging due to the presence of scattering point migration caused by target rotation in both directions.

2. Keystone transform algorithm

The Keystone transform algorithm is a scale-changing algorithm that first performs a Fourier transform in the range direction for the range-compressed data, so that the echo image is transformed to the $f - t_{\mathrm{m}}$ domain. Next, a scale-changing operation is performed in the azimuthal direction, taking

$$\tau_{\mathrm{m}} = \frac{f_{\mathrm{c}}}{f_{\mathrm{c}} + f} t_{\mathrm{m}} \tag{4.74}$$

where t_{m} and τ_{m} are the azimuthal coordinates before and after the transformation, respectively; f_{c} and f are the center frequency of the system and the additional frequency generated by the FM signal, respectively. The Keystone transform is shown in Fig. 4.9. The wedge shape of this scale-changing operation is similar to the Keystone shape, so it is named the Keystone transformation. It can be seen that the Keystone transform stretches the t_{m}-axis with respect to the frequency in the $f - t_{\mathrm{m}}$ domain, compressing the low frequency signal and stretching the high frequency signal.

Fig. 4.9 Schematic diagram of Keystone transform and labeling operation

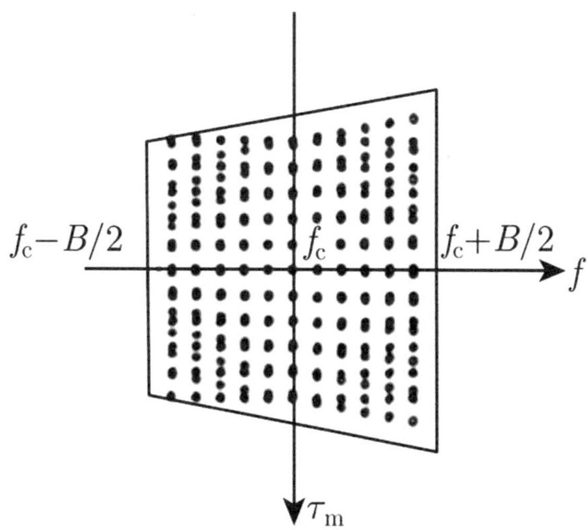

After the scale-changing operation, the transformed result is subjected to IFFT in the range direction to bring the image back to the $t_k - t_m$ domain, completing the Keystone scale-changing transform.

The Keystone transform improves the imaging resolution because the echo data undergoes a range direction Fourier transform, which is essentially a Fourier transform of a chirp signal of one pulse width in duration with a fixed phase attached, part of which is caused by the delay of the echo and part of which is caused by the carrier frequency; the essence of the Keystone transform is to rearrange the column numbers (the corresponding frequencies) of the stored matrix elements in the azimuthal direction according to the range direction through the Fourier transform., and the principle of rearrangement is to eliminate as much as possible the changes in the spectral components and distribution caused by the target motion.

The essence of distance and Doppler independence of the echo data is to calculate the virtual azimuthal moment τ assuming no change in the target distance (all, $f_c + f$ compressed or extended to f_c). The coupling of the radially moving target distance f and Doppler information is mainly caused by the target motion, and the main effect of the target radial motion on the echo is the change in distance. Therefore, it is very effective to calculate the virtual azimuthal moments assuming that the target is not moving or that the target motion has no effect on the relative distance.

The Keystone transform corrects the echoes originally located at different range cells to the same range cell, compensating for the range migration. After the correction of range migration, the inter-envelope variation of each echo in the image is significantly weakened, which leads to a better result of envelope alignment. The degree of focus of the image after coherent accumulation is improved. However, the Keystone transform algorithm only compensates for the range migration of the target motion, but not for the azimuthal migration of the target.

In order to implement the Keystone transform, the most direct method is to perform interpolation in the azimuthal direction after the Fourier transform in the range direction according to the variational formula to achieve the variational scale in the azimuthal direction. However, the computational complexity of this direct calculation is high. The Chirp-Z transform can be used here to compute any set of equally spaced samples of the discrete Fourier transform on the unit circle.

Let $x[n]$ denote a sequence of N points and $X(e^{j\omega})$ denote its Fourier transform, taking its M samples on the unit circle that are at the frequencies shown in the following equation:

$$\omega_k = \omega_0 + k\Delta\omega, k = 0, 1, \cdots, M - 1 \tag{4.75}$$

where the starting frequency ω_0 and the frequency increment $\Delta\omega$ can be chosen arbitrarily. For the special case of the discrete Fourier transform (DFT), $\omega_0 = 0$, $M = N$, and $\Delta\omega = 2\pi/N$.

The Fourier transform corresponding to the above equation is as follows:

$$X(e^{j\omega}) = \sum_{n=0}^{N-1} x[n]e^{-j\omega_k n}, k = 0, 1, \cdots, M-1 \tag{4.76}$$

Definition

$$W = e^{-j\Delta\omega} \tag{4.77}$$

then you can get

$$X(e^{j\omega}) = \sum_{n=0}^{N-1} x[n]e^{-j\omega_0 n} W^{nk}, k = 0, 1, \cdots, M-1 \tag{4.78}$$

In order to represent the above equation as a convolution, using the equation

$$nk = \frac{1}{2}[n^2 + k^2 - (k-n)^2] \tag{4.79}$$

Equation (4.78) can be expressed as

$$X(e^{j\omega}) = \sum_{n=0}^{N-1} x[n]e^{-j\omega_0 n} W^{\frac{n^2}{2}} W^{\frac{k^2}{2}} W^{-\frac{(k-n)^2}{2}}, k = 0, 1, \cdots, M-1 \tag{4.80}$$

make

$$g[n] = x[n]e^{-j\omega_0 n} W^{\frac{n^2}{2}} \tag{4.81}$$

You can get

$$X(e^{j\omega}) = W^{\frac{k^2}{2}} \sum_{n=0}^{N-1} g[n]e^{-j\omega_0 n} W^{-\frac{(k-n)^2}{2}}, k = 0, 1, \cdots, M-1 \tag{4.82}$$

To obtain the more familiar form, interchange n and k to obtain

$$X(e^{j\omega}) = W^{\frac{n^2}{2}} \sum_{n=0}^{N-1} g[k]e^{-j\omega_0 k} W^{-\frac{(n-k)^2}{2}}, n = 0, 1, \cdots, M-1 \tag{4.83}$$

From the above equation, $X(e^{j\omega})$ corresponds to the result of the convolution of the sequence $g[n]$ with the sequence $W^{-\frac{n^2}{2}}$ multiplied by the sequence $W^{\frac{n^2}{2}}$.

Thus, to implement the Keystone transform, the azimuthal signal can be transformed to the frequency domain using the Chirp-Z transform, according to the calculated scalar ratio, and then inverse transformed using the inverse Fourier transform.

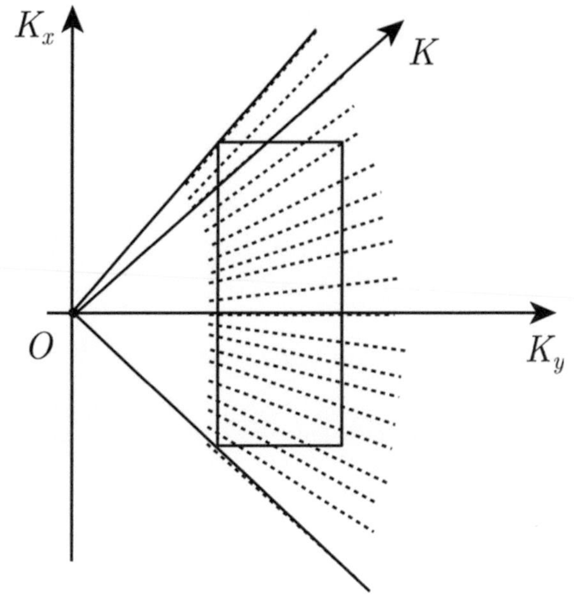

Fig. 4.10 Schematic diagram of PFA support domain

In this way, the computational complexity can be significantly reduced based on the fast Fourier transform.

3. PFA

The PFAis an imaging method for rotating targets and has been successfully applied to ISAR imaging, which has greatly improved the range of ISAR focus imaging and has been used quite extensively in turntable imaging.

For the echo signal, it is represented as a sector on the support domain in the two-dimensional frequency domain. If the coordinates in the range direction is transformed to K_R and the coordinates in azimuthal direction is transformed to K_a, the distribution of the echo points in the two-dimensional frequency domain can be obtained as shown in Fig. 4.10. The direct azimuthal FFT processing of the echoes is the RD processing algorithm, and the interpolation to make the support domain of the echoes rectangular is the PFA algorithm.

The PFA uses two interpolations, the first for the difference in the K_R direction, which changes the support domain from a sector to a trapezoid. The second interpolation in the K_a direction changes the trapezoid into a rectangle. Although PFA can obtain better focused imaging for a target with stable rotation speed, a more accurate rotation speed of the target is required in the calculation of the interpolation point position. At the same time, because PFA imaging requires interpolation in both directions, the computational complexity is high and will consume a long time in W-band ISAR imaging processing.

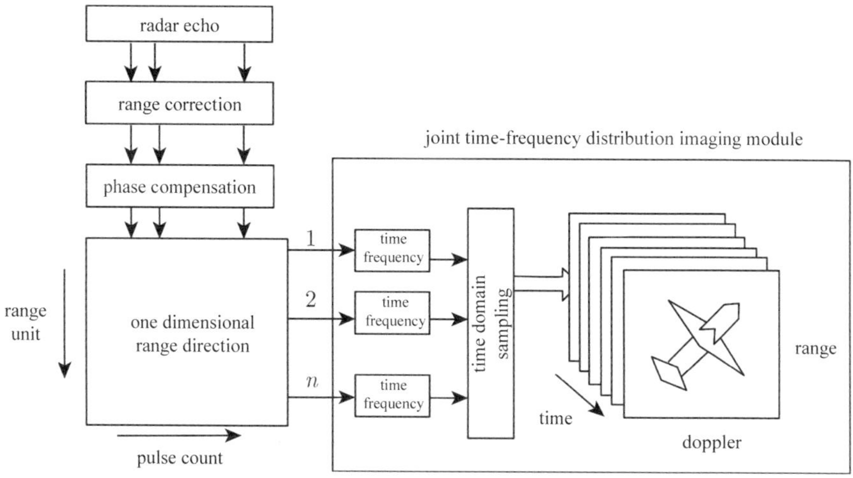

Fig. 4.11 Schematic diagram of time–frequency algorithm principle

4. Time–frequency algorithm

Time–frequency analysis is mainly developed for non-smooth signals whose spectrum varies with time, and it usesa joint function of time and frequency to describe the local performance of the signal at different times and frequencies simultaneously [11]. The time–frequency algorithm, that is, the joint time–frequency analysis is used instead of the Fourier transform for the echo signal in the process of ISAR imaging, so that the original two-dimensional range-Doppler matrix becomes a three-dimensional time-range-Doppler matrix. By sampling in the time dimension, a two-dimensional range-Doppler image can be obtained at each time sampling point. The imaging schematic of the time–frequency method is shown in Fig. 4.11.

Compared with conventional imaging methods, the time–frequency imaging method enables focused imaging for targets with varying rotational speed by using time–frequency analysis. Also, by generating time-varying imaging results, the motion of the target during the imaging time can be analyzed. However, since time–frequency analysis is equivalent to using only a portion of the echoes for imaging during the imaging process, it can bring a serious decrease in azimuthal resolution.

4.4.2 Rotational Speed Stabilization Target Imaging Method

The stable rotation target is the basic ISAR imaging target. For a stable rotational speed target, the point of maximum amplitude at a scattering point after motion compensation and the phase of that point are shown below:

$$t_{\rm k} = \frac{2r}{c} - \frac{2a\theta(t_{\rm m}) + r\theta^2(t_{\rm m})}{c} = \frac{2r}{c} - \frac{2a\omega t_{\rm m} + r(\omega t_{\rm m})^2}{c} \tag{4.84}$$

$$\varphi(t_\mathrm{m}) = 2\pi f_0\left(\frac{2r}{c} - \frac{2a\theta(t_\mathrm{m})}{c} - \frac{r\theta^2(t_\mathrm{m})}{c}\right) = 2\pi f_0\left(\frac{2r}{c} - \frac{2a\omega t_\mathrm{m}}{c} - \frac{r(\omega t_\mathrm{m})^2}{c}\right) \tag{4.85}$$

It can be seen that the range migration of the target is mainly generated by the $\frac{2a\omega t_\mathrm{m}}{c}$ term in the range direction and in the azimuthal direction, mainly by the $2\pi f_0\left(r(\omega t_\mathrm{m})^2/c\right)$ term. Meanwhile, the two parts of the migratory momentum are independent of each other. Therefore, this book proposes a two-dimensional migration compensated echo imaging algorithm that uses the Keystone transform instead of the azimuthal interpolation in the PFA, and uses a phase-compensated approach instead of the range interpolation in the PFA, reducing the computational complexity of the imaging method.

1. Two-dimensional migration compensation algorithm

The PFA uses a two-dimensional interpolation method to transform the two-dimensional support domain of the echo from a sector to a rectangle, which compensates for the migration of the target. The interpolation of the PFA in the azimuthal direction is very similar to the Keystone transform, which is a variable scale interpolation operation in the frequency domain for the azimuthal direction of the echoes, so the Keystone transform algorithm can be used instead, while the processing of this direction is independent of the rotation speed of the target.

In order to substitute for the PFA method range interpolation, that is, azimuthal migration compensation, the third term in Eq. (4.85) can be further expanded by substituting $t_\mathrm{m} = a/\mathrm{PRF}$, $R = \rho_\mathrm{r} r$, and $\rho_\mathrm{r} = c/2B$. where a, r, ρ_r are the number of azimuthal points, the number of range points, and the range resolution of the image after the Keystone transform, respectively. We can obtain

$$2\pi f_0\frac{r(\omega t_\mathrm{m})^2}{c} = \frac{\pi f_\mathrm{c}}{B(\mathrm{PRF})^2} \times a^2 r \times \omega^2 \tag{4.86}$$

It is assumed that the relative bandwidth of the transmitted signal is small and can be neglected. Equation (4.86) shows that the phase difference due to azimuthal migration is a function of the rotation speed of the target at a certain point of the image, so the azimuthal migration though resolution cell (MTRC) can be compensated if a more accurate angular velocity of the target is obtained and the phase calculated by Eq. (4.86) is superimposed on each point of the image.

In the actual imaging process, in many cases, the actual angular velocity of rotation of the target is more difficult to obtain, so here the angular velocity of rotation of the target can be estimated by substituting different angular velocities of rotation by a search method based on the quality of the obtained imaging results.

If the PFA is used for imaging, then each search cell has to compensate for the echo signal by interpolation in two dimensions, which is computationally intensive. Since the range migration compensation of the Keystone transform is independent of the angular velocity of the target, if the two-dimensional compensation algorithm is used, only the results of the Keystone transform need to be compensated and imaged, and each search cell only needs to superimpose a phase determined by the search angular velocity at each point of the echo signal, which greatly reduces the complexity of the operation.

In the search process of angular velocity, the entropyof the image can be used as the evaluation criterion for the quality of imaging results. The search for angular velocity can be performed in stages, using a larger search step for a larger range of angular velocity, and then gradually reducing the step size and range of angular velocity search, which can reduce the total number of searches.

In the process of azimuthal MTRC compensation, the angular velocity of rotation of the target can be estimated, so that this angular velocity can be used for azimuthal calibration of the target:

$$\rho_{\mathrm{a}} = \frac{\lambda}{2\theta} = \frac{\lambda}{2\omega\tau} \tag{4.87}$$

In order to make the azimuthal direction of the image and the range direction of each cell have the same scale, the azimuthal direction of the image can be scaled. The scale of the scaling can be calculated as follows:

$$\mu = \frac{\rho_{\mathrm{a}}'}{\rho_{\mathrm{a}}} = \frac{c}{2B} \times \frac{2\omega\tau}{\lambda} = \frac{f_{\mathrm{c}}}{B} \times \tau \times \omega \tag{4.88}$$

After the scaling, the resolution of the image is $\rho = c/(2B)$ in both the azimuthal and range directions. The scaling can be achieved using the Chirp Z transformation (CZT) method.

2. Algorithm processing flow

Based on the above analysis, the processing flow of the Two-dimensional migration compensation algorithm can be summarized as follows.

Step 1: Keystone the motion compensated results to compensate for the range migration of the scattering points.

Step 2: Based on the set initial search range and search compensation, the current set angular velocity value is obtained.

Step 3: Use the current angular velocity to calculate the phase to be compensated in the image, and then compensate the image for imaging.

Step 4: Calculate the entropyof the imaging result image.

Step 5: Repeat steps 2 to 4 until all angular velocities in the current search range are searched, take the angular velocity corresponding to the smallest entropy in the current range, set the area around this angular velocity as the next search range, and reduce the search compensation accordingly.

Step 6: Repeat steps 2 to 5 until all searches are completed, and calculate the minimum entropy of the image and its corresponding angular velocity for all angular velocities.

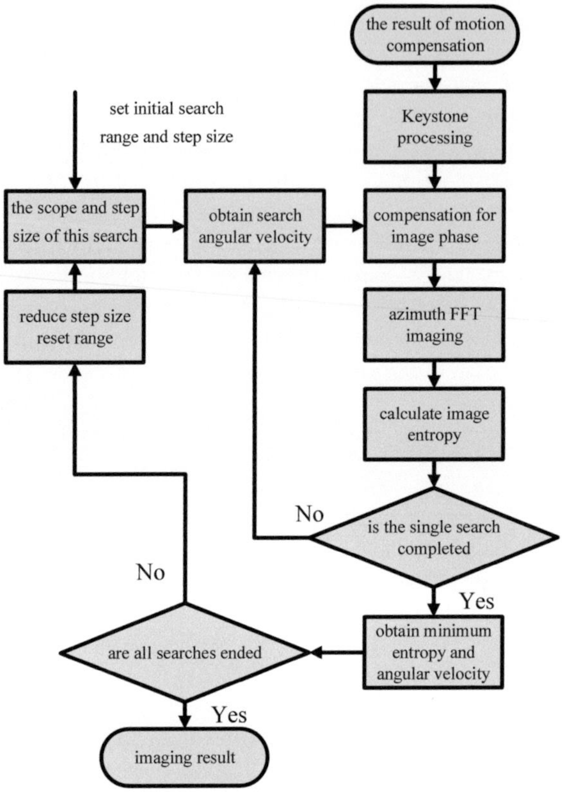

Fig. 4.12 Two-dimensional migration compensation algorithm imaging flowchart

Step 7: Use the angular velocity value obtained in step 6 to calculate the phase to be compensated and compensate the original data for imaging.

Figure 4.12 gives the detailed processing flow.

3. Algorithm error analysis

Assuming an error of $\Delta\omega$ in the estimation of the angular velocity and neglecting the relative bandwidth of the signal and the effect of Keystone transform on the time dimension, the azimuthal phase is affected as shown in Eq. (4.89), where τ is the imaging time.

$$\varphi_{\mathrm{a}}^{*} = \frac{4\pi f_0}{c} A\omega t_{\mathrm{m}} + \frac{2\pi f_0}{c} R[\omega^2 t_{\mathrm{m}}^2 - (\omega + \Delta\omega)^2 t_{\mathrm{m}}^2] \tag{4.89}$$

If full compensation is performed, the azimuthal imaging phase is

$$\varphi_{\mathrm{a}} = \frac{4\pi f_0}{c} A\omega t_{\mathrm{m}} \tag{4.90}$$

This gives the phase that causes the error in azimuth, as shown in Eq. (4.91). Here it is assumed that $\Delta\omega$ is a small quantity with respect to ω.

$$\Delta\varphi_{\mathrm{a}} = \frac{2\pi f_0}{c} R[\omega^2 t_{\mathrm{m}}^2 - (\omega + \Delta\omega)^2 t_{\mathrm{m}}^2] \approx \frac{4\pi f_0 R\omega\Delta\omega t_{\mathrm{m}}^2}{c} \tag{4.91}$$

The target point offset caused by the phase error after imaging is shown in Eq. (4.92), where N is the total number of azimuthal points in the imaging.

$$\Delta N = \frac{\Delta\varphi_{\mathrm{a}}/N}{2\pi/N} = \frac{\Delta\varphi_{\mathrm{a}}}{2\pi} = \frac{2R\omega\Delta\omega t_{\mathrm{m}}^2}{\lambda} \tag{4.92}$$

It can be seen that this offset increases as the target distance increases toward the distance from the center of rotation, and also increases with time. Assuming that ΔN can achieve the highest 1/2-grid compensation accuracy for the largest offset after the minimum entropy search process, the error in the angular velocity estimation is shown in Eq. (4.93), where D is the maximum distance from the target distance to the center of rotation.

$$\Delta\omega = \frac{\lambda}{4D\omega\tau^2} \tag{4.93}$$

The accuracy of the calibration can be obtained at the same time:

$$\Delta\rho_{\mathrm{a}} = \frac{\lambda\Delta\omega}{2\omega^2\tau} = \frac{\lambda^2}{8D\omega^3\tau^3} \tag{4.94}$$

When the maximum tolerable relative estimation error is assumed to be k%, the minimum estimable angular velocity can be obtained by substituting $\Delta\omega = k\omega/100$ into Eq. (4.93):

$$\omega_{\min} = 5\sqrt{\frac{\lambda}{kD\omega\tau^2}} \tag{4.95}$$

From Eqs. (4.94) and (4.95), it can be seen that the larger azimuthal migration caused by the increased resolution in the W-band imaging brings a higher accuracy in the estimation of the angular velocity, as well as an improved accuracy in the azimuthal calibration using the azimuthal migration in a single image.

4.4.3 Imaging Method for Targets with Varying Rotational Speed

Section 4.4.2 describes how to calculate the angular velocity and compensate for a rotational speed stable target, however, in practical applications, the angular velocity of most targets will change to varying degrees. Therefore, it is necessary to consider the effect of target angular velocity variation on the results [12–19]. This book proposes an improved sub-aperture method to reduce the azimuthal migration caused by the target rotational speed variation.

1. Improved sub-aperture imaging method

The time–frequency algorithm is the more commonly used method for imaging the rotational speed variation target; however, the time–frequency algorithm does not utilize the information of all echoes when imaging a certain point of time, and at the same time, there is no connection between the imaging of each point of time, and it does not generate an image with a high resolution. Therefore, the method of dividing sub-apertures to image separately and coherently superimposing each sub-aperture is considered here to improve the imaging results of the target with varying rotational speed.

However, in the W-band imaging process, the range and azimuthal migrations of the target are large. If the sub-aperture is divided directly, there are three factors that affect the scattering point coordinates in each sub-aperture: the range migration of the scattering point, the azimuthal migration of the scattering point, and the migration caused by the change of rotational speed. If the sub-apertures are combined, these three factors need to be compensated separately to obtain better results. Therefore, this book proposes an improved sub-aperture method by first compensating the signal according to the migration method of the stable speed target, and then dividing the aperture by the compensation result. After the migratory compensation process, the change of coordinates of the scattering points between sub-apertures will be mainly caused by the change of the target speed. It is easier to compensate only this one component. In the following, the improved sub-aperture method is derived and analyzed.

Assuming that the angle as a function of time is $\omega(t)$, the above compensation process for the uniform angular velocity target is first performed for the target with varying angular velocity. Assuming that the estimated angular velocity is ϖ, the phase error caused by the angular velocity variation at the point (A, R) can be obtained as

$$\Delta\varphi = \varphi[\omega(t)] - \varphi[\varpi] = \frac{4\pi f_0}{c} A[\omega(t)\tau - \varpi\tau] - \frac{2\pi f_0}{c} R\left[\int^2 \omega(t)dt - (\varpi\tau)^2\right] \tag{4.96}$$

The former term in Eq. (4.96) represents the azimuthal shift caused by the change in the angular velocity of the target and needs to be further compensated. The latter term in Eq. (4.96) can be compensated by the previously performed phase compensation process if the average angular velocity of the target can be estimated more accurately.

If there is no change in the angular velocity of the target and the average angular velocity is $\overline{\omega}$, the azimuthal coordinate of the point (A, R) on the target in the imaging image can be obtained from Eq. (4.97). And if the angular velocity of the target is $\omega(t)$, then at each slow time point, the azimuthal coordinate point of the target is

$$N = \frac{\varphi}{2\pi} = \frac{2f_0 A\varpi\tau}{c} = \frac{2A\varpi\tau}{\lambda} \tag{4.97}$$

$$N' = \frac{\varphi'}{2\pi} = \frac{2f_0 A\omega(\tau)\tau}{c} = \frac{2A\omega(\tau)\tau}{\lambda} \tag{4.98}$$

Since the angular velocity of the target varies with time, the azimuthal orientation of the target is continuously shifted throughout the synthetic aperture time. Here, assuming that the angular velocity of the target is continuous, the change in the angular velocity of the target will be smaller for a small period of time, that is to say, if only a small section of the aperture is imaged, the migration of the target in the azimuthal direction caused by the change in angular velocity will also be smaller. Thus, the azimuthal migration due to angular velocity variation can be reduced by dividing the echo data into several sub-apertures for imaging and resynthesis. Assuming that the average angular velocity of the kth sub-aperture is ϖ_k, the azimuthal coordinates of the point (A, R) on the target can be obtained as follows

$$N_k = \frac{\varphi_k}{2\pi} = \frac{2f_0 A\varpi_k\tau}{c} = \frac{2A\varpi_k\tau}{\lambda} \tag{4.99}$$

Multiplying the left and right sides of Eq. (4.99) by ϖ/ϖ_k we get

$$N_k\frac{\varpi}{\varpi_k} = \frac{2A\varpi_k\tau}{\lambda}\frac{\varpi}{\varpi_k} = \frac{2A\varpi\tau}{\lambda} = N \tag{4.100}$$

From Eq. (4.100), it can be seen that only azimuthal scaling of each sub-aperture image is required to obtain a sub-image with identical coordinates of each target point; after obtaining the sub-image, it is only necessary to convert each sub-image back to the time domain, superimpose it and then re-image it to obtain a synthetic image; the effect of angular velocity variation can be effectively reduced by processing the sub-aperture.

In the actual processing, due to the large error in the estimation of the average angular velocity of each sub-aperture, the method of selecting a certain sub-aperture as a benchmark and aligning the other sub-apertures can be carried out. In order to obtain the stretching multiplier of a certain sub-aperture relative to the reference aperture, the method can be carried out by setting different stretching multipliers, imaging two apertures synthetically, and searching according to the entropy of the resulting image.

The average angular velocity of the selected reference aperture, that is, the relationship between the average angular velocity corresponding to the image imaged by the above method and the estimated average angular velocity of the image, can be calculated from the scaling factor of each sub-aperture according to the following equation

$$\varpi' = \varpi \frac{\sum T_i}{K} \tag{4.101}$$

where T_i is the expansion multiple of the i-th sub-aperture; K is the number of sub-apertures.

By using the improved sub-aperture imaging method, the azimuthal migration caused by the change of the target angular velocity can be reduced. This method can be applied to arbitrary changes in angular velocity within a certain magnitude; moreover, this method is parameter selection from the imaging results, which has strong anti-interference performance; also, this method has a lower operational complexity and is less time-consuming.

2. Algorithm processing flow

With the above analysis, the processing of the improved sub-aperture imaging method can be divided into the following steps.

Step 1: Perform two-dimensional compensation processing on the motion compensation results according to the stable speed target, and compensate for the distance migration and the azimuthal migration due to the average rotation angular velocity.

Step 2: The result of step 1 is divided into apertures in the signal domain according to the given number of apertures, and aperture 1 is considered as the aligned aperture and aperture 2 is considered as the aperture to be aligned.

Step 3: Stretch the aperture to be aligned according to the current search stretch multiplier.

Step 4: Combine the apertures to be aligned with the aligned apertures and image.

Step 5: Calculate and record the entropy value of the imaging results.

Step 6: Repeat steps 3~5 until the search range of aperture stretching is completed, and the stretching multiplier of the aperture to be aligned is calculated according to the minimum entropy value.

Step 7: Stretch the apertures to be aligned according to the stretching multiplier and synthesize them with the aligned apertures as the aligned apertures, and use the subsequent apertures as the apertures to be aligned, and repeat steps 3~7 until all the apertures are aligned and the final imaging results are obtained.

Figure 4.13 gives the processing flow for merging the two sub-apertures. Figure 4.14 shows the detailed processing flow for the improved sub-aperture imaging method. The flow chart shows the processing of the 2 apertures, and when there are

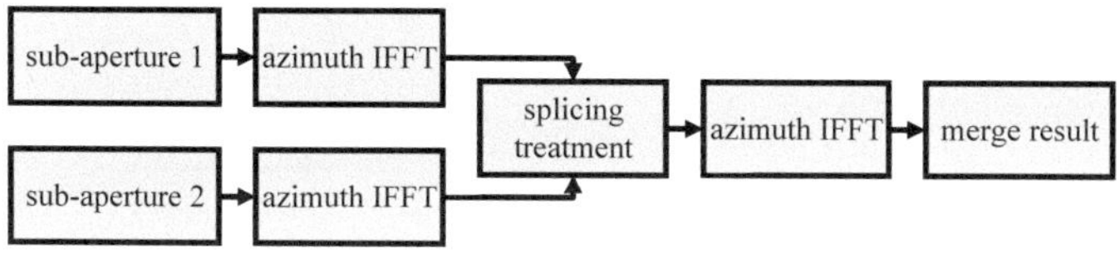

Fig. 4.13 Flow chart of sub-aperture synthesis processing

more apertures, the subsequent apertures are processed in the same way as the 2nd aperture.

3. Algorithm imaging effect analysis

The sub-aperture imaging method is a method of coherently superimposing the echo signal by dividing it into a number of small apertures, which are estimated and compensated for the average angular velocity separately. Therefore, the imaging error caused by the sub-aperture algorithm is mainly due to the error between the average angular velocity estimation and compensation and the actual angular velocity of the target.

The angular velocity of the target is expanded against time to give

$$\omega = \omega_0 + \frac{\mathrm{d}\omega}{\mathrm{d}t}t + \frac{1}{2}\frac{\mathrm{d}^2\omega}{\mathrm{d}t^2}t^2 + \frac{1}{3}\frac{\mathrm{d}^3\omega}{\mathrm{d}t^3}t^3 + \cdots \tag{4.102}$$

From this, the error in angular velocity can be obtained as

$$\begin{aligned}\overline{\omega} - \omega = & \frac{\int\left(\omega_0 + \frac{\mathrm{d}\omega}{\mathrm{d}t}t + \frac{1}{2}\frac{\mathrm{d}^2\omega}{\mathrm{d}t^2}t^2 + \frac{1}{3}\frac{\mathrm{d}^3\omega}{\mathrm{d}t^3}t^3 + \cdots\right)\mathrm{d}t}{T} \\ & - \omega_0 + \frac{\mathrm{d}\omega}{\mathrm{d}t}t + \frac{1}{2}\frac{\mathrm{d}^2\omega}{\mathrm{d}t^2}t^2 + \frac{1}{3}\frac{\mathrm{d}^3\omega}{\mathrm{d}t^3}t^3 + \cdots \\ = & \frac{\int \omega_0 \mathrm{d}t}{T} - \omega_0 + \frac{\int \frac{\mathrm{d}\omega}{\mathrm{d}t} t \mathrm{d}t}{T} - \frac{\mathrm{d}\omega}{\mathrm{d}t}t + \frac{1}{2}\left(\frac{\int \frac{\mathrm{d}^2\omega}{\mathrm{d}t^2} t^2 \mathrm{d}t}{T} - \frac{\mathrm{d}^2\omega}{\mathrm{d}t^2}t^2\right) + \cdots\end{aligned} \tag{4.103}$$

In order to estimate the results obtained from the above equation, the overall error is estimated by taking the cumulative sum S of the errors at each echo moment:

$$S = \sum_{t_\mathrm{m}} |\overline{\omega} - \omega| \tag{4.104}$$

set up

$$S_n = \frac{1}{n}\sum\left|\frac{\int \frac{\mathrm{d}^n\omega}{\mathrm{d}t^n} t^n \mathrm{d}t}{T} - \frac{\mathrm{d}^n\omega}{\mathrm{d}t^n}t^n\right| \tag{4.105}$$

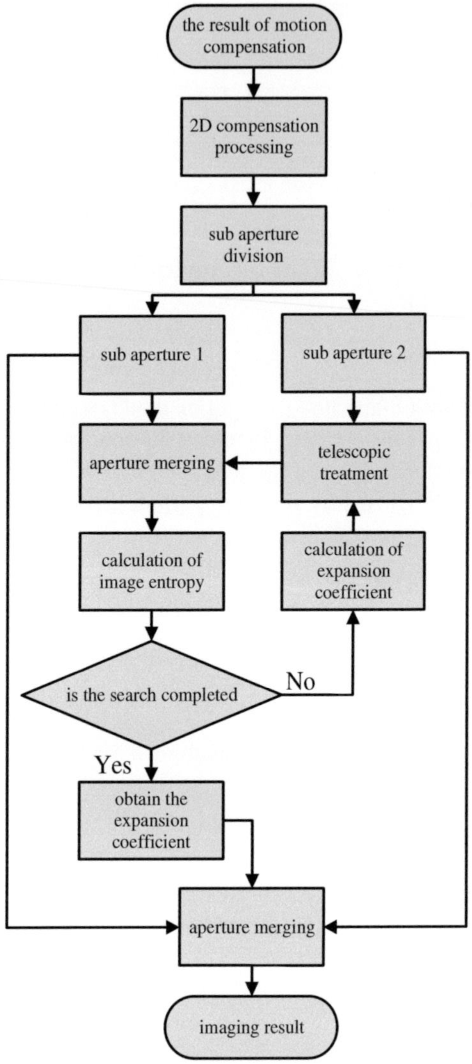

Fig. 4.14 Flow chart of improved sub aperture imaging method

then we can get

$$S = S_0 + S_1 + S_2 + \cdots \tag{4.106}$$

The sub-aperture treatment is equivalent to changing the summation interval of the mean to the original 1/K. Consider the following integration function:

$$S(f, x) = \int_0^1 \left| f(x) - \int_0^1 f(x)\mathrm{d}x \right| \mathrm{d}x \tag{4.107}$$

Table 4.2 Numerical calculation table of integral function

	x^2	x^3	x^4
1	0.2566	0.2362	0.2140
2	0.1261	0.1235	0.1209
3	0.0837	0.0829	0.0821
4	0.0627	0.0623	0.0620
8	0.0313	0.0312	0.0312
16	0.0156	0.0156	0.0156

If the error after sub-aperture processing can be expressed as

$$S_K(f,x)=\int_0^{\frac{1}{K}}\left|f(x)-K\int_0^{\frac{1}{K}}f(x)\mathrm{d}x\right|\mathrm{d}x+\int_{\frac{1}{K}}^{\frac{2}{K}}\left|f(x)-K\int_{\frac{1}{K}}^{\frac{2}{K}}f(x)\mathrm{d}x\right|\mathrm{d}x+\cdots$$
$$+\int_{\frac{K-1}{K}}^{1}\left|f(x)-K\int_{\frac{K-1}{K}}^{1}f(x)\mathrm{d}x\right|\mathrm{d}x \tag{4.108}$$

In order to estimate the error improvement after the sub-aperture treatment, the above equation can be integrated numerically, and Table 4.2 shows the obtained results.

From the table, it can be seen that after taking K integration intervals, the total error is approximately 1/K of the original one for each sub-function:

$$\breve{S}_K(f,x)\approx\frac{1}{K}S(f,x) \tag{4.109}$$

From this, the total angular velocity error is approximately 1/K of that without the sub-aperture treatment when K sub-aperture treatments are used.

According to Eq. (4.98), the change in the number of imaging points caused by the angular velocity error can be obtained as follows

$$\Delta N=|N_\omega-N_{\overline{\omega}}|=\frac{2A\Delta\omega\tau}{\lambda} \tag{4.110}$$

After sub-aperture treatment:

$$\Delta N_K=\frac{2A\frac{\Delta\omega}{K}\tau}{\lambda}=\frac{1}{K}\frac{2A\Delta\omega\tau}{\lambda} \tag{4.111}$$

It can be seen that if K sub-apertures are used for the imaging process, the degree of defocussing caused by the change in the angular velocity of the target can be reduced to 1/K of the original.

Further expanding Eq. (4.111) by substituting $\tau=\frac{\Delta\theta}{\omega}$, $\Delta\theta=\frac{\lambda}{2\rho_a}$, we get

$$\Delta N_K = \frac{1}{K}\frac{2A\Delta\omega}{\lambda}\frac{1}{\omega}\frac{\lambda}{2\rho_a} = \frac{1}{K}\frac{\Delta\omega}{\omega}\frac{A}{\rho_a} \tag{4.112}$$

To keep the image from defocusing, the offset is required to be less than N. Substituting this conditionwill give the equation for the proportional change in the angular velocity of the target when the target is not defocused:

$$\frac{\Delta\omega}{\omega} \leq \frac{\rho_a}{A}KN \tag{4.113}$$

From Eq. (4.113), it can be seen that the sub-aperture imaging method increases the proportion of target angular velocity changes that can be allowed without defocusing the image by K time.

4.5 W-Band ISAR Actual Measurement Data Processing

In this section, the phase coherent processing proposed, the envelope alignment technique based on envelope regularization, and the three commonly used phase correction techniques in Sect. 4.3, and the targets with stable rotational speed and targets with variable rotational speed imaging methods proposed in Sect. 4.4 are simulated and the simulation results are analyzed, respectively.

4.5.1 Simulation System and Target Parameters

Table 4.3 shows the parameters used in the simulation system. Here, the echo generation and simulation are performed for a simulated aircraft target consisting of 37 points with a maximum scale of ± 4 m. The residual velocity of the target after tracking is 4 m/s and the acceleration is -4 m/s^2. A random vibration with an amplitude of 0.1 m is also added to the target.

Table 4.3 System simulation parameters table

Simulation parameters	Value
Center frequency	94 GHz
Signal bandwidth	5 GHz
Sampling rate	6 GHz
Pulse repetition frequency	4000
Pulse duration	100 us
Measurement duration	0.5 s
Distance to target	5000 m
Angular velocity of the target	0.1 rad/s

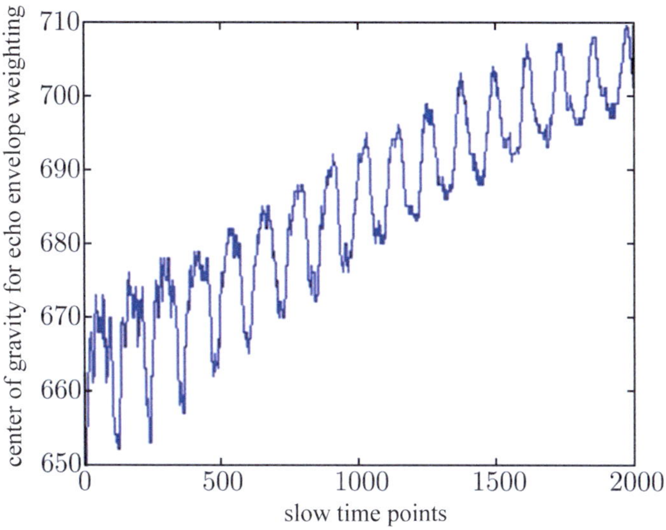

Fig. 4.15 Figure of the position of the target center of gravity

4.5.2 Coherent Processing

Figure 4.15 shows the estimated center of gravity of the target in the echo, and it can be seen that the center of gravity of the echo has a more drastic change due to the interference between the scattering points.

Figure 4.16 shows the results of fitting the displacement of the target's center of gravity, and it can be seen that the fitted results can reflect the trend of the target's motion more accurately.

By aligning the short window lengths, the vibration amplitude of the target is obtained and superimposed into the target center of gravity displacement fitting curve, Fig. 4.17 shows the actual offset value of the target center of gravity obtained.

Using the target offset values obtained in Fig. 4.17, the phase required for compensation can be calculated and the echo signal can be coherently processed.

4.5.3 Envelope Regularization Processing

In this subsection, the pre-regulation and post-regulation envelopes are processed using the correlation method, the 1st order distance method and the minimum entropy method, respectively, and the results obtained are shown below.

Figure 4.18a shows the result of the range compression of the echo signal obtained by simulation, while Fig. 4.18b shows the result of the envelope rectification of (a), which shows that the echo envelope changes more obviously with the change of

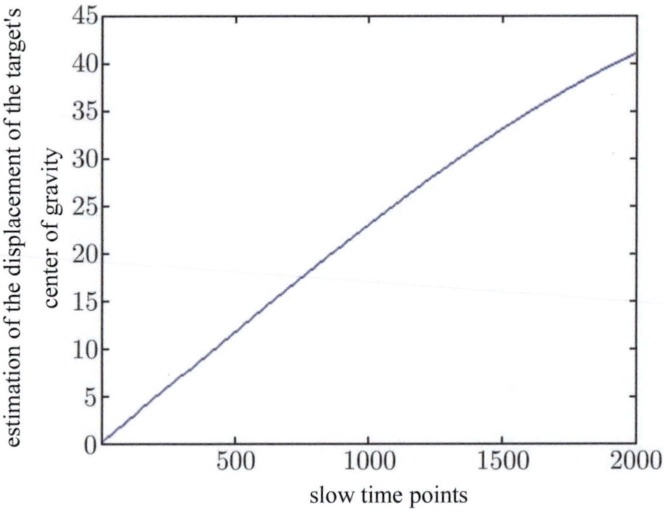

Fig. 4.16 Fitting curve of target center of gravity movement

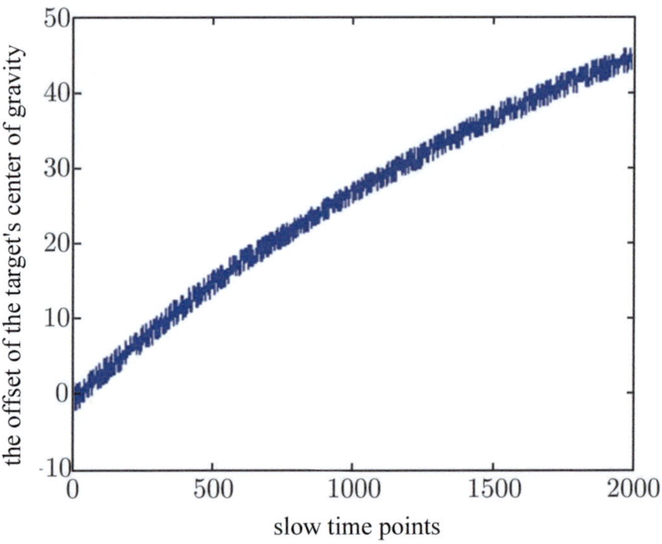

Fig. 4.17 Offset of the target center of gravity

azimuth (vertical direction) time in (a), while the envelope of (b) basically does not change after rectification.

Figure 4.19 shows the search curves of the three envelope alignment methods when aligning the 12th echo. It can be seen that all three methods have good estimation accuracy and can obtain the offset of the envelope more accurately.

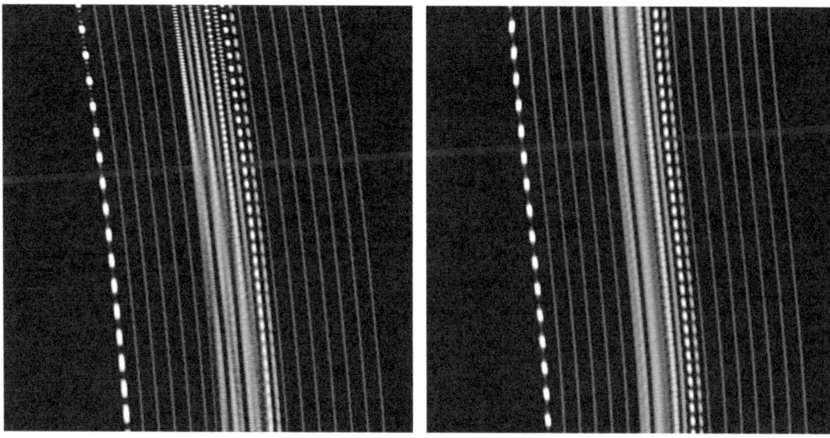

(a) without envelope regularization (b) after envelope regularization

Fig. 4.18 Imaging results of range compression

Figures 4.20, 4.21, 4.22, 4.23, 4.24 and 4.25 shows the alignment results of the correlation method, the1st order distance method, and the minimum entropy method before and after envelope regularization, and the corresponding imaging results after phase correction by the PGA method, respectively.

From the alignment results of the above figures, it can be seen that the direct envelope alignment of the range compressed results produces a bending phenomenon due to the change of the envelope, while the results of the processing of the envelope-regulated images are more accurately aligned. It can be seen that the imaging results without envelope alignment have obvious scattering in the range direction, which has a more serious impact on the imaging resolution, while the imaging results after envelope alignment are very well focused and can achieve a high resolution.

In order to evaluate the imaging results, the correlation method, the 1st order distance method and the minimum entropy method were evaluated using three parameters before and after envelope regularization: the intensity of the extreme point of the image, the entropy of the image, and the contrast ratio of the image, and the results obtained are shown in Table 4.4.

Fig. 4.19 Search curves for adjacent envelopes using various envelope alignment methods

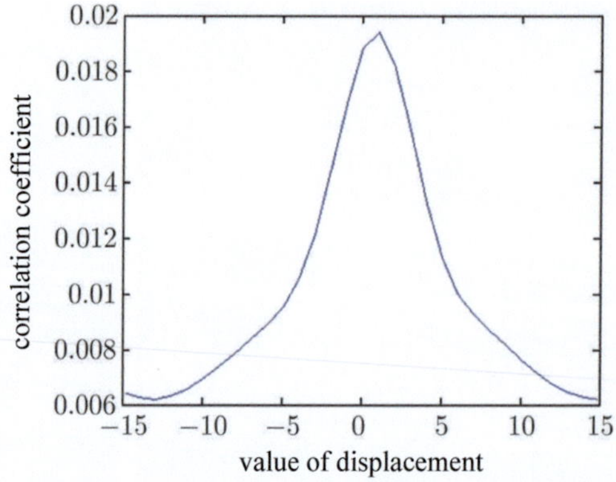

(a) the search curve for correlation method

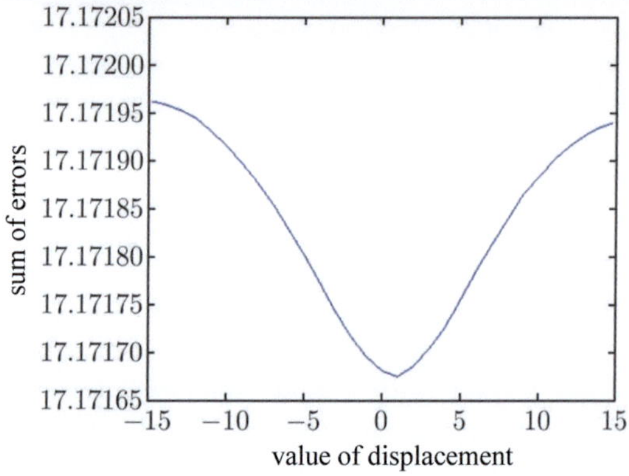

(b) the search curve for first order distance method

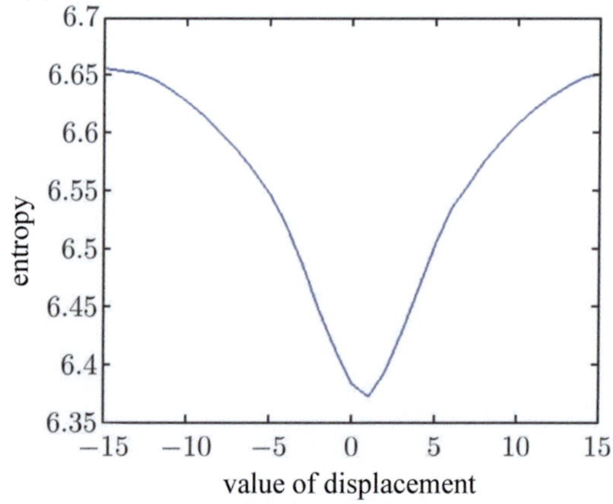

(c) the search curve of the minimum entropy method

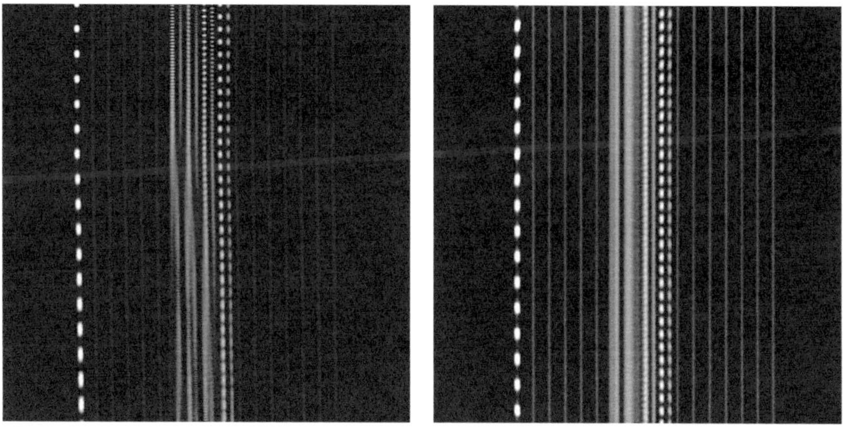

(a) without envelope regularization (b) after envelope regularization

Fig. 4.20 Imaging results of alignment using correlation method

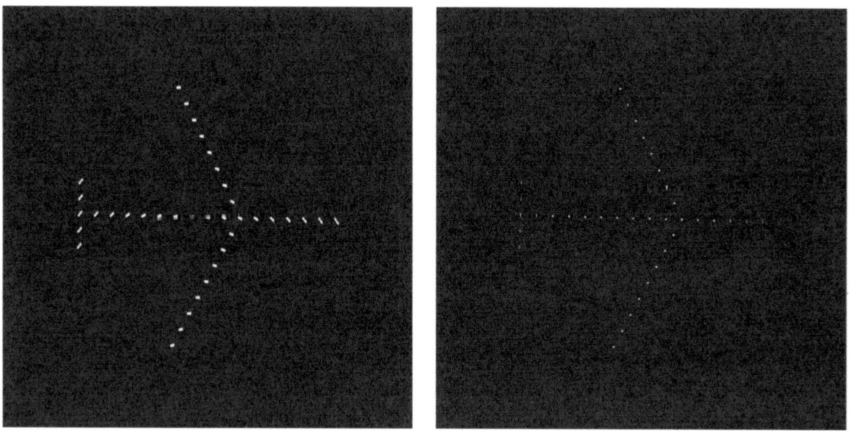

(a) without envelope regularization (b) after envelope regularization

Fig. 4.21 Imaging results of correlation method

Through comparison, it can be seen that through envelope regularization processing, the intensity of extreme points in the imaging results of each alignment algorithm has significantly improved, reaching approximately 6 dB. The overall image entropy is decreasing, indicating that the image is sharpened. The contrast ratio of the images is also improved to a large extent.

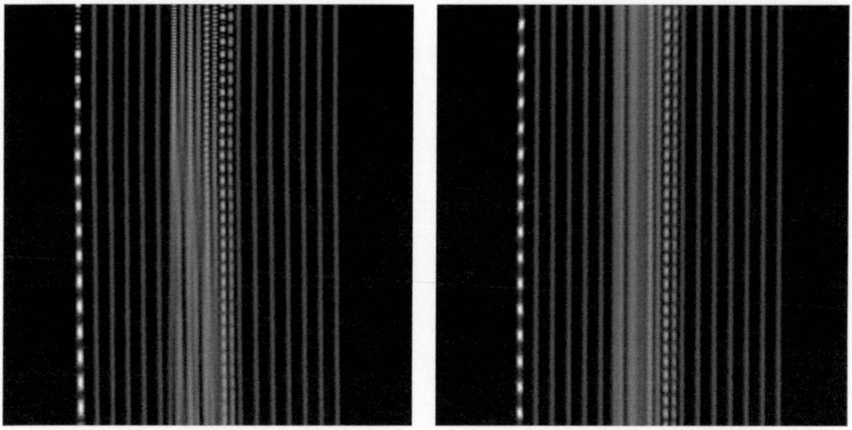

(a) without envelope regularization (b) after envelope regularization

Fig. 4.22 Imaging results of alignment using the first order distance method

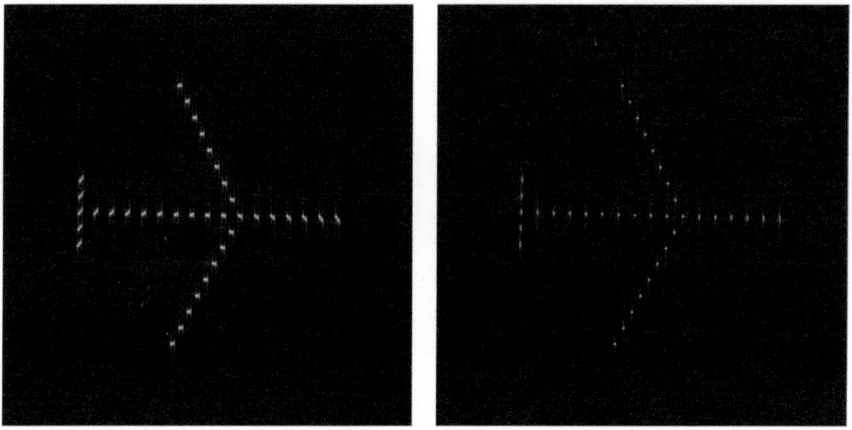

(a) without envelope regularization (b) after envelope regularization

Fig. 4.23 Imaging results of the first order distance method

4.5.4 Phase Correction Simulation Experiment

After the envelope alignment process, the results need to be phase corrected. Figure 4.26 shows the imaging results of imaging directly without phase correction, and it can be seen that the presence of the error phase produces severe scatter in the azimuthal direction.

After the envelope alignment, Fig. 4.27 shows the phase correction results using the Prominent Point method, Doppler centroid tracking method, and PGA method,

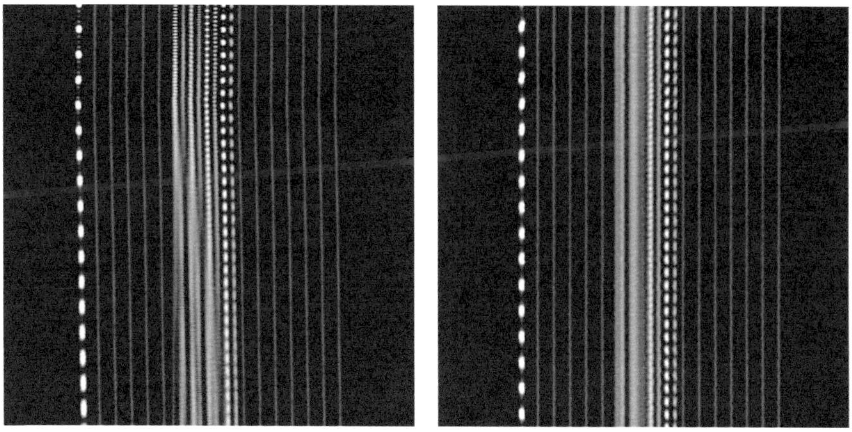

(a) without envelope regularization (b) after envelope regularization

Fig. 4.24 Imaging results of minimum entropy alignment

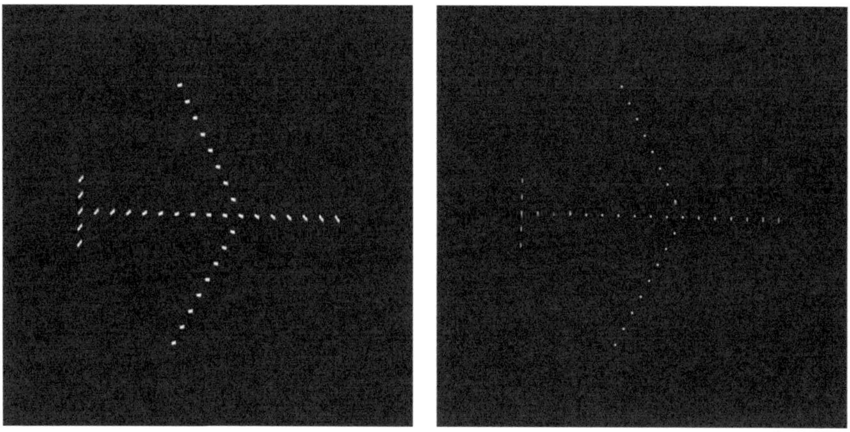

(a) without envelope regularization (b) after envelope regularization

Fig. 4.25 Imaging results of minimum entropy method

Table 4.4 Comparison of envelope regularization effects

Methods	Has the envelope been regularized	The intensity of the extreme point of the image /dB	The entropy of the image	The contrast ratio of the image
Correlation method	Yes	122.5	8.59	16.7
	No	116.0	8.69	13.3
1st order distance method	Yes	120.3	8.69	16.3
	No	112.9	9.13	13.0
Minimum entropy method	Yes	122.6	8.60	16.8
	No	116.2	8.69	13.4

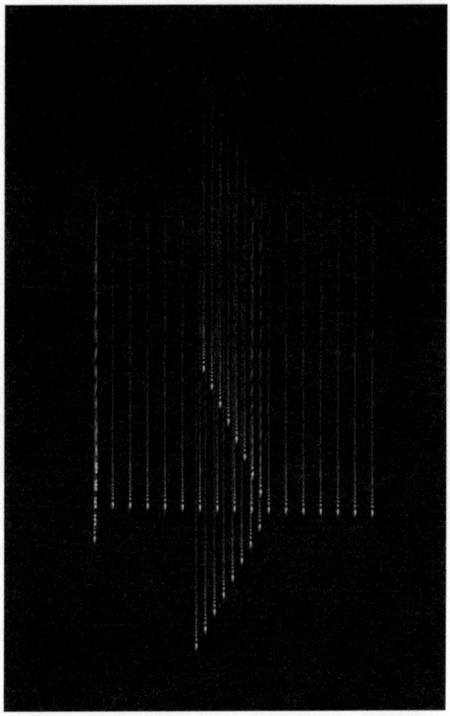

Fig. 4.26 Imaging result without phase correction

respectively. It can be seen that each phase correction method can accurately compensate for the error phase and achieve good focusing results.

Evaluate the imaging results of the above three methods using the intensity of the extreme point of the image, the entropy of the image and the contrast ratio of the image. Table 4.5 shows the results of the evaluation. It can be seen that the the intensity of the extreme point of all three methods has increased significantly, by about 9 dB, while the entropy of the image and the contrast ratio of the image have improved significantly.

4.5.5 Target Imaging with Stable Rotational Speed

The system parameters used in this section of the simulation are the same as those in Table 4.3. The target is a 4 $\times$ 9 array of dots, the distance between two adjacent dots is 1 m, and the maximum position of the target is $\pm$12 m.

The results obtained using the RD imaging algorithm are shown in Fig. 4.28a. The resolution is high in the W-band, and the ambiguities in both the azimuthal and range directions of the target is very obvious, which requires further image processing algorithms. After the range compression, the results of the range compensation using

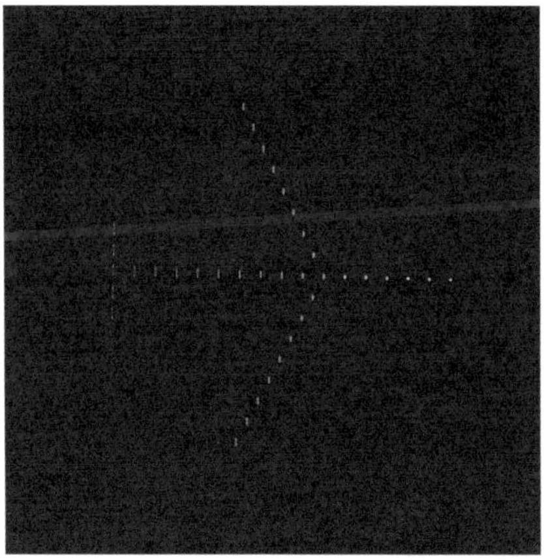

(a) prominent point method

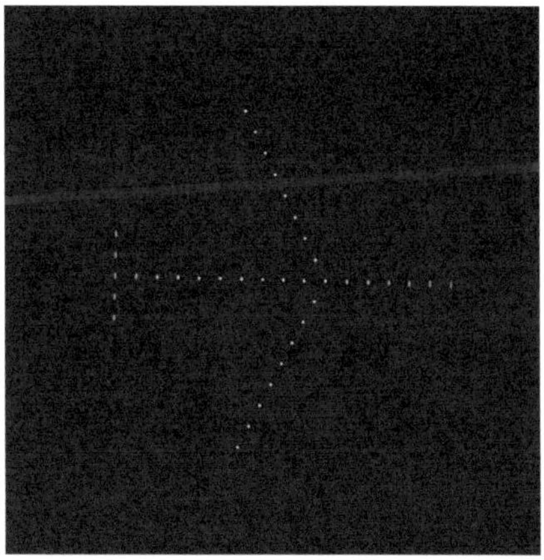

(b) doppler centroid tracking method

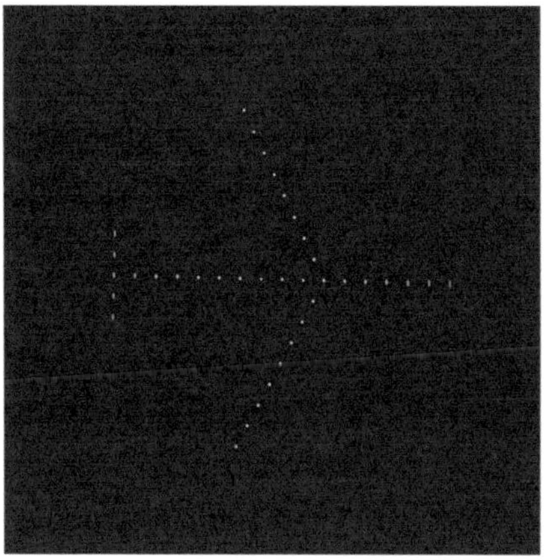

(c) PGA

Fig. 4.27 Imaging results after compensation by various phase correction methods

Table 4.5 Comparison of imaging results using phase correction methods

Methods	The intensity of the extreme point of the image /dB	The entropy of the image	The contrast ratio of the image
Original image	113.10	9.52	8.18
Prominent point method	122.67	8.77	15.84
Doppler centroid tracking method	122.68	8.61	16.52
PGA method	122.64	8.60	16.76

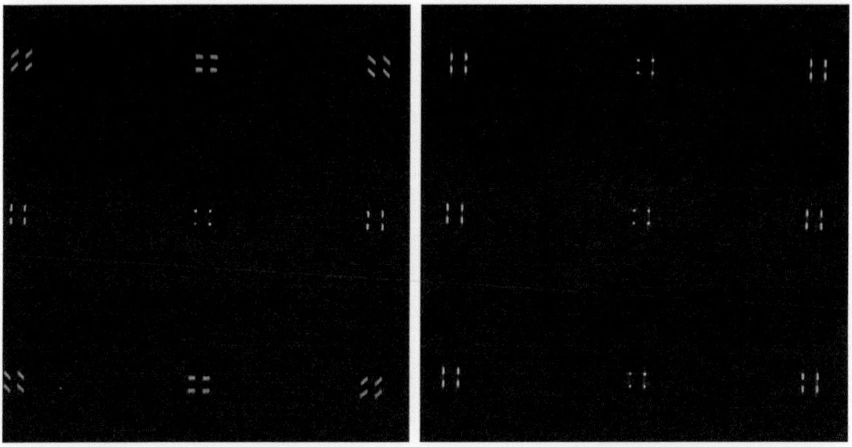

(a) imaging result of RD (b) imaging result of Keystone transform

Fig. 4.28 Figure of the compensation effect of the Keystone method

the Keystone method are shown in Fig. 4.28b, which shows that the migration of the left target in the range direction has been corrected by the Keystone algorithm, but there is still a large migration in the azimuth direction.

Next, the azimuthal compensation is performed. The search curve for the angular velocity estimation is shown in Fig. 4.29. It can be seen that after several searches, the entropy of the image decreases as the search step as well as the search range decreases during the search process.

Figure 4.30 shows the imaging results after two-dimensional compensation, and it can be seen that the image quality has been significantly improved by two-dimensional compensation processing.

Table 4.6 shows the numerical evaluation of the results of the above three methods. It can be seen that the overall entropy of the image decreases significantly after Keystone transform imaging and two-dimensional compensation imaging, while the contrast ratio of the image also improves significantly. The small change in the intensity of the extreme point of the image is due to the less migration of the target center point and the resulting less scattering, so the improvement in the alignment of these two imaging methods is also small.

4.5.6 Target Imaging with Changes in Rotational Speed

When the angular velocity of the target changes, it causes a change in Doppler frequency, which in turn causes a shift in the azimuthal imaging position and causes azimuthal ambiguities of the imaging. In this subsection, the angular velocity of the target is set to have second-order and third-order angular acceleration, and it is

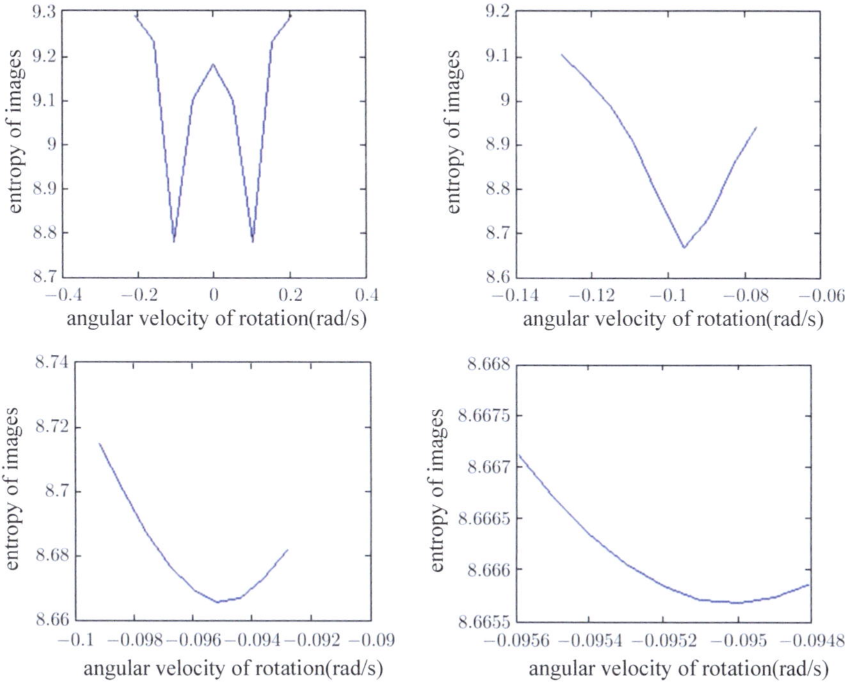

Fig. 4.29 Search curve for angular velocity

Fig. 4.30 Figure of the effect of minimum entropy phase compensation

Table 4.6 Comparison table of imaging results of various methods

Methods	The intensity of the extreme point of the image /dB	The entropy of the image	The contrast ratio of the image
RD imaging algorithm	120.39	9.71	12.43
Keystone transform algorithm	120.90	9.18	21.44
Two-dimensional migration compensation algorithm	120.97	8.66	27.66

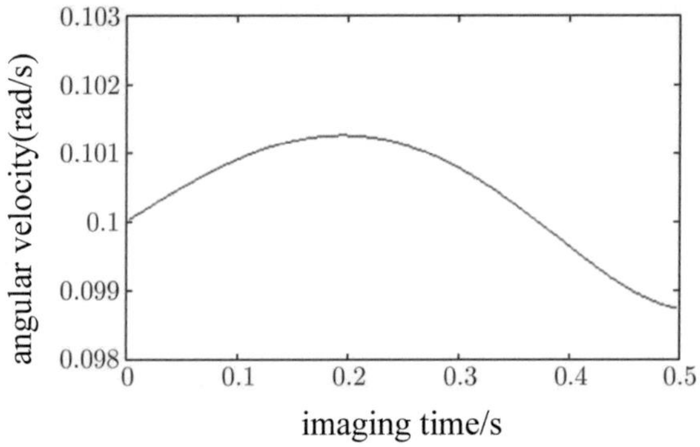

Fig. 4.31 The variation of the angular velocity of the target with time

imaged using a improved sub-aperture method. Figure 4.31 shows the variation of the angular velocity of the target with time.

Figure 4.32a shows the results obtained using the rotational speed stabilization imaging algorithm for 2, 4, and 8 sub-apertures, respectively. Table 4.7 shows the scaling factors for each aperture processing. From the table, it can be seen that the more the number of apertures, the closer the stretching coefficients of each sub-aperture are to the true rotational angular velocity profile of the target.

Using the improved sub-aperture method, Fig. 4.32 shows the results obtained. It can be seen from the figure that the imaging quality of the improved sub-aperture method, although somewhat degraded compared to that without angular acceleration, reduces the azimuthal ambiguities caused by angular acceleration and yields better imaging results compared to Fig. 4.32a.

Table 4.8 shows the results of the numerical analysis of the results obtained by the imaging methods for each number of apertures. It can be seen from the table that the improved sub-aperture method has a significant improvement in the image entropy and contrast ratio indexes, while the higher the number of apertures, the better the image quality. There is a small decrease in the intensity of the extreme

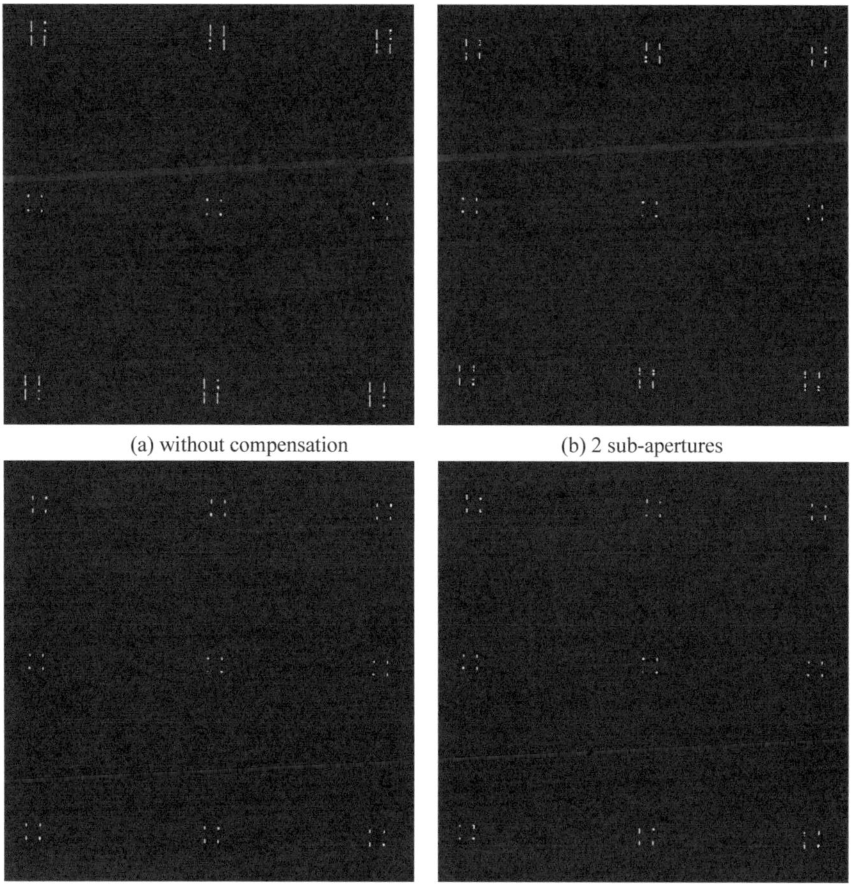

(a) without compensation (b) 2 sub-apertures (c) 4 sub-apertures (d) 8 sub-apertures

Fig. 4.32 Figure of the effect of sub-aperture compensation

Table 4.7 Table of aperture expansion coefficient for improved sub-aperture method

Number of apertures	1	2	3	4	5	6	7	8
2 sub-apertures	1.000	0.980						
4 sub-apertures	1.000	1.000	0.980	0.960				
8 sub-apertures	1.000	1.010	1.005	1.010	0.985	0.970	0.960	0.950

point of the image, which is mainly caused by the sub-aperture scaling and the errors in the synthesis process.

Table 4.8 Comparison of imaging effects of improved sub-aperture methods

Methods	The intensity of the extreme point of the image /dB	The entropy of the image	The contrast ratio of the image
Two-dimensional migration compensation algorithm	120.21	9.15	19.63
2 sub-apertures	119.20	8.99	22.02
4 sub-apertures	119.08	8.92	23.27
8 sub-apertures	119.05	8.78	24.02

References

1. Bao M, Zhou P, Li YC, et al. ISAR imaging of complex moving targets based on product high-order phase function [J]. System Engineering and Electronics, 2011, 33(5): 1018-1022.
2. Berger T, Hamran S E. An efficient scaled maximum likelihood algorithm for translational motion estimation in ISAR imaging[C]. IEEE Radar Conference 2010, 2010:75–80.
3. Berger T, Tollisen S, Hamran S E. Comparison of two methods for automatic range alignment in ISAR imaging[C]// IEEE Radar Conference 2011, 2011: 6–11.
4. Bucciarelli M, Pastina D. Multi-grazing ISAR for side-view imaging with improved cross-range resolution [J]. IEEE Radar Conference, Kansas, 2011: 939–944.
5. Cao P, Xing M D, Sun G C, et al. Minimum entropy via subspace for ISAR autofocus[J]. IEEE Trans. Geosci. Remote Sens. Lett., 2010, 7(1): 205-209.
6. Ender J H G. On compressive sensing applied to radar [J]. Signal Processing, 2010, 90(5): 1402-1414.
7. Martorella M, Acito N, Berizzi F. Statistical CLEAN technique for ISAR imaging [J]. IEEE Transactions on Geoscience and Remote Sensing, 2007, 45(11): 3552-3560.
8. Munoz-Ferreras J M, Perez-Martinez F, Datcu M. Generalisation of inverse synthetic aperture radar autofocusing methods based on the minimisation of the Renyi entropy[J]. IET Radar Sonar Navig., 2010, 4(4): 586-594.
9. Peng SB, Xu J, Xiang JB, et al. A Novel ISAR Imaging Time Selection Method for Maneuvering Target Based on Phase Linear Degree [J]. Journal of Electronics & Information Technology, 2010, 32(12): 2795-2801.
10. Peng S B, Xu J, Peng Y N, et al. Parametric inverse synthetic aperture radar manoeuvring target motion compensation based on particle swarm optimiser [J]. IET Radar Sonar Navig., 2011, 5(3): 305-314.
11. Qiu XH, Zhao Y, Cheng A H W, et al. Consistency Study of Minimum Entropy Autofocusing with Phase Compensation in ISAR Imaging [J]. Journal of Electronics & Information Technology, 2007, 29(8): 1799-1801.
12. Wang Y, Jiang Y C. Inverse synthetic aperture radar imaging of maneuvering target based on the product generalized cubic phase function[J]. IEEE Geoscience and Remote Sensing Letters, 2011, 8(5): 958-962.
13. Wang J F, Kasilingam D. Global range alignment for ISAR [J]. IEEE Trans. on AES, 2003, 39(1): 351-357.
14. Wang L, Zhu D Y, Zhu Z D. Range alignment for ISAR using genetic algorithms[C]. IGARSS'05, 2005: 4666–4669.
15. Xing M, Wu R, Li Y, et al. New ISAR imaging algorithm based on modified WignerVille distribution[J]. IET Radar, Sonar and Navig., 2009, 3(1): 70-80.

16. Yang J, Xu RC, Bao QL, et al. The Implemental Methods of High Velocity Compensation in ISAR Imaging Based on the Entropy Minimization[J]. Signal Processing, 2009, 25(12): 1861-1866.
17. Zhang L, Xing M D, Qiu C W, et al. Achieving higher resolution ISAR imaging with limited pulses via compressed sampling[J]. IEEE GRS Lett., 2009, 6(3): 567-571.
18. Zhang L, Xing M D, Qiu C W, et al. Resolution enhancement for inversed synthetic aperture radar imaging under low SNR via improved compressive sensing [J]. IEEE Trans. on GRS, 2010, 48(10): 3824-3838.
19. Zhu D Y, Wang L, Yu Y S, et al. Robust ISAR range alignment via minimizing the entroy of the average range profile [J]. IEEE Geoscience and Remote Sensing Letters, 2009, 6(2): 204-208.

Chapter 5
Ka-Band Synthetic Aperture Radar Technology for Continuous Wave Regimes

5.1 Overview

Multi-satellite network observation is an important development direction of satellite-based SAR technology, which can meet the needs of national economic construction applications in different aspects by integrating and complementing each other's functions through the cooperative work of multiple SAR satellites. The development of satellite-based SAR system to miniaturization is the way to overcome the problems of long development period of traditional Earth observation satellite system, complicated and expensive design of satellite platform and launch tool.

As described in Chap. 2, the nearly 100% transmit duty cycle of continuous wave SAR can significantly reduce the number of amplifier components, which is particularly advantageous in reducing cost when applied to satellite-based systems. At the same time, the FMCW SAR system adopts the de-modulation processing method, which can effectively reduce the bandwidth of the echo signal, reduce the hardware requirements of the digital receiving system, and effectively reduce the data volume of the echo signal, which is of high value for relieving the huge pressure of the satellite-ground data transmission of high-resolution SAR. Further, the on-board FMCW SAR system uses separate platforms for transceiver and transmitter operation, resulting in a simplified architecture with small receiving satellite energy requirements. The combination of Ka band image- optic-like and small antenna size can effectively control the cost of the networking system while acquiring highly interpretable SAR images.

On the one hand, the prerequisite for the proper imaging of transceiver-separated FMCW SAR is to ensure the space–time frequency phase synchronization between the binary satellites, and the gapless operation of the continuous-wave system makes the traditional method of binary phase synchronization no longer applicable. On the other hand, the separation of radar transceiver antennas in dual-satellite mode leads to different ranges of the same target from the radar transceiver and transmitter platforms, which makes the spatial geometry relationship more complicated. Moreover,

H. Wang, *The Millimeter Wave Synthetic Aperture Radar Technology*,
https://doi.org/10.1007/978-981-97-1044-7_5

the movement of the transceiver platform during the long pulse time of FMCW SAR cannot be neglected, and these factors require the development of targeted algorithms for the imaging processing of dual-satellite FMCW SAR.

This chapter firstly introduces the synchronization method of the space–time frequency phase with the operating characteristics of the Ka-band binary FMCW SAR, and then carries out the echo modeling analysis of the binary FMCW SAR on this basis, and introduces the imaging method of the binary FMCW SAR for the strip and sliding cluster modes in a targeted manner.

5.2 Ka-Band FM Continuous Wave SAR Double Satellite Synchronization

The transmitting and receiving of radar signals in the continuous wave system are simultaneous; therefore, the isolation between the transmitting and receiving ends is the main problem that plagues the satellites of the continuous wave system, and the isolation of the transmitting and receiving ends is solved by using the spatial propagation attenuation of electromagnetic waves placed on different satellites. To ensure the effective amplitude of the imaging, the transmitter and receiver of the dual-satellite FMCW SAR system should have uniform beam coverage, which requires spatial synchronization at the transmitter and receiver. In order to make the whole system units work in a coordinated manner, the transmitter and receiver of the dual-satellite FMCW SAR system need to have a uniform time standard, which requires time synchronization at both transmitter and receiver stations. For the FM continuous wave radar, the receiving station must adjust the local oscillation to make it coherent with the transmit signal in order to realize the differential beat decoupling with the return signal, which is based on the time, frequency and phase synchronization of the transmitting and receiving satellites.

5.2.1 Ka-Band SAR Long-Range Spatial Synchronization Method

The Ka-band FMCW SAR carrier adopts a binary split transceiver mode due to the constraint of simultaneous transceiver operation in the continuous wave regime, and the Ka-band has the characteristics of short wavelength and small beamwidth. The receiving antenna cannot obtain all the scattered signals of the target, resulting in the loss of the effective observation width similar to Fig. 5.1, or even a complete inability to work effectively.

In order to meet the working requirements of the Ka-band FM continuous wave SAR load, a suitable attitude cooperative control scheme for the transmitting and receiving satellites is needed, as shown in Fig. 5.2. In order to ensure the ground

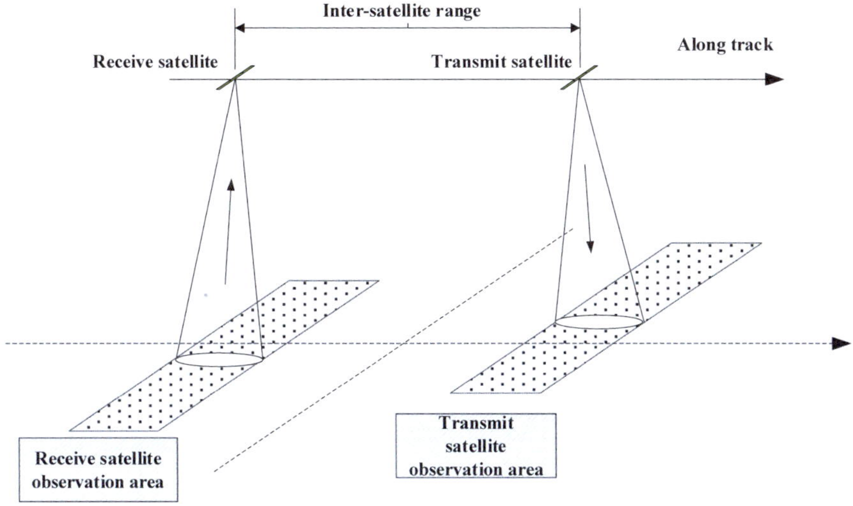

Fig. 5.1 The receiving and transmitting satellite space is not synchronized

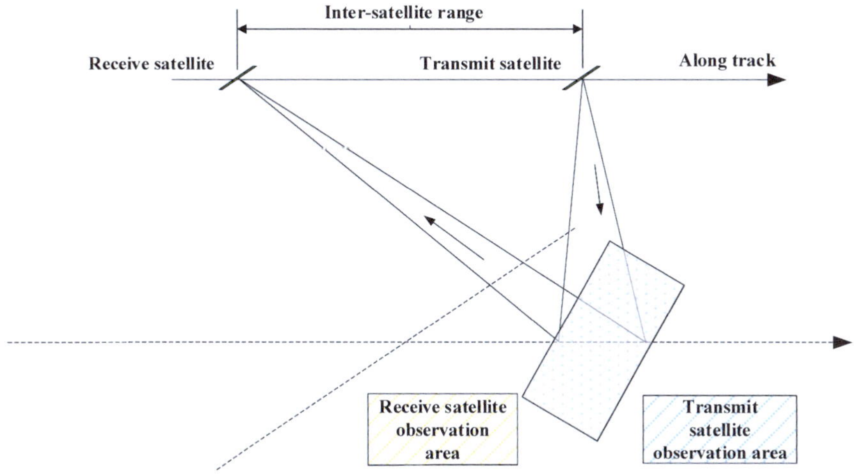

Fig. 5.2 Satellite space synchronization

beam coincidence of the primary and secondary loads, the beam pointing of the receiving satellites is changed to make the observation area of the transmitting and receiving satellites consistent, so as to complete the spatial synchronization of the Ka-band continuous wave load.

However, the following technical difficulties exist in Ka-band continuous-wave spatial synchronization:

(1) The traditional guidance law cannot be applied to the binary at the same time, because the Ka-band beam is narrow, the binary spacing is large, and the binary beams do not overlap at all, so a new guidance law or improvement on the original guidance law is needed to make the ground beam pointing coincide.
(2) The accuracy of attitude control for light and small satellites requires high precision. When the attitude control error is large, the double-satellite beam pointing will have large deviation, which affects the beam reception effect, and this part of the work puts high requirements on the attitude control scheme of the satellite, which will not be discussed in this section.

1. Launch satellite full zero Doppler guidance technology

Let the position and velocity of the satellite with respect to the center of the earth be R_s and v_s, the position and velocity of the ground target with respect to the center of the earth be R_t and v_t, and the angular velocity of the earth's rotation be ω_e, then the Doppler center frequency of the satellite is obtained as

$$\begin{aligned} f &= \frac{-2}{\lambda}\cdot\frac{\boldsymbol{R}\cdot\boldsymbol{R}}{R} = \frac{-2}{\lambda}\frac{\boldsymbol{R}}{R}(v_s - v_t) \\ &= \frac{-2}{\lambda}\frac{\boldsymbol{R}}{R}[v_s - \boldsymbol{\omega}_e \times (\boldsymbol{R}_s - \boldsymbol{R})] = \frac{-2}{\lambda}\frac{\boldsymbol{R}}{R}(v_s + \boldsymbol{R}_s \times \boldsymbol{\omega}_e) \end{aligned} \tag{5.1}$$

To calculate the attitude guidance angle, the satellite body coordinate system is defined, and the satellite body coordinate system is set to coincide with the orbital coordinate system at the initial moment: the origin is at the satellite center of mass; the Z-axis points from the satellite center of mass to the center of the earth; the X-axis is in the orbital plane, perpendicular to the Z-axis, and points to the satellite velocity direction as positive (because there is a trajectory angle, the X-axis direction is not exactly the same as the satellite velocity direction); the Y-axis is determined by the right-hand rule. In addition, ix, iy and iz are set as the unit vectors of X, Y and Z axes respectively; X, Y and Z axes are called roll axis, pitch axis and yaw axis respectively [1].

The satellite is not attitude guided and has in the body coordinate system

$$\boldsymbol{R}_s = R_s i_z = -\frac{a(1-e^2)}{1+e\,cosf} i_z \tag{5.2}$$

$$v_s = v_x i_x + v_z i_z = \sqrt{\frac{\mu}{a(1-e^2)}}(1+ecosf)i_x - \sqrt{\frac{\mu}{a(1-e^2)}} e i_z \sin f \tag{5.3}$$

$$\omega_e = \omega_e(i_x \cos u \sin i - i_y \cos i - i_z \sin u \sin i) \tag{5.4}$$

where u is the latitudinal amplitude angle. At this point, there is

$$v_s + R_s \times \omega_e = \begin{bmatrix} \cos\theta & 0 & -\sin\theta \\ 0 & 1 & 0 \\ \sin\theta & 0 & \cos\theta \end{bmatrix} \times \begin{bmatrix} \cos\psi & \sin\psi & 0 \\ -\sin\psi & \cos\psi & 0 \\ 0 & 0 & 1 \end{bmatrix} \times \begin{bmatrix} v_x + R_s\omega_e \cos i \\ R_s\omega_e \cos u \sin i \\ v_z \end{bmatrix}$$
$$= [a_1\ a_2\ a_3]^T \tag{5.5}$$

where: $a_1 = \cos\theta\cos\psi(v_x + R_s\omega_e \cos_i) + \cos\theta \times \sin\psi R_s\omega_e \cos u \sin i - v_z \sin\theta$;
$a_2 = -\sin\psi(v_x + R_s\omega_e \cos_i) + \cos\psi R_s\omega_e \cos u \sin i$;
$a_3 = \sin\theta \times \cos\psi(v_x + v_x) + \sin\theta\sin\psi R_s\omega_e \cos u \sin i + v_z \cos\theta$.

The position vector R between the SAR antenna center loudness and the target point is represented in the satellite system only with respect to the SAR antenna installation. It is assumed that the SAR antenna is installed in the YOZ plane of the system with the orientation perpendicular to the X-axis and the downward angle of view α (i.e., the angle between the antenna center and the Z-axis is α). At this time, after the yaw and pitch guidance, there is

$$\boldsymbol{R}/R = [0\ \sin\alpha\ \cos\alpha\ ]^T \tag{5.6}$$

Then all variables in Eq. (5.6) are represented in the satellite principal system after performing the two-dimensional bootstrap, with

$$f = -\frac{2}{\lambda}\begin{bmatrix} 0 \\ \sin\alpha \\ \cos\alpha \end{bmatrix} \cdot \begin{bmatrix} a_1 \\ a_2 \\ a_3 \end{bmatrix} = a_4 + a_5 \tag{5.7}$$

where: $a_4 = -\frac{2}{\lambda}\sin\alpha[-\sin\psi(v_x + R_s\omega_e \cos i) + R_s w_e \cos\psi \cos u \sin i]$;

$$a_5 = -\frac{2}{\lambda}\cos\alpha[\sin\theta\cos\psi(v_x + R_s\omega_e \cos i) + R_s w_e \sin\theta\sin\psi\cos u \sin i + v_z \cos\theta]$$

Analyzing Eq. (5.7), in order to make the full zero Doppler guidance independent of the downward viewing angle, and the antenna center in the SAR antenna range direction always pointing to the zero Doppler line, it is only necessary to satisfy

$$-\sin\psi(v_x + R_s w_e \cos i) + R_s w_e \cos\psi \cos u \sin i = 0 \tag{5.8}$$

$$\sin\theta\cos\psi(v_x + R_s w_e \cos i) + R_s w_e \sin\theta \times \sin\psi \cos u \sin i + v_z \cos\theta = 0 \tag{5.9}$$

Solving Eqs. (5.8) and (5.9), the yaw guidance angle is obtained as

$$\psi = \arctan\frac{R_s\omega_e \cos u \sin i}{v_x + R_s\omega_e \cos i} \tag{5.10}$$

Pitch guidance angle:

$$\theta = \arctan\left(-\frac{v_z}{\cos\psi(v_x + R_s\omega_e \cos i)} + R_s\omega_e \sin\psi \cos u \sin i \right) \tag{5.11}$$

where

$$R_s = -\frac{a(1-e^2)}{1+e\cos f} \tag{5.12}$$

$$v_x = \sqrt{\frac{u}{a(1-e^2)}}(1+e\cos f) \tag{5.13}$$

$$v_z = -\sqrt{\frac{u}{a(1-e^2)}} e \sin f \tag{5.14}$$

From the above analysis, it can be seen that if the yaw and pitch angles of the launcher are guided in two dimensions according to the above distribution, the center of the SAR antenna always points to the zero Doppler line in the azimuthal direction, and there is no Doppler residual, and zero Doppler attitude guidance can be achieved.

To simplify the guide operation, we can take e = 0 and the elliptical orbit becomes a circular orbit, and Eq. (5.10) can be simplified as

$$\psi = -\arctan\frac{\cos u \sin i}{\omega_s/\omega_e - \cos i} \tag{5.15}$$

where ω_s is the angular rate of the satellite in real time, and $\omega_s = (\mu/a)^{1/2}$

2. Long-range receiving satellite following the guidance technology

The receiver follow-on guidance is shown in Fig. 5.3, and the transmitter uses full zero Doppler guidance to achieve zero azimuthal Doppler, eliminating the effect of Earth rotation on SAR imaging. The receiving satellite uses roll and pitch attitude maneuvers to point at the ground target point of the primary satellite.

The position of the remote sensing point on the ground is first calculated from the attitude of the primary satellite, and then the three-axis attitude of the secondary satellite can be calculated from the relative position of the secondary satellite and the remote sensing point on the ground of the primary satellite, where the roll and pitch angles determine the pointing of the secondary satellite load, and the yaw angle is the rotation of the load in the direction of the sensitive axis. In this simulation, with an orbit altitude of 500 km and a phase deviation of 0.1° between the transmitting and receiving satellites, the roll and pitch angles of the receiving satellite are shown in Fig. 5.4 and Fig. 5.5, respectively.

The deviation of the aiming point range between the two satellites on the ground and the Doppler center frequencies of the transmitting and receiving satellites of the continuous system are solved, as shown in Fig. 5.6 and Fig. 5.7, respectively.

It can be seen from Fig. 5.6 that the receiving satellite follow-up technique can eliminate the range and azimuthal bias of remote sensing points, but the Doppler center frequency of the Ka-band binary system is still large, and the Doppler center

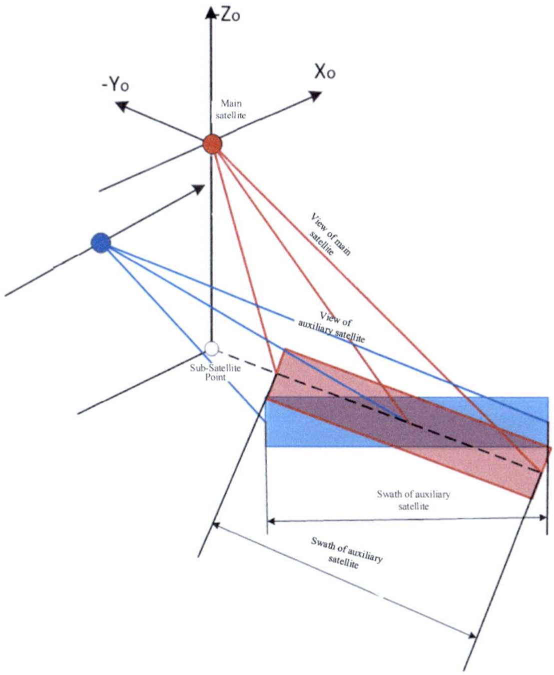

Fig. 5.3 Principle of tracking and aiming scheme

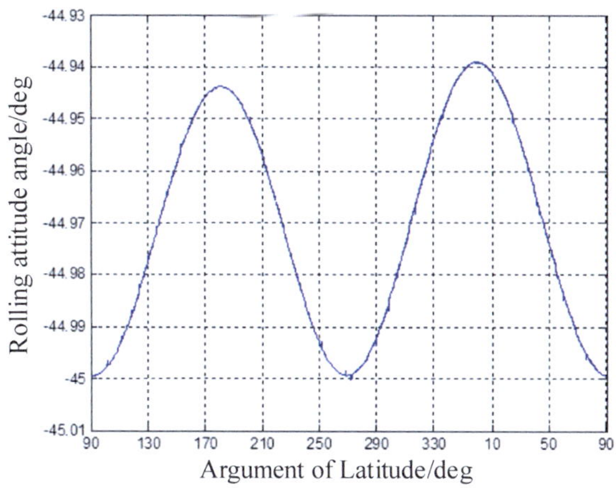

Fig. 5.4 Receiving satellite rolling angle

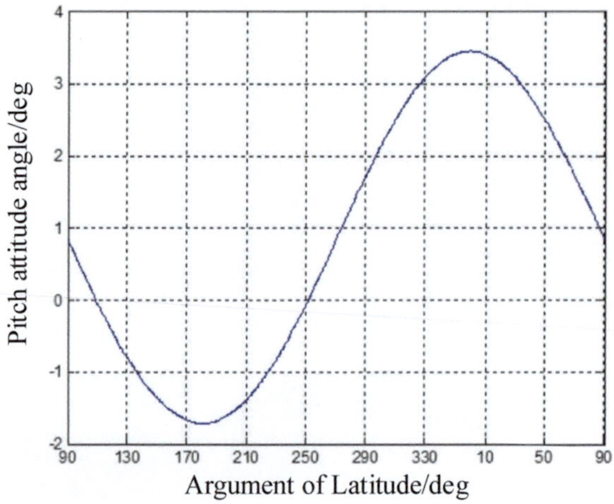

Fig. 5.5 Receiving satellite altitude

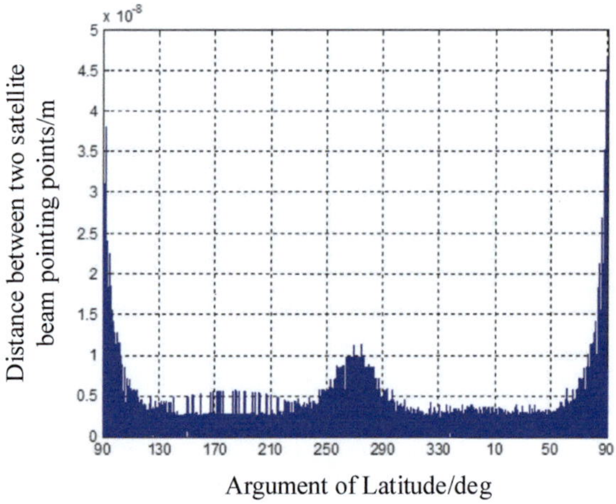

Fig. 5.6 Deviation of the aiming point of the transmitting and receiving satellites

frequency of the combined binary or single receiving satellite is in the order of 104 Hz.

Considering the engineering reality, there are certain attitude control errors for both the transmitting and receiving satellites at the transceiver end of the continuous wave SAR system, and the control accuracy of the satellite platform is taken as 0.01°. Considering the actual situation, an attitude error of 0.01° is added to each of the three axes of the primary satellite and −0.01° is added to each of the three axes of the secondary satellite, and Figs. 5.8 and 5.9 show the range deviation of the ground remote sensing points and the Doppler center frequency of the two satellites after adding the control error, respectively.

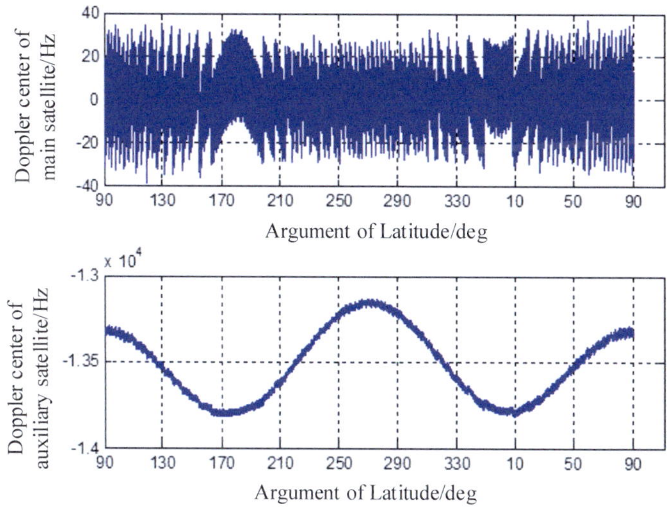

Fig. 5.7 Doppler center frequencies of transmitting and receiving satellites

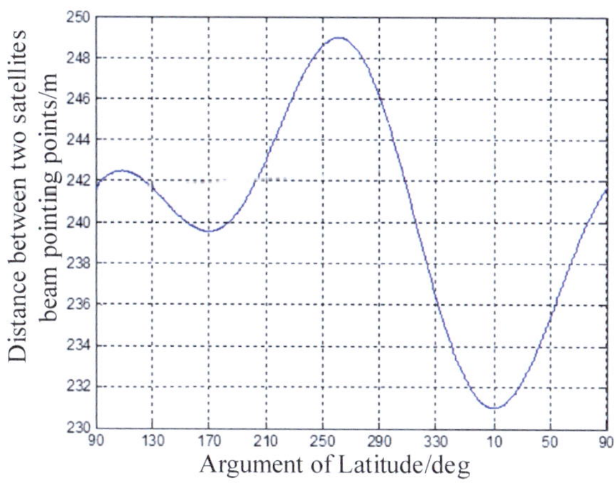

Fig. 5.8 Deviation of ground aiming point distance between two satellites

Comparing Fig. 5.9 with Fig. 5.7, it is obtained that there is a deviation in the Doppler center frequency between the guidance method considering the control error and the guidance method without considering the control error, and the deviation value is shown in Fig. 5.10.

Considering that the azimuthal beamwidth of the antenna is about 0.09°, the maximum receivable Doppler deviation is about 1360 Hz. The control accuracy can guarantee the echo reception.

Considering that the beam pointing accuracy of the actual satellite system is not only limited by the capability of the platform, but also constrained by the antenna

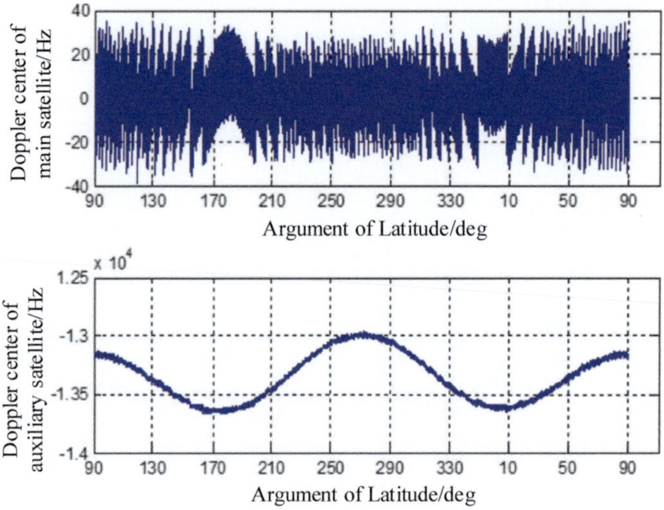

Fig. 5.9 Two satellite doppler center frequencies

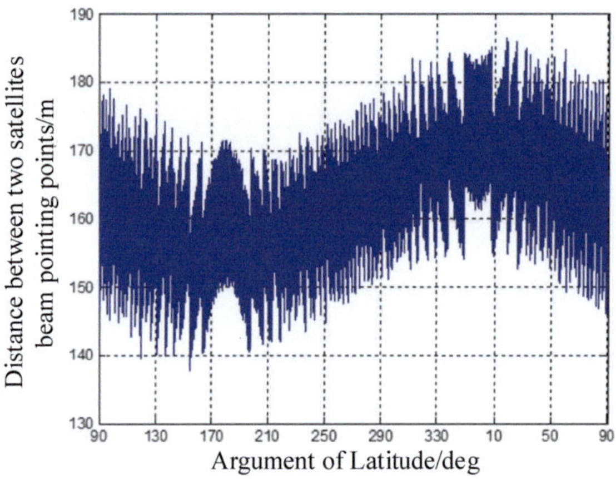

Fig. 5.10 Doppler center frequency deviation before and after adding control error

beam pointing accuracy, the current on-board antenna beam pointing accuracy can be better than 0.01° (0.006° is taken as the analysis in the following of this section), and the introduced Doppler deviation is about 220 Hz, combining the platform maneuver and the Doppler deviation introduced by the antenna beam pointing, which will introduce the azimuthal antenna equivalent Therefore, the echo-based long-range receiving satellite follow-up method is adopted here.

The echo-based long-range receiving satellite follow-up method is divided into the following two steps:

Estimate the Doppler deviation fdc value from the echo and find its difference from the ideal f_{dc};
The antenna azimuthal beam scan is used to compensate for the multi-sourced fdc.

Based on the above two steps, it can be derived that the final azimuthal Doppler deviation is divided into two parts: the echo estimation error f_{dc} and the azimuthal beam pointing error f_{dc} of the antenna.

Considering the real time processing in orbit, it is difficult to take more points, for example, 4096 points are taken at PRF = 4000 Hz and divided into 64 sub-blocks. fdc is estimated to take the maximum Doppler spectrum amplitude, considering the accuracy of calculation, the upper limit of error is about one sub-block, i.e. 64 points, which is about 62.5 Hz. fdc is mainly from the azimuthal antenna beam pointing error, which is about 94.8 Hz, and the integrated Doppler deviation is about 157.3 Hz. Therefore, the angle to be adjusted by the receiving satellite is solved by the integrated Doppler deviation, and the angle adjustment is done by the beam scan of the receiving satellite antenna direction map, and the integrated Doppler deviation is about 5%. Considering the control accuracy of current light and small satellite platforms and the beam pointing accuracy of phased array antennas, the echo-based long-range receiving satellite tracking method can achieve better indexes.

5.2.2 *Time–Frequency Synchronization Method Based on Signal-To-Transmission*

1. Time synchronization

High-precision time synchronization program is usually used to tame the local clock, two-way transponder timing and other time synchronization methods, the above two time synchronization methods can meet the high precision time synchronization application requirements. However, taking into account the time synchronization accuracy, anti-interference performance and other factors, it can be achieved by using the microwave two-way answer time synchronization method, the accuracy of the method can reach 10 ns, or even within 5 ns.

The inter-satellite synchronization using two-way answer timing is achieved by synchronizing two terminals on two transmitting and two receiving satellites based on two-way data chain round-trip timing (RTT) messages. Compared with the GPS/BeiDou co-viewing timing method, the RTT synchronization method has the characteristics of good stability, high accuracy, and strong anti-interference capability, and its basic principle is shown in Fig. 5.11.

Where, TOA_I is the time when the interrogation message arrives at the response end determined by the responding station (master); TOA_R is the time when the RTT response message arrives determined by the interrogating station (slave); t_d is the duration of the interrogation message; t_p is the propagation time of the RTT message;

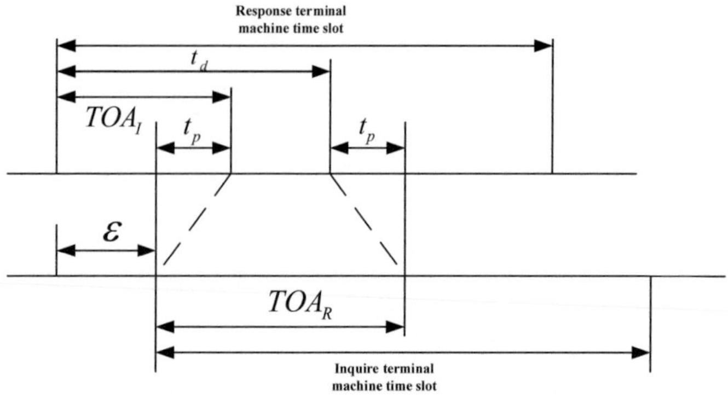

Fig. 5.11 Schematic diagram of RTT synchronization method principle

ε is the initial clock difference between the two stations. With a small relative motion speed between the two stations and the same round-trip transmission range, tp = $TOA_I - \varepsilon = TOA_R + \varepsilon - t_d$, the clock difference of the interrogating station with respect to the master station is:

$$\varepsilon = \frac{TOA_I - TOA_R + t_d}{2} \tag{5.16}$$

The interrogation station adjusts its own clock according to ε to achieve accurate time synchronization. The key to achieve high precision time synchronization by RTT synchronization method lies in the accurate measurement of TOA_I and TOA_R. In order to quickly achieve high precision timing, the interrogation message and the answer message are loaded onto a pseudo-random sequence, and the receiver can accurately measure the arrival moment of the interrogation or answer message by using the pseudo-code phase measurement technique. Theoretically, the time resolution measurement accuracy of this method can reach one hundredth of the code slice width without considering the existence of relative motion between stations. However, the timing accuracy is affected by round-trip path asymmetry, crystal stability, temperature, and Doppler of pseudocode due to relative motion.

(1) Impact of asymmetry in the round-trip link

A major factor affecting the measurement accuracy is the asymmetry in the transmission delay of the uplink and downlink due to the relative motion between stations. In the RTT synchronization principle, the upstream and downstream transmission delay tp is assumed to be equal. In fact, due to the time difference between the arrival of the interrogation and reply messages, in the presence of relative motion between the binary satellites, there is a time difference in the uplink and downlink transmission delay tε whose magnitude depends on the speed of the relative motion and the interval between the sending of the interrogation and reply messages. In this case, the calculation formula of the clock difference is modified as $\varepsilon = (TOA_I - TOA_R + t\varepsilon + td)/2$. The time interval between the master receiving the interrogation message and sending the reply message (td − TOA_I) is around 3 ms, and the duration of the whole timing cycle is calculated as 10 ms, assuming that the relative motion speed is 1 m/s, the asymmetric error introduced by the relative motion in the upstream and downstream The asymmetric error introduced by the relative motion in the link is $t_\varepsilon/2 = (10\ \text{ms} \times 1\ \text{m/s}) / (2 - c) = 0.02$ ns. This error is fixed and can be compensated if the information of the relative motion can be obtained from the differential positioning module.

(2) Effect of Doppler frequency bias

Another major factor affecting the measurement accuracy is the pseudo-code period variation caused by the Doppler shift, so when there is a Doppler shift, the arrival time of the slave to extract the master's transmit sequence will be affected. Assuming a relative speed of 1 m/s and a code slice rate of 10.23 MHz, the Doppler shift generated at the code slice frequency is

$$f = v \cdot \frac{10.23 \times 10^6\ \text{MHz}}{3 \times 10^8 \text{m/s}} = 0.08\ \text{Hz} \tag{5.17}$$

The pseudocode Doppler shift caused by the relative motion is small, and the time measurement error within the time range of one pseudocode cycle is

$$T_\varepsilon = \left(\frac{1}{10.23\text{MHz}} - \frac{1}{10.23\text{MHz} + 1.36\text{Hz}}\right) \times 1023 = 0.22\text{ns} \tag{5.18}$$

The RTT synchronization is performed every second and the error does not accumulate; the main determinant of the error is the duration of the answer message. This error is also a fixed error and can be compensated for if the relative speed of motion is available as well.

(3) Estimation of total time synchronization error

Combining the above factors, the accuracy of the time synchronization system is better than 5 ns when there is a relative motion of 1 m/s, and the time synchronization accuracy can be further improved under the corrected condition. In order to further meet the requirements of high precision time synchronization, the time of two-way calibration needs to be reduced, and frequency domain parallel code correlation is the key to achieve time–frequency synchronization. The capture process requires the input signal and the locally generated pseudocode to perform the correlation operation, and the correlation operation in the time domain is converted into the phase multiplication operation in the frequency domain by FFT, and then the correlation value of each code phase in the time domain is found by IFFT, so that the frequency domain parallel code phase search calculates the correlation value of all the code phases at a certain local oscillation frequency.

The frequency domain parallel code phase capture algorithm is calculated by FFTing and conjugating the received IF signal. The local pseudo-noise (PN) code is modulated with the carrier, and the modulated signal is FFTed and conjugated.

The result of the IFFT is the correlation value between the input IF signal and the locally generated signal, and the maximum value of the correlation value is found and compared with the predefined threshold. The arrival time of the signal can be calculated by the position of the peak, and then the synchronization error can be calculated.

The simulation of the time synchronization accuracy at different code speeds is shown in Fig. 5.12.

The simulation results of peak position capture at a code rate of 20.46 Mcps with different signal-to-noise ratios are shown in Fig. 5.12a, and the standard deviation of synchronization accuracy is 0.8 ns, 1.35 ns and 2.38 ns for 25 dB, 20 dB and

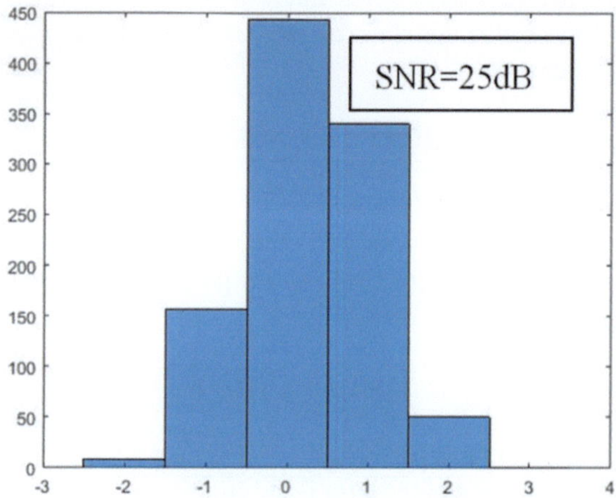

(a) At a bit rate of 20. 46 Mcps

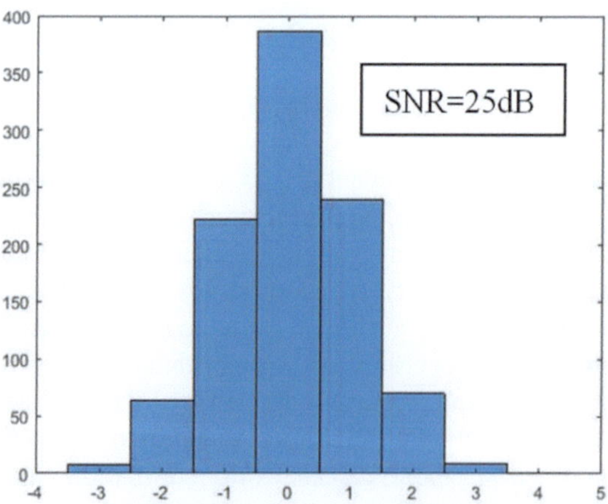

(b) At a bit rate of 10. 23Mcps

Fig. 5.12 Time synchronization accuracy at rates of 20.46 Mcps and 10.23 Mcps

15 dB, respectively, and the simulation results at a code rate of 10.23Mcps are shown in Fig. 5.12b, and the standard deviation of synchronization accuracy is 1.1 ns, 1.89 ns and 3.0 ns for 25 dB, 20 dB and 15 dB, respectively. Since the pseudocode and digital transmission are transmitted in the same channel and the calibration is spread spectrum, it can ensure a high signal-to-noise ratio and meet the time accuracy requirements of the SAR system on the transmitting and receiving satellites.

2. Frequency synchronization

The ultimate goal of frequency synchronization is to provide frequency and phase reference signals for the Ka FMCW SAR transmitter and receiver. The reference frequency source for the transmitter and receiver is a tame crystal/rubidium clock, and since the accuracy of the highly stable crystal/rubidium clock used on the satellite is above the 10−11 level, the radar carrier frequency difference between the transmitter and receiver is small enough after taming.

The disciplined oscillator combines the excellent short-term stability characteristics of a quartz crystal oscillator with the good long-term stability of the signal. The principle of the disciplined oscillator is shown in Fig. 5.13. The disciplining circuit is essentially a phase-locked loop (PLL). The output of the disciplined oscillator, after frequency division and shaping, generates a pulse per second (PPS) signal. This signal is compared in phase with the PPS pulse output from the receiver to obtain a phase difference. This phase difference is then digitally filtered and, after digital-to-analog (D/A) conversion, results in an error voltage. This error voltage controls the voltage-controlled port of the disciplined oscillator to achieve frequency adjustment.

The frequency error measurement can be better performed by using the ratio-time method, in which the number of clock periods N within the 1PPS signal interval of the local time uniform is measured multiple times from the station, then according to the calculation equation

$$\frac{\Delta T}{\tau} = \frac{\Delta f}{f_0} \tag{5.19}$$

The frequency error of the slave station with respect to the master can be obtained. Since the number of clock cycles experienced by the master during this time is an integer 109, $\Delta T = (N \cdot 10^{-8} - \tau)s$, $\tau = 100$ s, and f0 = 100 MHz. It can be seen that the accuracy of N determines the estimation accuracy of the frequency error.

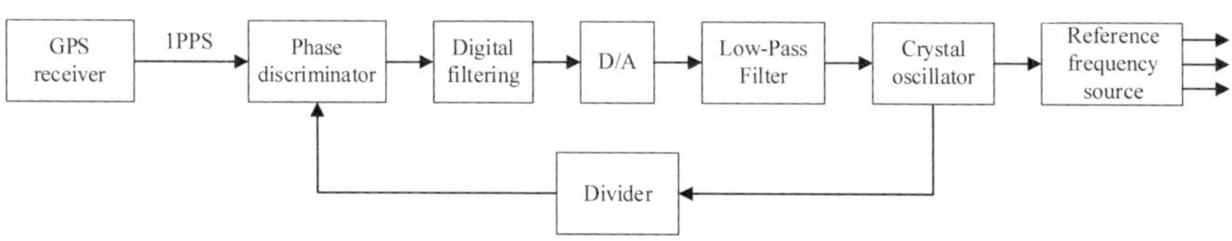

Fig. 5.13 Schematic diagram of synchronization based on global positioning system

The accuracy of N depends on the timing accuracy of the 1PPS signal at the beginning and end of the timing. Depending on the time-synchronization technique used in the time synchronization subsection, the 1PPS accuracy can reach 1 ns (1σ), and the frequency error estimation accuracy can reach 3×10^{-11} for a timing interval $\tau = 100$ s.

The correction of the frequency source can be achieved by adjusting the voltage value loaded on the frequency control pin of the thermostatic crystal. This is done by converting the estimated frequency error Δf and feeding it to a digital-to-analog converter (DAC), which in turn controls the resistance of the digital potentiometer, thereby changing the voltage value loaded on the frequency control pin.

The frequency domain synchronization is designed to ensure that the phase rotation of the transmit and receive signals is less than 2π within 1 s. At a clock frequency of 100 MHz, the synchronization accuracy of Ka-band frequency domain can reach 3×10^{-11}, which makes the frequency difference of the main oscillation signal generated by the final frequency synthesis better than 1 Hz, meeting the working requirements of the continuous wave transmitter and receiver end loads.

Figure 5.14 shows the radar carrier frequency difference after taming.

The analysis of the frequency synchronization accuracy is described below.

(1) Crystal characteristics determine its accuracy with time will have a large drift, the reference clock using "high-precision atomic clock + clock taming" method, through high-precision synchronization pulses to complete frequency adjustment, and ultimately maintain a more constant frequency state. This is achieved

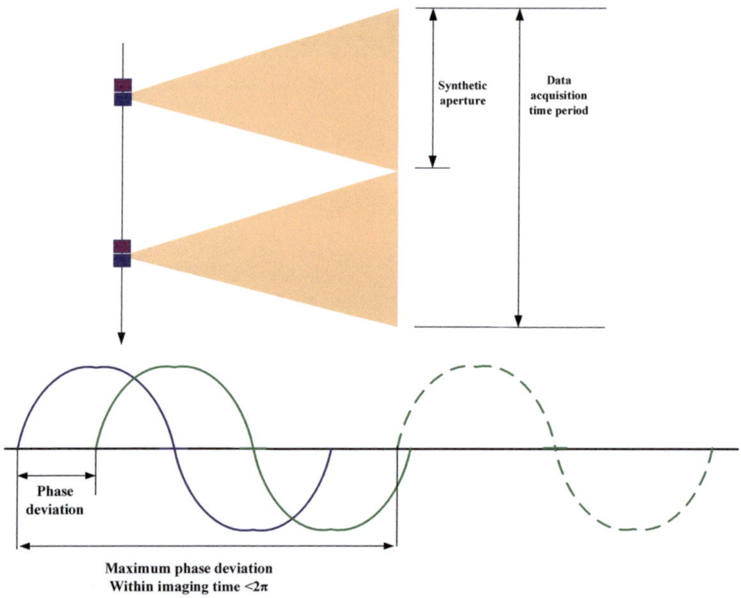

Fig. 5.14 Radar carrier frequency difference after domestication

by using high-precision synchronization pulses to obtain the reference frequency source of the master and slave, so that the frequency difference between the master and slave is small enough after taming.

(2) The tamed clock can combine the excellent short-term stability of the quartz crystal oscillator and the good long-term stability of the atomic clock signal, the taming circuit is actually a phase-locked loop, the tamed crystal output is sent to the phase-locked loop after frequency division and shaping, the tamed crystal output is divided and shaped to generate the PPS signal, which is compared with the PPS pulse output of the data chain to obtain the phase difference, the phase difference is digitally filtered, and the error voltage is obtained after D/A conversion to control the voltage control port of the tamed crystal to achieve frequency adjustment. The phase difference is digitally filtered and the error voltage is obtained after D/A conversion to control the voltage control port of the tamed crystal to achieve frequency adjustment.

Based on the above mentioned high precision synchronization pulses and taming circuits, the final frequency synchronization accuracy is guaranteed to be better than 3×10^{-11}.

5.2.3 Continuous Wave SAR Heterodyne Phase Synchronization Method

In the dual-satellite system, the reference frequency difference of the dual satellites, phase noise, phase jitter of the transceiver channel, phase error caused by receiver noise, and phase variation caused by the Doppler effect of the relative motion of the dual satellites all cause phase deviation, which affects the imaging quality. In the process of measuring the phase synchronization error, the sampling frequency is much smaller than the radar carrier frequency difference between the two satellites, which causes ambiguity in the phase error extraction process and increases the difficulty of phase error extraction and compensation, and the accuracy is low. The traditional pulse system SAR system uses the non-transmitting and non-receiving time slots in the cycle to realize the phase synchronization signal pair transmission, and extracts the synchronization phase based on the synchronization signal to realize the synchronization between the transmitting and receiving satellites. However, the continuous wave system has no transmitting and receiving time slots due to its large signal time width, and only a separate phase synchronization channel can be used to transmit and receive the synchronization signal. In order to avoid electromagnetic interference between the phase synchronization signal and the echo signal, a heterodyne signal with a certain frequency difference from the main oscillation signal needs to be used [2–4].

The factors causing the phase variation include: the difference between the reference frequencies Δf_i at the transmitting and receiving ends, the phase noise $n_{\varphi i}$, the phase jitter φ_{sysTi} and φ_{sysRi} at the transmitting and receiving channels, the phase variation φ_{anti} introduced by the phase direction diagram of the synchronized horn antenna, the phase error φSRNi caused by the receiver noise, the phase variation $2\pi\Delta d/\lambda$ caused by the Doppler effect of the relative motion of the transmitting and receiving satellites, etc., where $i = 1$ or 2, respectively. represent the transmitting or receiving satellites, respectively. If the radar carrier frequencies of the transmitting and receiving satellites are $f_i = f_0 + \Delta f_i$, where f_0 is the nominal frequency and Δf_i is the frequency deviation of satellite i, then the radar carrier phase of satellite i at time t is

$$\varphi_i(t) = 2\pi \int_0^t (f_0 + \Delta f_i)dt + \varphi_{0i} + n_{\varphi_i}(t) \tag{5.20}$$

where φ_{0i} is the radar carrier initial phase of satellite i; $n_{\varphi i}$ is the carrier phase noise.

With φ_{ji} denoting the demodulated phase obtained by satellite i transmitting satellite j receiving, the demodulated phase obtained by the transmitting satellite transmitting a synchronous pulse at time t and received by the receiving satellite at time $t + \tau_{12}$ is $\varphi_{21}(t + \tau_{12})$, and the compensated phase is

$$\varphi_0(t) = \frac{1}{2}\left[\varphi_{21}(t + \tau_{12}) + \varphi_{12}(t + \tau_{sys} + \tau_{21})\right] \tag{5.21}$$

The purpose of extracting the compensation phase is to obtain the phase synchronization error caused by the frequency deviation of the transmitting and receiving satellites and phase noise, and the other phases introduced in the process will become interference phases that affect the phase synchronization performance.

The phase difference caused between binary satellites can be summarized as follows:

(1) Phase difference formed by frequency deviation
(2) The phase difference formed by phase noise;
(3) Phase errors introduced by receiver noise;
(4) Phase errors introduced by phase jitter in the radar transceiver channel
(5) The phase error introduced by the directional diagram of the synchronized horn antenna due to the change of the relative positions of the transmitting and receiving satellites during one phase-synchronous pulse pair transmission;
(6) The phase error introduced by the Doppler effect of the relative motion of the satellites.

Items (1) and (2) are the components needed to compensate the phase of the receiving satellite echo and achieve phase synchronization. The other terms are additional phase errors introduced during the transmission of bidirectional synchronized pulses, which will form the residual phase error after phase synchronization. According to engineering experience, the phase jitter of radar transceiver channel can be generally controlled within 1°, which also has a small impact on the phase synchronization performance; when the speed of relative motion of satellites is stable or does not change much, the phase introduced by Doppler effect is approximately constant, which has a small impact on the phase synchronization performance, so the main factor affecting the phase synchronization performance will be the receiver noise, which can be extracted from the transceiver binary by means of carrier pair transmission The phase information of the transmitting and receiving binary can be extracted by the carrier pair transmission method, and high precision phase synchronization can be achieved by the sliding window accumulation.

The phase synchronization factor extraction process is as follows:

(1) The transmitting frequency synthesis generates Ka-band synchronized signals under highly stable clock excitation provided by the time–frequency unit, which are switched to synchronized transmitting antenna radiation through the circulator;
(2) Synchronous receiving antenna receives the other party's radiation signal, which is converted and gain adjusted by the RF receiving component and output to the digital receiver;
(3) The digital receiver is sampled and processed to obtain the phase factor, which is used for imaging acquisition data compensation.

The real-time processing component uses the phase difference method for instantaneous frequency measurement. After the quadrature processing, the inverse tangent algorithm is used to extract the instantaneous phase from the quadrature baseband signal, and then the phase difference of fixed interval points is accumulated by the sliding window, and finally the low-pass filter is used to low-pass filter the phase difference accumulation result, whose workflow is shown in Fig. 5.15.

s_Q (n), s_I (n) denote the in-phase component and the quadrature component of the signal s(n), respectively. Then the phase θ(n) of the signal s(n) at time n can be estimated by taking the arctangent of $s_Q(n)/s_I(n)$, i.e.

$$\hat{\theta}(n) = \arctan\frac{s_Q(n)}{s_I(n)} = 2\pi f_d n t_s + \phi \tag{5.22}$$

Then the phase difference Δθ(k) of the signal at two different moments with sampling interval k is estimated as

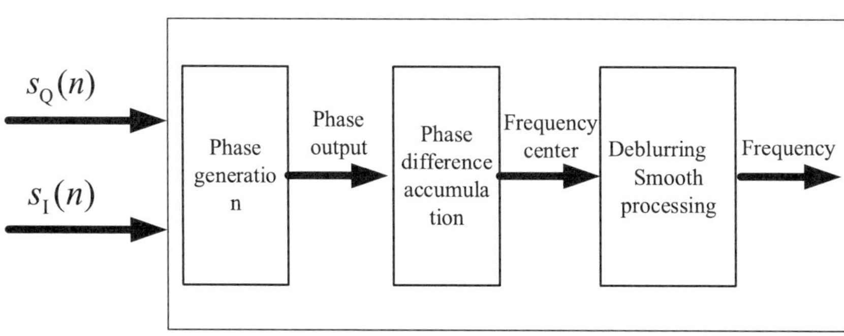

Fig. 5.15 Flow chart of direct phase difference algorithm

$$\Delta\hat{\theta}(k) = \hat{\theta}(n+k) - \hat{\theta}(n) = 2\pi f_d k t_s \tag{5.23}$$

Therefore, fd can be estimated by the following equation:

$$\hat{f}_d = \frac{\Delta\hat{\theta}(k)}{2\pi k t_s} = \frac{\Delta\hat{\theta}(k)}{2\pi k} f_s \tag{5.24}$$

The phase of the signal varies linearly in the range [−π, π), so there are phase mutation points (from π to −π) in each cycle. Here, the defuzzification is performed by adding an offset of −2π/2π at the phase mutation point according to the direction of the phase mutation, so that the phase remains linear for a period of time after the mutation, and setting the selector so that the normal phase output is used before the mutation point, and the offset phase output is used for k (k is the sampling interval) points after the mutation point. after the offset phase output.

The algorithm simulation can achieve better than 10° phase synchronization accuracy with better than 20 dB SNR, thus ensuring the imaging quality.

5.3 Double Satellite FM Continuous Wave SAR Strip Imaging

This section introduces the dual-base FMCW SAR imaging algorithm. In dual-base FMCW SAR, the transmitting and receiving antennas are separated on two separate satellite platforms, which can effectively overcome the problem of excessive signal coupling between transmitting and receiving, avoid signal leakage, and improve the range of operation. The dual-base mode operation can also achieve higher mobility, stealth, and better anti-interference and anti-interception performance of SAR satellites [5].

Here, the geometric configuration of the satellite-based dual-base SAR operation is first analyzed in detail. A detailed analysis of the slant range history of the electromagnetic waves during the transmitting and receiving of the satellite-based FMCW SAR reveals that the radar antenna branches in the dual-base mode, resulting in completely different ranges from the radar receiving and transmitting platforms for the same target, and the complexity of the spatial geometric relationship leads to the slant range history of the satellite-based dual-base radar containing a double square root (DSR) term, which is expressed as a flat top hyperbola (FTH). This makes the development of the dual-base imaging algorithm difficult [6, 7]. Based on the detailed analysis of the slant range history, a model of the dual-base FMCW SAR de-frequency echo signal under certain approximation conditions is constructed here to derive the mathematical equation of the point target echo to accurately represent the effect of the instantaneous slant range variation on the echo signal during the sweep cycle; because the receiver and transmitter are located on different satellite platforms and the range between the two satellites is too large, it cannot be simply equated to a single-base FMCW SAR system. Because the receiver and transmitter are located on different satellite platforms and the binary spacing is too large, they cannot be simply equated to a single-base FMCW SAR system and imaged using the conventional satellite-based single-base FMCW SAR imaging algorithm [8]. In the following paper, based on the traditional frequency scaling (FS) algorithm and the characteristics of the dual-base geometry, we propose two imaging algorithms for dual-base FMCW SAR.

5.3.1 Dual Satellite FM Continuous Wave SAR Echo Modeling

To explain the principle of satellite-based dual-base FMCW SAR imaging, it is necessary to first construct a dual-satellite operation system, calculate the oblique range variation of the transmitting and receiving satellite radar platforms relative to the scene target in the mapping belt according to the dual-satellite operation system, establish a satellite-based dual-base FMCW SAR echo model based on the de-frequency reception regime, and then complete the imaging process by the imaging algorithm. As shown in Fig. 5.16, the transmitting and receiving satellite radar platform adopts the widely used co-orbital following configuration, i.e., the transmitting and receiving satellites operate in the same orbit and maintain a relatively fixed range. The gray shaded area represents the target imaging mapping belt, and the transmitting satellite maintains a backward oblique view with an oblique angle of θt0 and the receiving satellite maintains a forward oblique view with an oblique angle of θ_{r0} with respect to the target imaging mapping belt, and the center line of their beams intersects at the center of the scene.

Here, the dual-base FMCW SAR imaging geometry model is constructed from the dual-satellite operation system in Fig. 5.16, as shown in Fig. 5.17.

If SAR transmits an ideal linear FM continuous wave signal, the echo of the target P at one point in the scene is analyzed, and the closest range (or vertical range) from this point to the orbit of the transmitting satellite T_x is R_{t0}, and the closest range (or vertical range) to the orbit of the transmitting satellite R_x is R_{r0}. The conventional pulsed system SAR, because the duty cycle of the transmitting pulse signal is much less than 1, has a very short pulse duration, which is generally in the microsecond range. The conventional pulse system SAR has a duty cycle much less than 1 and a very short pulse duration, usually in the microsecond range, so it can be processed with the "stop–go-stop" assumption. As the duty cycle of FMCW SAR is equal to 1, the instantaneous slope range is related to the fast time t_r and the slow time t_a when the relative displacement of the target and the antenna phase center caused by the range to the fast time within the signal modulation period is not negligible. At any moment tr + ta the sum of the slope range from the phase center of the transmitting and receiving satellite radar antenna to P is R(tr,ta), and let the complex envelope of the radar transmitting signal be $S_t(t_r) = a_r(t_r)\exp(j\pi\gamma t_r^2)$, γ is the

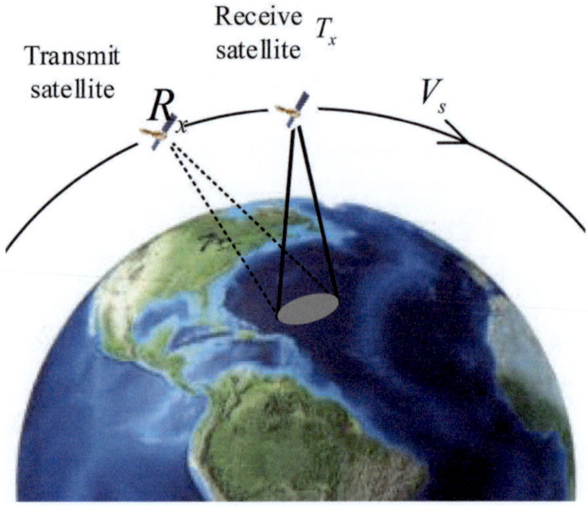

Fig. 5.16 Dual satellite operating system

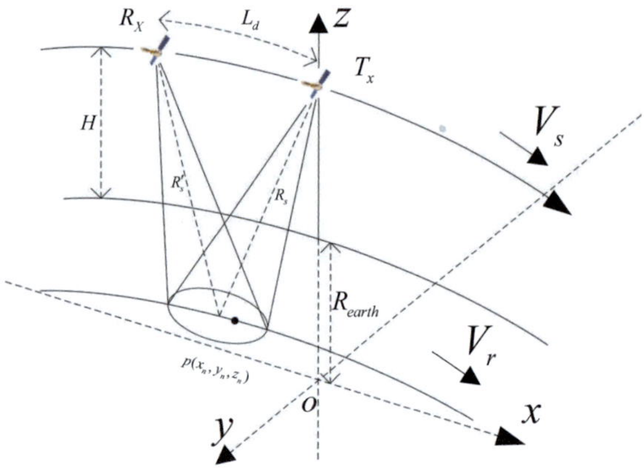

Fig. 5.17 Geometric modeling of spaceborne bistatic FMCW SAR imaging

modulation frequency of the transmitted linear frequency modulation (LFM) signal, and its received fundamental frequency signal of the above point target echo in the range to fast time—bearing slow time domain (tr − ta domain) can be written as

$$s(t_r, t_a) = w_a(t_a)w_\mathrm{r}\left(t_r - \frac{2R(t_r, t_a)}{\mathrm{c}}\right) \cdot \exp\left[j\pi\gamma\left(t_r - \frac{2R(t_r, t_a)}{\mathrm{c}}\right)^2\right] \cdot \exp\left[-\mathrm{j}\frac{4\pi}{\lambda}R(t_r, t_a)\right] \tag{5.25}$$

where $w_r(-)$ and $w_a(-)$ are the window function and azimuth window function of the radar linear FM signal, respectively, the former is a rectangular window when unweighted and the latter is related to the antenna beam shape in addition to filtering weighting; $\lambda = c/f_c$ is the wavelength corresponding to the carrier frequency.

The received point target echo can be expressed as the differential beat output after the decoupling frequency modulation process

$$\begin{aligned} s(t_r, t_a) = {} & w_a(t_a) \cdot w_r\left(t_r - \frac{2R(t_r, t_a)}{c}\right) \\ & \cdot \exp\left\{ -j\frac{4\pi}{c}\gamma\left[R(t_r, t_a) - R_{ref}\right]t - \frac{2R_{ref}}{c})\right\} \\ & \cdot \exp\left[-j\frac{4\pi}{\lambda}R(t_r, t_a) - R_{ref}\right] \cdot \exp\left\{-j\frac{4\pi\gamma}{c^2}\left[R(t_r, t_a) - R_{ref}\right]^2\right\} \end{aligned} \quad (5.26)$$

where γ is the range-oriented tuning frequency; R_{ref} is the reference range for the solved line frequency tuning process. The first exponential term in Eq. (5.26) is the range-oriented phase course, the second exponential term is the azimuthal signal, and the third exponential term is the RVP (residual video phase), which, if not removed, will have an impact on the azimuthal energy focus.

5.3.2 *RD Imaging Methods*

The RD algorithm has been used since the beginning of the development of SAR imaging because it is the simplest and easiest to understand of the many radar imaging algorithms. The main idea of the RD algorithm is to decompose the two-dimensional received signal into two one-dimensional signals cascaded in the azimuth and range directions, and then perform signal processing on the two one-dimensional signals separately. Since the received signal has been pre-processed with de-modulation, the range compression is completed by doing range FFT in the range direction; in the azimuthal processing, the range and azimuthal signals are coupled due to the existence of range migration and cannot be compressed directly by FFT, so the range migration correction must be performed before the azimuthal pulse compression [9]. In the RD algorithm, the range migration correction is usually performed by interpolation, followed by the corresponding matching filtering, and finally the azimuthal pulse compression of the pulse. Since the "stop–go-stop" model is used in the research of pulsed SAR RD algorithm, in the RD algorithm of FMCW SAR, the range migration will change accordingly, and the RVP term will be generated in the process of receiving signal de-modulation. Therefore, this term needs to be eliminated before simulation.

1. The basic principle of range Doppler algorithm

Taking the fast time into account, the instantaneous slope between the transmitter and the point target is shown in Eq. (5.27), and by expanding the instantaneous slope at the fast time in a Taylor series and neglecting the effect of the third term later, we obtain

$$\begin{aligned} R_t(t_r, t_a) &= \sqrt{V_r^2(t_r + t_a)^2 + R_{t0}^2 - 2R_{t0}V_r(t_r + t_a)\sin\theta_{t0}} \\ &\approx \sqrt{V_r^2t_a^2 + R_{t0}^2 - 2R_{t0}V_rt_a\sin\theta_{t0}} + \frac{V_r^2t_a - R_{t0}V_r\sin\theta_{t0}}{\sqrt{V_r^2t_a^2 + R_{t0}^2 - 2R_{t0}V_rt_a\sin\theta_{t0}}}t_r \end{aligned} \tag{5.27}$$

The Taylor series expansion is obtained by omitting four and higher terms.

$$\begin{aligned} R_t(t_a) &= \sqrt{V_rt_a^2 + R_{t0}^2 - 2R_{t0}V_rt_a\sin\theta_{t0}} \\ &\approx R_{t0} - V_rt_a\sin\theta_{t0} + \frac{V_r^2\cos^2\theta_{t0}}{2R_{t0}}t_a^2 + \frac{V_r^3\sin\theta_{t0}\cos^2\theta_{t0}}{2R_{t0}^2}t_a^3 \end{aligned} \tag{5.28}$$

where $-\mathrm{V_rt_asin\theta_{t0}}$ is called the linear walk component, denoted by $\Delta\mathrm{R_t(t_a)}$, which can cause a shift in the radar Doppler center. During the imaging process, this term is corrected separately using a correction factor to eliminate the influence of the linear walk component of the transmitter and receiver, the corresponding correction factor is given in Eq. (5.34), and after correcting this term we ignore the effect of the cubic term, which can be approximated as

$$\begin{aligned} R_t(t_a) = R_t(t_a) - \Delta R_t(t_a) = R_{t0} + \tfrac{V_r^2\cos^2\theta_{t0}}{2R_{t0}}t_a^2 \\ \approx \sqrt{R_{t0}^2 + V_r^2t_a^2\cos^2\theta_{t0}} \end{aligned} \tag{5.29}$$

Substituting Eq. (5.29) into (5.28), the instantaneous slope range between the radar transmitter and the point target is obtained as

$$R_t(t_r, t_a) \approx R_t(t_a) + \frac{V_r^2t_a - R_{t0}V_r\sin\theta_{t0}}{R_t(t_a)}t_r \tag{5.30}$$

Similarly, the instantaneous slope range between the radar receiver and the point target can be derived as

$$R_r(t_r, t_a) \approx R_r(t_a) + \frac{V_r^2t_a - R_{r0}V_r\sin\theta_{r0}}{R_r(t_a)}t_r \tag{5.31}$$

where, $R_r(t_r) = \sqrt{R_{r0}^2 + V_r^2t_a^2\cos^2\theta_{r0}}$, the total instantaneous slope range expression from the transceiver platform to the point target is

$$R(t_r, t_a) = R_t(t_a) + R_r(t_a) + \left[\frac{V_r^2 t_a - R_{t0}\sin\theta_{t0}}{R_t(t_a)} + \frac{V_r^2 t_a - R_{r0}\sin\theta_{r0}}{R_r(t_a)}\right]t_r \quad (5.32)$$

Substituting Eq. (5.32) into (5.26), the echo signal can be rewritten as

$$\begin{aligned} s(t_r, t_a) = &\exp\left\{-j\frac{4\pi}{\lambda}\left[\frac{V_r^2 t_a - R_{t0}\sin\theta_{t0}}{R_t(t_a)} + \frac{V_r^2 t_a - R_{r0}\sin\theta_{r0}}{R_r(t_a)}\right]t_r\right\} \\ &\cdot \exp\left\{-j\frac{4\pi}{\lambda}[R_t(t_a) + R_r(t_a)]\right\} \\ &\cdot \exp\left\{-j\frac{4\pi}{\lambda}\gamma\{R_t(t_a) + R_r(t_a)\right. \\ &\left. + \left[\frac{V_r^2 t_a - R_{t0}\sin\theta_{t0}}{R_t(t_a)} + \frac{V_r^2 t_a - R_{r0}\sin\theta_{r0}}{R_r(t_a)}\right]\left(t_r - \frac{2R_{ref}}{c}\right)\right\} \end{aligned} \quad (5.33)$$

The linear walk component mentioned above needs to be eliminated before doing the azimuthal FFT, taking into account the linear walk component of the receiver and transmitter from the original $\Delta R(t_a) = \Delta R_t(t_a) + \Delta R_r(t_a) = -(V_r t_a \sin\theta_{t0} + V_r t_a \sin\theta_{r0})$,then the corresponding correction function is

$$H_f = \exp\left\{j\frac{4\pi}{c}\gamma\Delta R(t_a)\left(t - \frac{2R_{ref}}{c}\right) + j\frac{4\pi}{\lambda}\Delta R(t_a)\right\} \quad (5.34)$$

By applying the azimuthal FFT of Eq. (5.33) to the range Doppler domain (range to time domain, azimuth to frequency domain) according to the standing phase principle, the total phase expression of the system is as follows

$$\theta(t_a) = \theta_t(t_a) + \theta_r(t_a) + 2\pi f_a t_a \quad (5.35)$$

The above equation is decomposed into the sum of the standing phases of the transmitter and receiver, where the standing phase principle of the transmitter is as follows:

$$\begin{aligned} \theta_t(t_a) &= -\frac{4\pi}{\lambda}\sqrt{R_{t0}^2 + V_r^2\cos^2\theta_{t0}t_a^2} \\ &\quad - j\frac{4\pi}{c}\gamma\sqrt{R_{t0}^2 + V_r^2\cos^2\theta_{t0}t_a^2}\left(t_r - \frac{2R_{ref}}{c}\right) + \pi f_a t_a \\ &= -\frac{4\pi}{\lambda}\sqrt{R_{t0}^2 + V_r^2\cos^2\theta_{t0}t_a^2}\left[1 + \frac{\gamma}{f_0}\left(t_r - \frac{2R_{ref}}{c}\right)\right] + \pi f_a t_a \end{aligned} \quad (5.36)$$

Taking the differentiation gives

$$t_{ta} = \frac{\lambda f_a R_{t0}}{2V_r^2\cos^2\theta_{t0}\sqrt{\left[1 + \frac{\gamma}{f_0}\left(t - \frac{2R_{ref}}{c}\right)\right]^2 - \frac{\lambda^2 f_a^2}{4V_r^2\cos^2\theta_{t0}}}} \quad (5.37)$$

Substituting Eq. (5.37) into (5.36) yields

$$\theta_t(f_a) = -\frac{4\pi}{\lambda}\left[1+\frac{\gamma}{f_0}\left(t-\frac{2R_{ref}}{c}\right)\right]\frac{R_{t0}\left[1+\frac{\gamma}{f_0}\left(t_r-\frac{2R_0}{c}\right)\right]}{\sqrt{\left[1+\frac{\gamma}{f_0}\left(t-\frac{2R_{ref}}{c}\right)\right]^2-\left(\frac{\lambda f_a}{2V_r\cos\theta_{t0}}\right)^2}}$$
$$+2\pi f_a\frac{\lambda f_a R_{t0}}{2V_r^2\cos^2\theta_{r0}\sqrt{\left[1+\frac{K}{f_0}\left(t-\frac{2R_{ref}}{c}\right)\right]^2-\left(\frac{\lambda f_a}{2v\cos\theta_{t0}}\right)^2}}$$
$$= -\frac{4\pi R_0}{\lambda}\sqrt{\left[1+\frac{K}{f_0}\left(t-\frac{2R_{ref}}{c}\right)\right]^2-\left(\frac{\lambda f_a}{2v\cos\theta_{t0}}\right)^2} \tag{5.38}$$

A Taylor series expansion at $t=\frac{2R_{\text{ref}}}{2}$ gives

$$\theta_t(f_a) = -\frac{4\pi R_{t0}}{\lambda}\left\{\sqrt{1-\left(\frac{\lambda f_a}{2V_r\cos\theta_{t0}}\right)^2}+\frac{\gamma}{f_0\sqrt{1-\left(\frac{\lambda f_a}{2V_r\cos\theta_{t0}}\right)^2}}\left(t-\frac{2R_{ref}}{c}\right)\right\}$$
$$-\frac{4\pi R_{t0}}{\lambda}\left\{\frac{\left(\frac{\gamma}{f_0}\right)^2\left[1-\left(1-\left(\frac{\lambda f_a}{2V_r\cos\theta_{t0}}\right)^2-1\right)\right]}{\left[1-\left(\frac{\lambda f_a}{2V_r\cos\theta_{t0}}\right)^2\right]^{\frac{3}{2}}}\left(t-\frac{2R_{ref}}{c}\right)^2\right\} \tag{5.39}$$

make $\beta_t(f_a)=\sqrt{1-\left(\frac{\lambda f_a}{2V_r\cos\theta_{t0}}\right)^2}$, then there are

$$\begin{aligned}\theta_t(f_a) = &-\frac{4\pi R_{t0}}{\lambda}\beta_t(f_a)-\frac{\gamma}{f_c\beta_t(f_a)}\frac{4\pi R_{t0}}{\lambda}\left(t_t-\frac{2R_{ref}}{c}\right)\\ &-\frac{4\pi R_{t0}\lambda\gamma^2}{c^2}\frac{\beta_t^2(f_a)-1}{\beta_t^3(f_a)}\left(t_r-\frac{2R_{ref}}{c}\right)^2\end{aligned} \tag{5.40}$$

Similarly, the phase of the receiver can be obtained, which can be expressed as

$$\begin{aligned}\theta_r(f_a) = &-\frac{4\pi R_{r0}}{\lambda}\beta_r(f_a)-\frac{\gamma}{f_c\beta_r(f_a)}\frac{4\pi R_{r0}}{\lambda}\left(t_r-\frac{2R_{ref}}{c}\right)\\ &-\frac{4\pi R_{r0}\lambda\gamma^2}{c^2}\frac{\beta_r^2(f_a)-1}{\beta_r^3(f_a)}\left(t_r-\frac{2R_{ref}}{c}\right)^2\end{aligned} \tag{5.41}$$

where, $\beta_t(f_a)=\sqrt{1-\left(\frac{\lambda f_a}{2V_r\cos\theta_{r0}}\right)^2}$. then the total phase expression of the same track dual-base FMCW SAR is

$$\begin{aligned}\theta(f_a) &= \theta_t(f_a) + \theta_r(f_a) \\ &= -\frac{4\pi R_{t0}}{\lambda}\left[\beta_t(f_a) + \beta_r(f_a)\right] \\ &\quad -\left[\frac{R_{t0}}{\beta_t(f_a)} + \frac{R_{r0}}{\beta_r(f_a)}\right]\frac{4\pi\gamma}{f_c\lambda}\left(t_r - \frac{2R_{ref}}{c}\right) \\ &\quad -\left[\frac{4\pi R_{t0}\lambda\gamma^2}{c^2}\frac{\beta_t^2(f_a)-1}{\beta_t^3(f_a)} + \frac{4\pi R_{r0}\lambda\gamma^2}{c^2}\frac{\beta_r^2(f_a)-1}{\beta_r^3(f_a)}\right]\left(t_r - \frac{2R_{ref}}{c}\right)^2\end{aligned} \tag{5.42}$$

Then the range Doppler spectrum of the echo signal can be expressed as

$$\begin{aligned}s(f_a, t_r) &= \exp\left\{-j\left[\frac{4\pi R_{t0}\beta_t(f_a)}{\lambda} + \frac{4\pi R_{r0}\beta_r(f_a)}{\lambda}\right]\right\} \\ &\quad \cdot \exp\left\{-j\frac{4\pi\gamma}{c}\left[\frac{R_{t0}}{\beta_t(f_a)} + \frac{R_{r0}}{\beta_r(f_a)} - R_{ref}\right]\left(t_r - \frac{2R_{ref}}{c}\right)\right\} \\ &\quad \cdot \exp\left\{-j\left[\frac{4\pi R_{t0}\lambda\gamma^2}{c^2}\frac{\beta_t^2(f_a)-1}{\beta_t^3(f_a)} + \frac{4\pi R_{r0}\lambda\gamma^2}{c^2}\frac{\beta_r^2(f_a)-1}{\beta_r^3(f_a)}\right]\right. \\ &\quad \left.\left(t_r - \frac{2R_{ref}}{c}\right)^2\right\}\end{aligned} \tag{5.43}$$

$\frac{R_{t0}}{\beta_t(f_a)} + \frac{R_{r0}}{\beta_r(f_a)}$ in the second exponential term in Eq. (5.43) can be approximated if the null variability of the range migration is neglected as

$$\frac{R_{t0}}{\beta_t(f_a)} + \frac{R_{r0}}{\beta_r(f_a)} \approx R_{t0} + R_{r0} + \left(\frac{R_{t0}\lambda^2 f_a^2}{8V_r^2\cos^2\theta_{t0}} + \frac{R_{r0}\lambda^2 f_a^2}{8V_r^2\cos^2\theta_{r0}}\right) \tag{5.44}$$

From Eq. (5.44), it can be seen that the expression for the range migratory momentum in the RD algorithm of the dual-base FMCW SAR is

$$\Delta R_{RCM} = \frac{R_{t0}\lambda^2 f_a^2}{8V_r^2\cos^2\theta_{t0}} + \frac{R_{r0}\lambda^2 f_a^2}{8V_r^2\cos^2\theta_{r0}} \tag{5.45}$$

The RD algorithm usually uses sinc interpolation to process the range migration momentum, and the RCM-corrected signal expression is

$$\Delta R_{RCM} = \frac{R_{t0}\lambda^2 f_a^2}{8V_r^2\cos^2\theta_{t0}} + \frac{R_{r0}\lambda^2 f_a^2}{8V_r^2\cos^2\theta_{r0}} \tag{5.46}$$

The third exponential term in Eq. (5.46) is a quadratic function of the range to the fast time and requires a quadratic range compression reference function to correct for:

$$H_{src} = \exp\left\{j\frac{4\pi(R_{t0}+R_{r0})\lambda\gamma^2}{c^2}\left[\frac{\beta_t^2(f_a)-1}{\beta_t^3(f_a)}+\frac{\beta_r^2(f_a)-1}{\beta_r^3(f_a)}\right]\left(t_r-\frac{2R_{ref}}{c}\right)^2\right\} \tag{5.47}$$

The azimuthal pulse compression is performed by means of matched filtering, and the corresponding matched filtering function is

$$H_{ac} = \exp\left\{j\frac{4\pi(R_{t0}+R_{r0})}{\lambda}[\beta_t(f_a)+\beta_r(f_a)]\right\} \tag{5.48}$$

At this point, the algorithm has finished the processing of the azimuthal direction, the signal in the azimuthal direction to do IFFT, the signal transformed to the two-dimensional time domain, then the echo function becomes

$$s(t_r, t_a) = \exp\left\{-j\frac{4\pi\gamma}{c}[R_{t0}+R_{r0}-R_{ref}]\left(t_r-\frac{2R_{ref}}{c}\right)\right\} \tag{5.49}$$

To do range-oriented compression on the echo signal, it can be seen from the previous section that only range-oriented FFT can be done on the processed echo signal. The expression of the signal after the range direction compression is as follows:

$$s(f_r, t_a) = \sin c\left\{\pi T_p\left[f_r+\frac{2\gamma}{c}(R_{t0}+R_{r0}-R_{ref})\right]\right\}\cdot\exp\left(-j\frac{4\pi R_{ref}}{c}f_r\right) \tag{5.50}$$

The second exponential term of Eq. (5.50) is brought about by the introduction of a reference range in the FMCW SAR during the de-modulation of the received signal, for which the compensation function is.

$$\text{The } H_{rvp} = \exp\left(j\frac{4\pi R_{ref}}{c}f_r\right) \tag{5.51}$$

RD imaging algorithm of the Double Satellite FMCW SAR is now in focus and ready for imaging, and the flow chart of the RD algorithm is shown in Fig. 5.18.

2. Simulation results and analysis

The following simulations are performed for a single point target of a dual-base FMCW SAR with the following parameters: signal bandwidth of 150 MHz, carrier frequency of 35 GHz, sweep period of 25 ms, range sampling frequency of 25 MHz, satellite orbit altitude of 500 km, and dual-satellite spacing of 1 km.

A stationary point target was placed on the centerline of the scene at the time of zero slow time and imaged using the method described in this section, and the results are shown in Fig. 5.19. Figure 5.19a and b shows the spatial response curves after range pulse compression without azimuthal pulse compression, where the vertical coordinate is the range frequency and the horizontal coordinate is the azimuthal time,

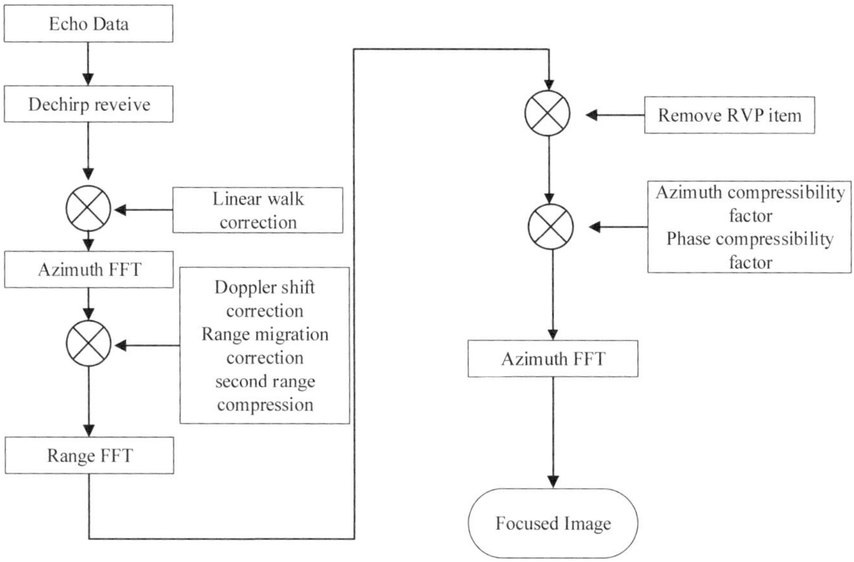

Fig. 5.18 Flow chart of dual satellite FMCW SAR RD algorithm

where Fig. 5.19a shows the case after uncompensated Doppler shift and Fig. 5.19b shows the case after compensated Doppler shift. It can be found that if the Doppler shift is not compensated, it will cause the migration of the echoes across the range unit and affect the focus of the image. In order to make the discrete data points continuous, the results are shown in Fig. 5.19c, its contour map in Fig. 5.19d, its range profile in Fig. 5.19e, and its azimuthal profile in Fig. 5.19f after using 16-fold interpolation. The performance metrics of the imaging results are given in Table 5.1, and it is found that both PSLR and ISLR are close to the theoretical values, which proves that the imaging method discussed in this section is effective.

5.3.3 Frequency Variable Scale Imaging Method

The most basic SAR imaging algorithm, the RD algorithm, which corrects the range migration term by interpolation, greatly increases the amount of system operations by interpolation, and the huge burden on the hardware system and many limitations in practice make it difficult to acquire images in real time.

The linear chirp scaling (CS) algorithm is a non-interpolation method for range migration correction, which is widely used in pulse SAR imaging because two linear FM signals are still linear FM signals when multiplied by each other. Mittermayer et al. proposed a frequency scaling (FS) algorithm for FMCW SAR imaging. algorithm for FMCW SAR imaging. In this section, we analyze the principle of FS algorithm for non-interpolated range migration correction, derive the required FM function for the FS algorithm process, and perform simulations to validate the FS algorithm for binary FMCW SAR systems based on the principle [10].

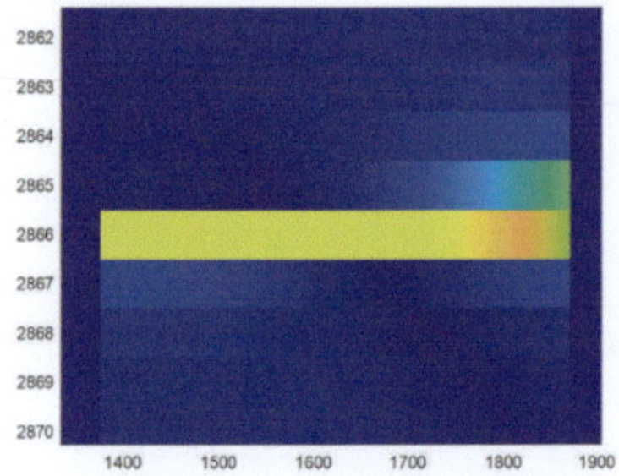

(a) Spatial response curve - Doppler frequency shift not compensated

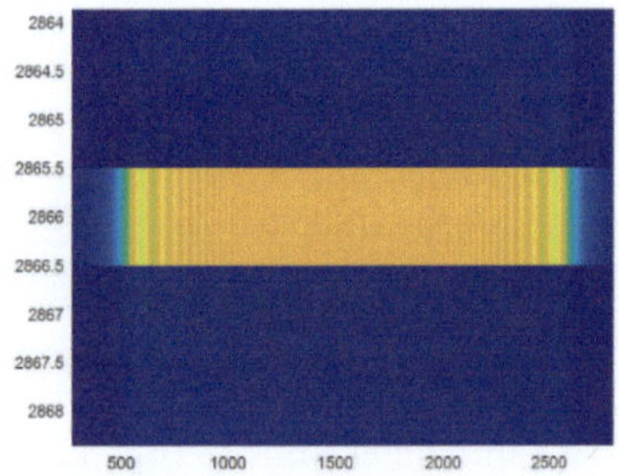

(b) Spatial response curve - after Doppler frequency shift compensation

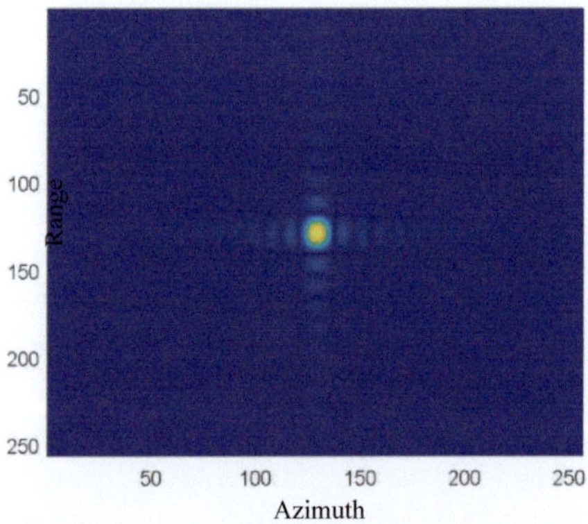

(c) 16x interpolation result

Fig. 5.19 Simulation results of dual satellite FMCW SAR RD algorithm

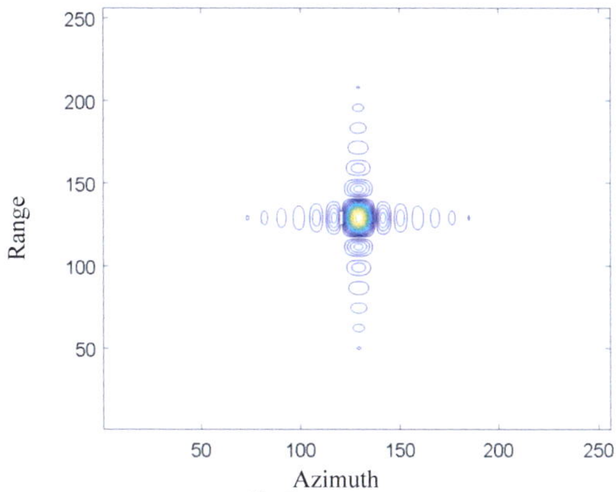

(d) Contour map

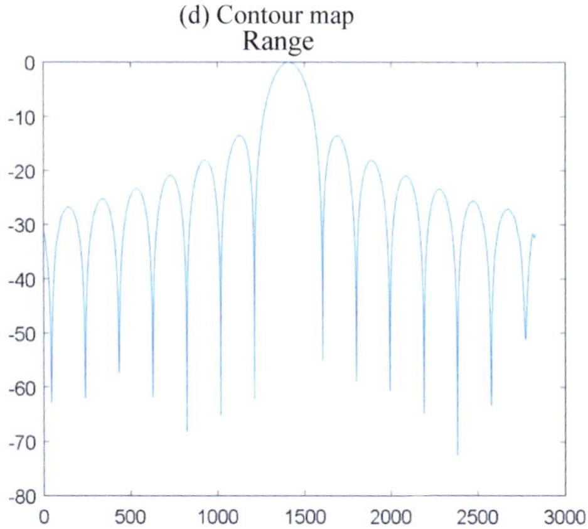

(e) Range profile

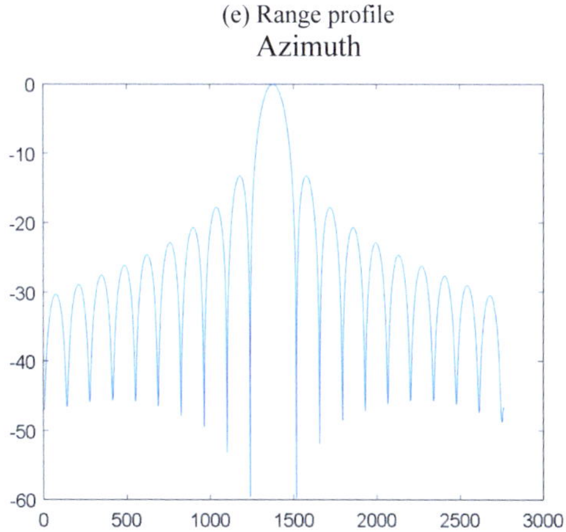

(f) Azimuth profile

Fig. 5.19 (continued)

Table 5.1 Performance index analysis-RD Algorithm

Performance index	PSLR (dB)		ISLR (dB)	
	Range	Azimuth	Range	Azimuth
Indicator value	−13.3768	−13.2621	−9.6932	−9.7218

1. Simulation results and analysis

The echo signal received by the receiver after the de-modulation process is rewritten after omitting the constant term as

$$\begin{aligned} s(t_r, t_a) = \exp\left[-\tfrac{j4\pi f_c}{c} R(t_r, t_a)\right] \exp\left\{-\tfrac{j4\pi\gamma}{c}\left[R(t_r, t_a) - R_{ref}\right]\left(t_r - \tfrac{2R_{ref}}{c}\right)\right\} \\ \cdot \exp\left\{\tfrac{j4\pi\gamma}{c^2}\left[R(t_r, t_a) - R_{ref}\right]^2\right\} \end{aligned} \tag{5.52}$$

Expanding the expression for the instantaneous slope range between the transceiver platform and the target to be measured by Taylor series at the fast time $t_r = 0$, we have

$$R(t_r, t_a) = R(t_a) + \frac{dR(t_r, t_a)}{dt} = R(t_a) - \frac{\lambda}{2} f_d t_r \tag{5.53}$$

$$R_t(t_r, t_a) = R_t(t_a) + \frac{V_r^2 t_a - R_{t0} V_r \sin\theta_{t0}}{R_t(t_a)} t_r \tag{5.54}$$

$$R_r(t_r, t_a) = R_r(t_a) + \frac{V_r^2 t_a - R_{r0} V_r \sin\theta_{r0}}{R_r(t_a)} t_r \tag{5.55}$$

$$R(t_a) = R_t(t_a) + R_r(t_a) \tag{5.56}$$

$$\begin{aligned} R(t_r, t_a) = R(t_a) + \frac{dR(t_r, t_a)}{dt_a} t_r = R(t_a) - \frac{\lambda}{2} f_d t_r \\ = R_t(t_a) + R_r(t_a) + \left[\frac{V_r^2 t_a - R_{t0} V_r \sin\theta_{t0}}{R_t(t_a)} + \frac{V_r^2 t_a - R_{r0} V_r \sin\theta_{r0}}{R_r(t_a)}\right] t_r \end{aligned} \tag{5.57}$$

where f_d is called the Doppler shift term, which is caused by the continuous motion of the carrier platform during the transmission and reception of the electromagnetic signal and can be disregarded in pulsed SAR. It is possible to obtain

$$f_d = -\frac{2}{\lambda}\left[\frac{V_r^2 t_a - R_{t0} V_r \sin\theta_{t0}}{R_t(t_a)} + \frac{V_r^2 t_a - R_{r0} V_r \sin\theta_{r0}}{R_r(t_a)}\right] \tag{5.58}$$

$$
\begin{aligned}
s_{IF}(t_r, t_a) = & \exp\left[-\frac{j4\pi f_c}{c}R(t_a)\right]\exp\left[-\frac{j4\pi\gamma}{c}\left(R(t_a) - R_{ref}\right)\left(t_r - \frac{2R_{ref}}{c}\right)\right] \\
& \cdot \exp\left[\frac{j4\pi\gamma}{c^2}\left(R(t_a) - R_{ref}\right)^2\right] \cdot \exp(-j2\pi f_d t_r)
\end{aligned}
\tag{5.59}
$$

Since the correction of the RCM term needs to be performed in the range Doppler domain, the frequency scalar process is also done in the range Doppler domain. The first step of the FS imaging algorithm is to perform azimuthal FFT on the received signal after de-modulation, so that it is transformed from the two-dimensional time domain to the range Doppler domain.

$$
\begin{aligned}
S(f_a, t_r) = & \exp\left\{-j\frac{4\pi R_{t0}}{\lambda}\sqrt{\left[1 + \frac{\gamma}{f_0}\left(t_r - \frac{2R_{ref}}{c}\right)\right] - \frac{\lambda^2 f_a^2}{4V_r^2\cos^2\theta_{t0}}}\right\} \\
& \cdot \exp\left\{-j\frac{4\pi R_{r0}}{\lambda}\sqrt{\left[1 + \frac{\gamma}{f_0}\left(t_r - \frac{2R_{ref}}{c}\right)\right] - \frac{\lambda^2 f_a^2}{4V_r^2\cos^2\theta_{r0}}}\right\} \\
& \cdot \exp(-j2\pi f_d t_r) \cdot \exp\left[-j\frac{4\pi\gamma}{c}R_{ref}\left(t_r - \frac{2R_{ref}}{c}\right)\right] \\
& \cdot \exp\left[-j\frac{4\pi\gamma}{c^2}R_{ref}\left(R(t_r, t_a) - \frac{2R_{ref}}{c}\right)\right]
\end{aligned}
\tag{5.60}
$$

make $\beta_t(f_a) = \sqrt{1 - \left(\frac{f_a\lambda}{2V_r\cos\theta_{t0}}\right)^2}$, $\beta_r(f_a) = \sqrt{1 - \left(\frac{f_a\lambda}{2V_r\cos\theta_{r0}}\right)^2}$, then

$$
\begin{aligned}
& \sqrt{\left[1 + \frac{K}{f_a}\left(t - \frac{2R_{ref})}{c}\right)\right]^2 - \frac{\lambda^2 f_a^2}{4V_r^2}} \\
& = \beta_t + \frac{\gamma}{f_0\beta_t}\left(t_r - \frac{2R_{ref}}{c}\right) - \frac{\gamma^2(1-\beta_t^2)}{2f_0^2\beta_t^3}\left(t_r - \frac{2R_{ref}}{c}\right)^2 \\
& \quad + \frac{\gamma^3(1-\beta_t^2)\left(t_r - \frac{2R_{ref}}{c}\right)^3}{2f_0^3\beta_t^5}
\end{aligned}
\tag{5.61}
$$

$$
\begin{aligned}
& \sqrt{\left[1 + \frac{\gamma}{f_a}\left(t_r - \frac{2R_{ref}}{c}\right)\right]^2 - \frac{\lambda^2 f_a^2}{4V_r^2\cos^2\theta_{r0}}} \\
& = \beta_r + \frac{\gamma}{f_0\beta_r}\left(t_r - \frac{2R_{ref}}{c}\right) - \frac{\gamma^2(1-\beta_r^2)}{2f_0^2\beta_r^3}\left(t_r - \frac{2R_{ref}}{c}\right)^2 \\
& \quad + \frac{\gamma^3(1-\beta_r^2)}{2f_0^3\beta_r^5}\left(t_r - \frac{2R_{ref}}{c}\right)^3
\end{aligned}
\tag{5.62}
$$

Then the transformed signal can be written as

$$
\begin{aligned}
S(f_a, t_r) = & \exp\left\{-j\left[\frac{4\pi R_{t0}\beta_t(f_a)}{\lambda} + \frac{4\pi R_{r0}\beta_r(f_a)}{\lambda}\right]\right\} \\
& \cdot \exp(-j2\pi f_d t_r) \cdot SRC(f_a, t_r) \cdot \exp\left\{j\frac{4\pi\gamma}{c^2}\left[R(t_r, t_a) - R_{ref}\right]^2\right\} \\
& \cdot \exp\left[-j\frac{4\pi\gamma}{c}\left(\frac{R_{t0}}{\beta_t(f_a)} + \frac{R_{r0}}{\beta_r(f_a)} - R_{ref}\right)\left(t_r - \frac{2R_{ref}}{c}\right)\right]
\end{aligned}
\tag{5.63}
$$

$$
\begin{aligned}
SRC(f_a, t_r) = & \exp\left\{-j\left[\frac{2\pi R_{t0}\gamma^2\lambda}{c^2}\frac{(\beta_t^2 - 1)}{\beta_t^3} + \frac{4\pi R_{r0}\gamma^2\lambda}{c^2}\frac{(\beta_r^2 - 1)}{\beta_r^3}\right]\right. \\
& \left.\left(t_r - \frac{2R_{ref}}{c}\right)^2\right\} \\
& \cdot \exp\left\{-j\left[\frac{2\pi R_{t0}\gamma^3\lambda^2}{c^3}\frac{(1 - \beta_t^2)}{\beta_t^5} + \frac{2\pi R_{r0}\gamma^3\lambda^2}{c^3}\frac{(1 - \beta_r^2)}{\beta_r^5}\right]\right. \\
& \left.\left(t_r - \frac{2R_{ref}}{c}\right)^3\right\}
\end{aligned}
\tag{5.64}
$$

Doing range-oriented FFT on the above equation and neglecting the effect of higher order terms yields

$$
\begin{aligned}
S(f_a, f_r) = & \sin c\left\{\pi T_p\left[f_r + \frac{2\gamma}{c}\left(\frac{R_{t0}}{\beta_t(f_a)} + \frac{R_{r0}}{\beta_r(f_a)} - R_{ref}\right)\right]\right\} \\
& \cdot \exp\left\{-j\left[\frac{4\pi R_{t0}\beta_t(f_a)}{\lambda} + \frac{4\pi R_{r0}\beta_r(f_a)}{\lambda}\right]\right\} \\
& \cdot \exp\left\{j\frac{4\pi\gamma}{c^2}\left[R(t_r, t_a) - R_{ref}\right]\right\}\exp\left[-j2\pi(f_r - f_d)\right]
\end{aligned}
\tag{5.65}
$$

At this point, the range migration equation can be obtained as

$$
RCM = \frac{2\gamma}{c}\left[\frac{R_{t0}}{\beta_t(f_a)} + \frac{R_{r0}}{\beta_r(f_a)} - R_{ref}\right] - f_d \tag{5.66}
$$

The second term of the above equation is compensated first, and after compensation the above equation becomes

$$
RCM' = \frac{2\gamma}{c}\left[\frac{R_{t0}}{\beta_t(f_a)} + \frac{R_{r0}}{\beta_r(f_a)} - R_{ref}\right] \tag{5.67}
$$

According to the above equation, the range migration correction is not only related to R_{t0} and R_{r0}, but also to β_r and β_t, then it is necessary to variably scale the range frequencies and normalize the scale factor as follows:

$$\begin{aligned}
RCM'' &= RCM1 + RCM2 + \Delta RCM \\
&= \frac{2\gamma}{c}\left[\frac{R_{t0}}{\beta_t(f_a)} - R_{ref}\right] + \frac{2\gamma}{c}\left[\frac{R_{r0}}{\beta_r(f_a)} - R_{ref}\right] + \frac{2\gamma R_{ref}}{c} \\
&= \frac{1}{\beta_t}\left[\frac{2\gamma}{c}\left(R_{t0} - \beta_t R_{ref}\right)\right] + \frac{1}{\beta_r}\left[\frac{2\gamma}{c}\left(R_{r0} - \beta_r R_{ref}\right)\right] + \frac{2\gamma R_{ref}}{c}
\end{aligned} \tag{5.68}$$

The last term of the above equation, ΔRCM, is compensated, and the compensated range migration term becomes

$$RCM''' = \frac{1}{\beta_t}\left[\frac{2\gamma}{c}\left(R_{t0} - \beta_t R_{ref}\right)\right] + \frac{1}{\beta_r}\left[\frac{2\gamma}{c}\left(R_{r0} - \beta_r R_{ref}\right)\right] \tag{5.69}$$

Two phase terms are introduced into the signal:

$$\begin{aligned}
S(t_a, f_r) = \sin c&\left\{\pi T_p\left(f_r + \frac{2\gamma}{c}\left[\frac{R_{t0}}{\beta_t(f_a)} + \frac{R_{r0}}{\beta_r(f_a)} - R_{ref}\right]\right)\right\} \\
&\cdot \exp\left\{-j\left[\frac{4\pi R_{t0}\beta_t(f_a)}{\lambda} + \frac{4\pi R_{r0}\beta_r(f_a)}{\lambda}\right]\right\} \\
&\cdot \exp\left\{j\frac{4\pi\gamma}{c^2}\left[R(t_a) - R_{ref}\right]\right\} \cdot \exp\left[-j2\pi(f_r - f_d)\right] \\
&\cdot \exp\left(-j\frac{\pi}{K}f_r^2\right)\exp\left(j\frac{\pi}{K}f_r^2\right)
\end{aligned} \tag{5.70}$$

By performing IFFT on the above equation in the range direction, the signal expression can be approximated as

$$\begin{aligned}
s_{IF}(t_r, t_r) = \exp&\left\{-j\left[\frac{4\pi R_{t0}\beta_t(f_a)}{\lambda} + \frac{4\pi R_{r0}\beta_r(f_a)}{\lambda}\right]\right\} \\
&\cdot \exp\left\{-j\frac{4\pi\gamma}{c}\left[\frac{\beta_r R_{t0} + \beta_t R_{r0}}{\beta_t\beta_r} - R_{ref}\right]\left(t_r - \frac{2R_{ref}}{c}\right)\right\} \\
&\cdot \exp(-j2\pi f_d t_r) \otimes \exp\left(-j\pi K t_r^2\right)
\end{aligned} \tag{5.71}$$

From Eq. (5.69), it can be seen that the range migration is related to R_{t0} and R_{r0} on the one hand, and when these two values are larger, the range migration value is also larger. At the same time, it is also related to the azimuthal frequency f_a. After the echo signal is dechirp, the range direction becomes a single-frequency signal, and the range frequency factor is scaled by $1/(\beta_r\ \beta_t)$, which corrects the complementary RCM, and the echo signal is multiplied or convolved with a series of functions to realize the scale-changing process. Figure 5.20 illustrates the frequency rescaling process of FS algorithm.

I(tr) is the system input function, which is convolved with the transform function $\exp(j\gamma_1 t^2{}_r)$ to obtain the first intermediate function with the transform function $T_1(t)$:

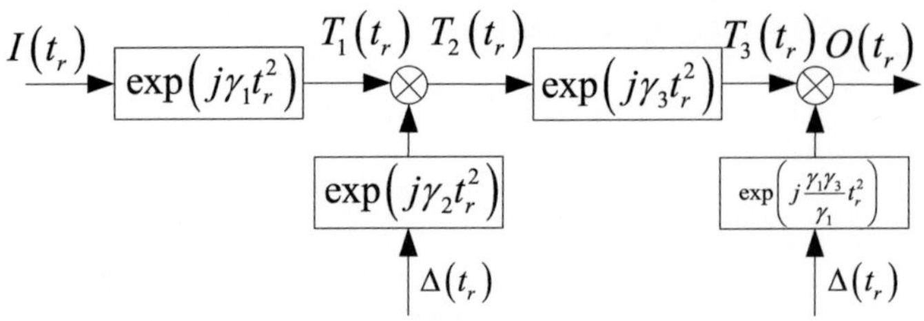

Fig. 5.20 Frequency scaling process

$$\begin{aligned} T_1(t_r) &= I(t) * \exp\left(j\gamma_1 t_r^2\right) \\ &= \exp\left(j\gamma_1 t_r^2\right) \int I(\tau)\exp\left(j\gamma_1\tau^2\right)\exp(-j2\gamma_1 t_r\tau)d\tau \end{aligned} \tag{5.72}$$

Multiplying T_1(tr) with the transformation function exp (jγ2 $t^2{}_r$) to obtain the second intermediate function $T_2(t_r)$, we have

$$\begin{aligned} T_2(t_r) &= T_1(t_r) * \exp\left(j\gamma_2 t_r^2\right) \\ &= \exp\left[j(\gamma_1+\gamma_2)t_r^2\right] \int I(\tau)\exp\left(j\gamma_1\tau^2\right)\exp(-j2\gamma_1 t_r\tau)d\tau \end{aligned} \tag{5.73}$$

Let $\gamma_1\tau = (\gamma_1+\gamma_2)\tau'$, then the above equation becomes

$$\begin{aligned} T_2(t_r) &= T_1(t) * \exp\left(j\gamma_2 t_r^2\right) \\ &= \exp\left[j(\gamma_1+\gamma_2)t_r^2\right] \int I(\tau)\exp\left(j\gamma_1\tau^2\right)\exp\left(-j2\gamma_1\tau^2\right)d\tau \\ &= \exp\left[j(\gamma_1+\gamma_2)t_r^2\right] \int I\left(\frac{\gamma_1+\gamma_2}{\gamma_1}\tau'\right) \\ &\qquad \exp\left[j\frac{(\gamma_1+\gamma_2)^2}{\gamma_1}\tau'^2\right]\exp\left[-j2(\gamma_1+\gamma_2)\tau'\right]d\tau' \\ &= \exp\left[j(\gamma_1+\gamma_2)t_r^2\right] \\ &\qquad \left[I\left(\frac{\gamma_1+\gamma_2}{\gamma_1}\tau'\right)\exp\left[j\frac{\gamma_2(\gamma_1+\gamma_2)}{\gamma_1}\tau'^2\right]\exp\left[j(\gamma_1+\gamma_2)\tau'^2\right]\right. \\ &\qquad \cdot\exp\left[-j2(\gamma_1+\gamma_2)t_r\tau'\right]d\tau' \\ &= \int I\left(\frac{\gamma_1+\gamma_2}{\gamma_1}\tau'\right)\exp\left[j\frac{\gamma_2(\gamma_1+\gamma_2)}{\gamma_1}\tau'^2\right] \\ &\qquad \exp\left[j(\gamma_1+\gamma_2)\left(\tau'-t\right)^2\right]d\tau \\ &= I\left(\frac{\gamma_1+\gamma_2}{\gamma_1}t_r\right)\exp\left[j\frac{\gamma_2(\gamma_1+\gamma_2)}{\gamma_1}t_r^2\right] \\ &\qquad \otimes\exp\left[j(\gamma_1+\gamma_2)t_r^2\right] \end{aligned} \tag{5.74}$$

Convolving T_2 (t_r) with the transformation function exp ($j\gamma 3\ t^2{}_r$) while letting $\gamma 1 + \gamma 2 + \gamma 3 = 0$, it can be expressed as

$$\begin{aligned} T_3(t_r) &= T_2(t_r) \otimes \exp\left(j\gamma_3 t_r^2\right) \\ &= I\left(\frac{\gamma_1+\gamma_2}{\gamma_1} t_r\right) \exp\left[j\frac{\gamma_2(\gamma_1+\gamma_2)}{\gamma_1} t_r^2\right] \\ &\quad * \exp\left[j(\gamma_1+\gamma_2) t_r^2\right] * \exp\left(j\gamma_3 t_r^2\right) \\ &= I\left(\frac{-\gamma_3}{\gamma_1} t_r\right) \exp\left(j\frac{-\gamma_2\gamma_3}{\gamma_1} t_r^2\right) * \exp\left(-j\gamma_3 t_r^2\right) * \exp\left(j\gamma_3 t_r^2\right) \\ &= I\left(\frac{-\gamma_3}{\gamma_1} t_r\right) \exp\left(j\frac{-\gamma_2\gamma_3}{\gamma_1} t_r^2\right) * \delta(t_r) \\ &= I\left(\frac{-\gamma_3}{\gamma_1} t_r\right) \exp\left(j\frac{-\gamma_2\gamma_3}{\gamma_1} t_r^2\right) \end{aligned} \tag{5.75}$$

The output function O (t_r) can be expressed as

$$O(t_r) = T_3(t) \cdot \exp\left(j\frac{\gamma_2\gamma_3}{\gamma_1} t_r^2\right) = I\left(-\frac{\gamma_3}{\gamma_1} t_r\right) \tag{5.76}$$

The analysis above shows that the input signal of I(t_r), after a series of transformations the final output $I\left(-\gamma_3/\gamma_1 t_r\right)$, realizes the scale transformation. The selection of parameters depends on the selection of the required transformation factor, and the selection of different parameters can realize the scale transformation between different factors, and the frequency variation scaling process required in this chapter can also be applied to the above system.

In the underlying analysis of FMCW SAR, T_1(tr) is equivalent to s_{IF}, , so the exp ($j\gamma_1\ t^2{}_r$) term is usually exp ($j\gamma t^2{}_r$) and the output signal O (tr) = I ($\beta_t\ \beta_r\ t_r$).

Then $\beta_t\ \beta_r = -\gamma 3 / \gamma 1$, $\gamma 2 = \pi\gamma\ (1 - \beta t\ \beta r)$, $\gamma 3\ \gamma 2 / \gamma 1 = \pi\gamma\beta t\ \beta r\ (\beta t\ \beta r - 1)$. Therefore, the frequency variation scale function is

$$H_{FS}(f_a) = \exp\left[j\pi K(1 - \beta_t\beta_r) t_r^2\right] \tag{5.77}$$

The echo received signal is mathematically transformed in the above steps to achieve a scale-to-scale transformation, and the effect of the RVP term in the transformation process is eliminated at the same time, introducing a frequency variation scale in the whole imaging algorithm.

As can be seen in Fig. 5.20, the FM system with parameter $\gamma 1$ generates an RVP term that rewrites the signal as

$$\begin{aligned} s_1(t_r, f_a) = &\exp\left\{-j\frac{4\pi\gamma}{c}\left[\frac{\beta_r R_{t0} + \beta_t R_{r0}}{\beta_t\beta_r} - R_{ref}\right]\left(\beta_t\beta_r t_r - \frac{2R_{ref}}{c}\right)\right\} \\ &\cdot \exp\left\{-j\left[\frac{4\pi R_{t0}\beta_t(f_a)}{\lambda} + \frac{4\pi R_{r0}\beta_r(f_a)}{\lambda}\right]\right\} \\ &\cdot \exp(-j2\pi f_d\beta_t\beta_r t_r) \cdot SRC(f_a, \beta_t\beta_r t_r) \otimes \exp\left(-j\pi K\beta_t\beta_r t_r^2\right) \end{aligned} \tag{5.78}$$

The signal is then multiplied with the FM function to obtain the intermediate function T_2 (t_r), with the expression

$$\begin{aligned} s_1(t_r, f_a) = &\exp\left\{-j\frac{4\pi\gamma}{c}\left[\frac{\beta_r R_{t0} + \beta_t R_{r0}}{\beta_t\beta_r} - R_{ref}\right]\left(\beta_t\beta_r t_r - \frac{2R_{ref}}{c}\right)\right\} \\ &\cdot \exp\left\{-j\left[\frac{4\pi R_{t0}\beta_t(f_a)}{\lambda} + \frac{4\pi R_{r0}\beta_r(f_a)}{\lambda}\right]\right\} \\ &\cdot \exp(-j2\pi f_d\beta_t\beta_r t_r) \cdot SRC(f_a, \beta_t\beta_r t_r) \\ &\otimes \exp\left(-j\pi K\beta_t\beta_r t_r^2\right) \end{aligned} \tag{5.79}$$

From Eq. (5.79), it can be seen that the fast time term before the convolution sign has completed the scale transformation of β. The fast time term after the convolution sign undergoes the scale transformation with the factor $\sqrt{\beta_t\beta_r}$. According to the step, T_2 (t_r) is convolved with the FM function with the parameter $\gamma 3$ to obtain T_3 (t_r), and the RVP term is eliminated, and the expression of the output signal is

$$\begin{aligned} O(t_r) = &\exp\left\{-j\frac{4\pi\gamma}{c}\left[\frac{\beta_r R_{t0} + \beta_t R_{r0}}{\beta_t\beta_r} - R_{ref}\right]\left(\beta_t\beta_r t_r - \frac{2R_{ref}}{c}\right)\right\} \\ &\cdot \exp(j2\pi f_d\beta_t\beta_r t_r) \cdot SRC(f_a, \beta_t\beta_r t_r) \\ &\cdot \exp\left\{-j\left[\frac{4\pi R_{t0}\beta_t(f_a)}{\lambda} + \frac{4\pi R_{r0}\beta_r(f_a)}{\lambda}\right]\right\} \\ &\cdot \exp\left[-j\pi\gamma\beta_t\beta_r(\beta_t\beta_r - 1)t_r^2\right] \end{aligned} \tag{5.80}$$

The last exponential term in the above equation is the redundant term generated after removing the RVP term, and it is eliminated with the correction function

$$H_{IFS} = \exp\left[-j\pi K\beta_t\beta_r(\beta_t\beta_r - 1)t_r^2\right] \tag{5.81}$$

In the FS algorithm, Eq. (5.81) is called the inverse frequency variation function, and the corrected signal expression is

$$\begin{aligned} O(t) &= \exp\left\{-j\frac{4\pi\gamma}{c}\left[\frac{\beta_r R_{t0}+\beta_t R_{r0}}{\beta_t\beta_r}-R_{ref}\right]\left(\beta_t\beta_r t_r-\frac{2R_{ref}}{c}\right)\right\} \\ &\cdot \exp(j2\pi f_d\beta_t\beta_r t_r)\cdot SRC(f_a,\beta_t\beta_r t_r) \\ &\cdot \exp\left\{-j\left[\frac{4\pi R_{t0}\beta_t(f_a)}{\lambda}+\frac{4\pi R_{r0}\beta_r(f_a)}{\lambda}\right]\right\} \end{aligned} \tag{5.82}$$

Performing a range-oriented FFT of Eq. (5.82) and observing the remaining RCM terms, we have

$$\begin{aligned} S(f_r,f_a) &= \sin c\left\{\frac{\pi T_p}{\beta_t\beta_r}\left[f_r+\frac{2K}{c}\left(\beta_r R_{t0}+\beta_t R_{r0}-\beta_t\beta_r R_{ref}\right)-\beta_t\beta_r f_d\right]\right\} \\ &\cdot \exp\left\{-j\left[\frac{4\pi R_{t0}\beta_t(f_a)}{\lambda}+\frac{4\pi R_{r0}\beta_r(f_a)}{\lambda}\right]\right\}\cdot SRC(f_a,f_r) \end{aligned} \tag{5.83}$$

From Eq. (5.83), the remaining RCM to be corrected is

$$RCM_2 = -\frac{2\gamma\beta_t\beta_r R_{ref}}{c} - \beta_t\beta_r f_d \tag{5.84}$$

The signal is FFTed in the azimuthal direction to transform the signal back to the range Doppler domain, and the RCM2 is corrected with the correction function

$$H_{RCM} = \exp\left\{-j4\pi\frac{\gamma\beta_t\beta_r R_{ref}}{c}\right\}\exp\{-2\pi\beta_t\beta_r f_d\} \tag{5.85}$$

The SRC correction function is

$$\begin{aligned} H_{SRC} &= \exp\left\{-j\left[\frac{2\pi R_{t0}\gamma^2\lambda}{c^2}\frac{\left(\beta_t^2-1\right)}{\beta_t^3}+\frac{4\pi R_{r0}\gamma^2\lambda}{c^2}\frac{\left(\beta_r^2-1\right)}{\beta_r^3}\right]\left(t_r-\frac{2R_{ref}}{c}\right)^2\right\} \\ &\cdot \exp\left\{j\left[\frac{2\pi R_{t0}\gamma^3\lambda^2}{c^3}\frac{\left(1-\beta_t^2\right)}{\beta_t^5}+\frac{2\pi R_{r0}\gamma^3\lambda^2}{c^3}\frac{\left(1-\beta_r^2\right)}{\beta_r^5}\right]\left(t_r-\frac{2R_{ref}}{c}\right)^3\right\} \end{aligned} \tag{5.86}$$

The expression after correcting the SRC is

$$\begin{aligned} S(f_r,f_a) &= \sin c\left\{\frac{\pi T_p}{\beta_t\beta_r}\left[f_r+\frac{2K}{c}\left(\beta_r R_{t0}+\beta_t R_{r0}-R_{ref}\right)\right]\right\} \\ &\cdot \exp\left\{-j\left[\frac{4\pi R_{t0}\beta_t(f_a)}{\lambda}+\frac{4\pi R_{r0}\beta_r(f_a)}{\lambda}\right]\right\} \end{aligned} \tag{5.87}$$

The matched filter for azimuthal compression is

$$H_a(f_a) = \exp\left\{-j\left[\frac{4\pi R_{t0}\beta_t(f_a)}{\lambda} + \frac{4\pi R_{r0}\beta_r(f_a)}{\lambda}\right]\right\} \tag{5.88}$$

The expression for the azimuthally compressed subsequent signal is

$$s(f_r, t_a) = \sin c\left\{\frac{\pi T_p}{\beta_t\beta_r}\left[f_r + \frac{2K}{c}\left(\beta_r R_{t0} + \beta_t R_{r0} - R_{ref}\right)\right]\right\} \cdot \sin c(\pi B_a t_a) \tag{5.89}$$

The RCM correction, SRC correction and azimuthal compression are done in the range Doppler domain, and the above correction can be done at the same time after the azimuthal FFT. The flow chart of the dual-satellite FMCW SAR FS algorithm is shown in Fig. 5.21.

2. Simulation results and analysis

The following MATLAB simulation is used to simulate the case of a single point target of a satellite-based dual-base FMCW SAR with the following simulation parameters: signal bandwidth 150 MHz, carrier frequency 35 GHz, sweep period 25 ms, range-to-sample frequency 25 MHz, satellite orbit altitude 500 km, and dual-satellite spacing 10 km.

A stationary point target is placed on the scene centerline at the moment of zero slow time, and the imaging process is performed using the method introduced in this section, and the results are shown in Fig. 5.22. Figure 5.22a and b show the

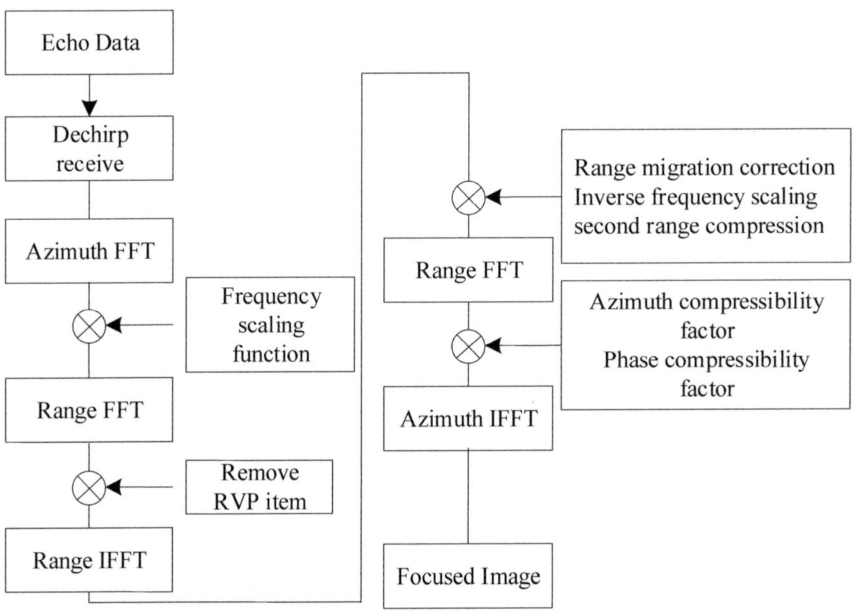

Fig. 5.21 Flow chart of dual satellite FMCW SAR FS algorithm

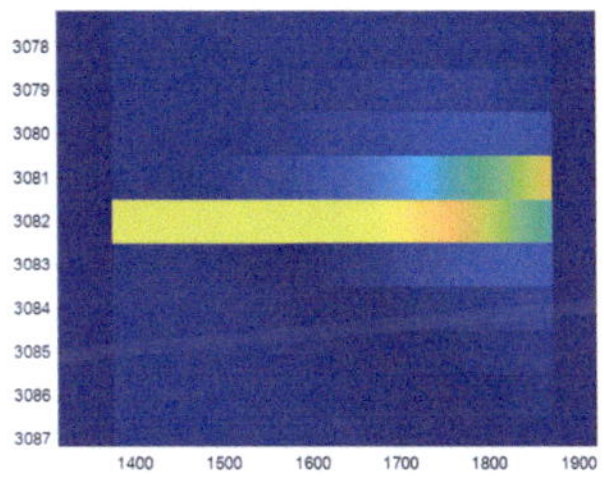

(a) Spatial response curve - Doppler frequency shift not compensated

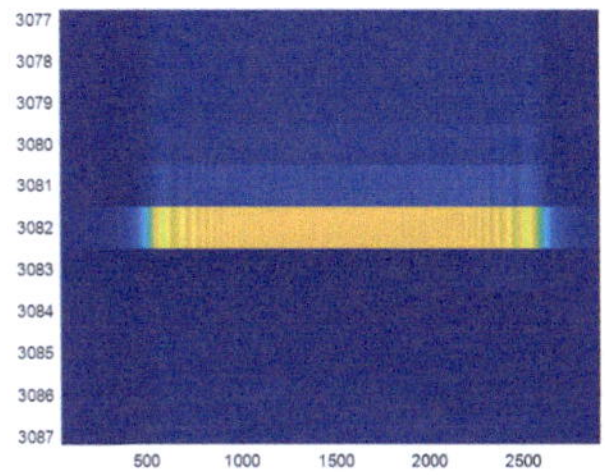

(b) Spatial response curve - after Doppler frequency shift compensation

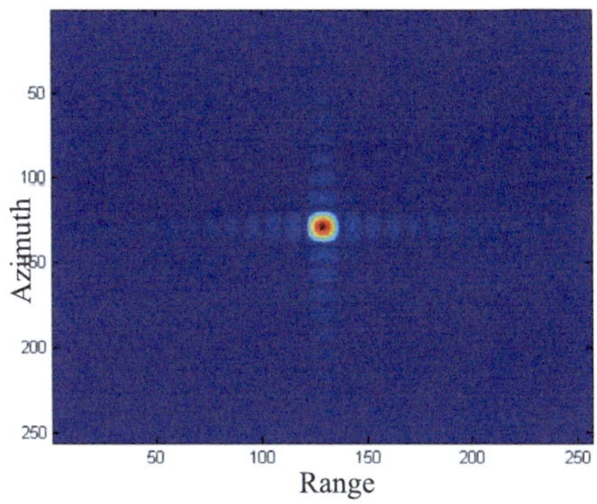

(c) 16x interpolation result

Fig. 5.22 Simulation results of dual satellite FMCW SAR FS Algorithm

spatial response curves after range impulse compression without azimuthal impulse compression, where the vertical coordinate is the range frequency and the horizontal coordinate is the azimuthal time, where Fig. 5.22a shows the case when the Doppler shift is not compensated and Fig. 5.22b shows the case after the Doppler shift is compensated, both figures are corrected for range migration. It can be found that if the Doppler shift is not compensated, it will cause the migration of the echoes across the range unit and affect the focus of the image. In order to make the discrete processed data points continuous, the results after 16-fold interpolation are shown in

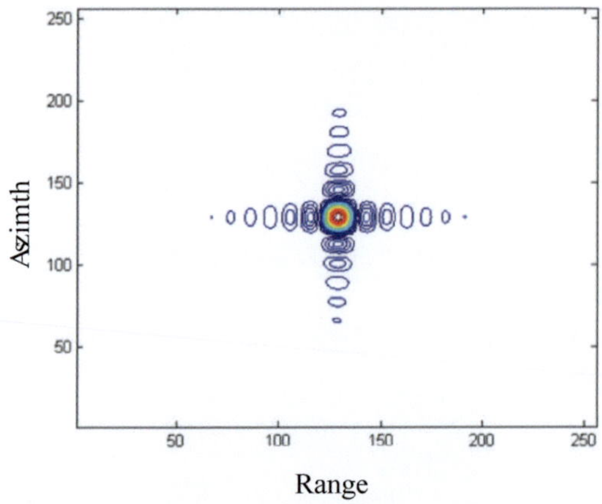

(d) Contour map

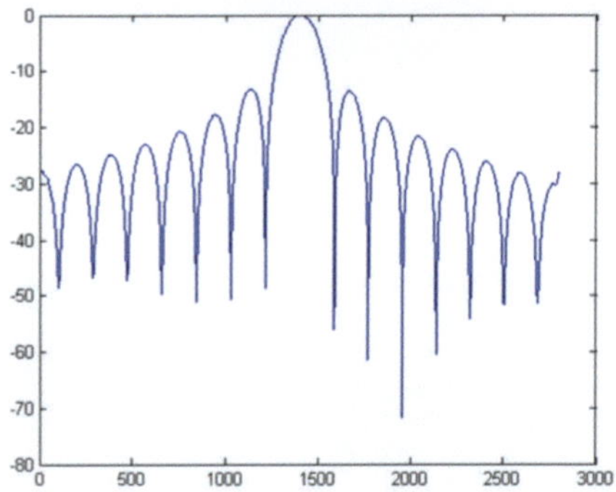

(e) Range profile

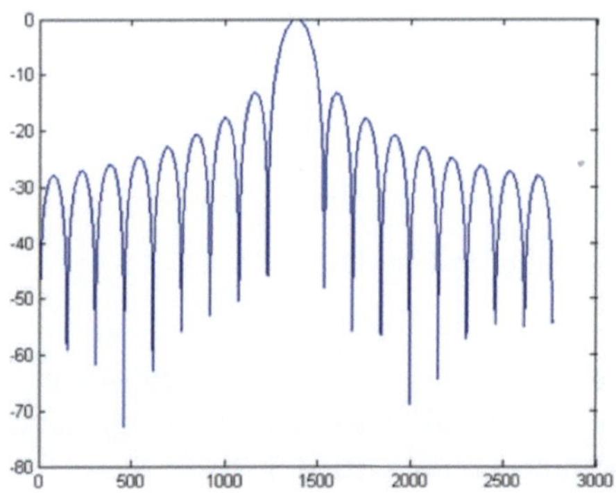

(f) Azimuth profile

Fig. 5.22 (continued)

Table 5.2 Performance index analysis-FS Algorithm

Performance index	PSLR (dB)		ISLR (dB)	
	Range	Azimuth	Range	Azimuth
Indicator value	−13.3638	−13.3114	−9.6386	−10.0087

Fig. 5.22c, its contour map in Fig. 5.22d, its range-oriented profile in Fig. 5.22e, and its azimuthal profile in Fig. 5.22f. The performance metrics of the imaging results are given in Table 5.2. It can be seen that both PSLR and ISLR are close to the theoretical values, which verifies the effectiveness of the frequency variant imaging method discussed in this section.

5.4 FM Continuous Wave SAR High-Resolution Imaging

Like the airborne continuous wave system SAR introduced in Chap. 2, the satellite-based continuous wave system SAR system also adopts differential reception, which can effectively reduce the bandwidth of the echo signal and the pressure on AD sampling by mixing the echo signal and the signal at the same frequency point as the echo signal, so it is a feasible technical means to effectively solve the problem of satellite-based ultra-high resolution imaging. At present, the high-resolution imaging on board adopts the clustering and sliding clustering mode, and the azimuthal resolution is improved by the azimuthal beam scanning of the antenna, which requires a scanning angle of more than 20°, and the serious problems of two-dimensional coupling, two-dimensional nulling and spectral folding introduced by the large oblique angle need to be studied. [11, 12].

5.4.1 Modeling the Interaction Mechanism Between Walking Correction and 2D Null-Variance Correction

To overcome the problems of spectral aliasing and severe two-dimensional coupling in satellite-based large oblique view, the walk correction method commonly used in airborne linear trajectory large oblique view needs to be introduced. However, in nonlinear trajectory imaging, the walk correction will change the original two-dimensional null variation characteristics of the signal, and its mechanism of action is expressed in the form of improving the azimuthal null variation or further deteriorating the azimuthal null variation [13]. First, the relative motion model of the large oblique nonlinear trajectory is established as R (t_a,r,X), where R (−) is the oblique range course from the satellite to the target, t_a is the azimuth time, and (r, X) is the two-dimensional position of the target. The echo signal can be expressed as $\mathrm{Echo}(t_a,r,X) = p(R(t_a,r,X))$, where p (−) denotes the echo signal model. Therefore, the echo signal is related to the two-dimensional position of the target, i.e., the echo signal is two-dimensional null-variant. Let the walk correction function be $f_{wc}(t_a) = k_{wc} - t_a$, when the walk correction function is introduced, the spectral aliasing can be suppressed, and also the target at different azimuthal positions at the reference range unit is changed, thus changing the original azimuthal null-variant characteristics of the signal. It can be conjectured that finding a suitable walk correction function can minimize the azimuthal null variation while ensuring that the signal spectrum does not overlap, and this conjecture can be equated to the following optimization problem with constraints:

$$\begin{aligned} &\hat{k}_{wc} = \arg\left\{ \min_{k_{wc}} \left| R\left(t_a, r_{ref}, X_l\right) - R\left(t_a, r_{ref}, X_{ref}\right) \right| \cap \min_{k_{wc}} \right. \\ &\qquad \left. \left| R\left(t_a, r_{ref}, X_r\right) - R\left(t_a, r_{ref}, X_{ref}\right) \right| \right\} \\ &s.t. \begin{cases} FFT_a\left\{Echo\left(t_a, r, X_{ref}\right)\right\} < PRF \\ FFT_a\{Echo(t_a, r, X_l)\} + FFT_a\{Echo(t_a, r, X_r)\} < PRF \end{cases} \end{aligned} \tag{5.90}$$

where $\hat{k}_{wc}$ is the slope of the optimally obtained walk correction function; (r_{ref},X_{ref}) is the two-dimensional position of the reference target; X_l and X_r are the azimuthal positions of the targets located at the leftmost and rightmost sides of the azimuthal scene, respectively; FFT_a (−) denotes the azimuthal spectrum of the signal; PRF denotes the width of the entire azimuthal spectrum.

For clarity, Fig. 5.23 gives a schematic representation of the interaction mechanism between walk correction and azimuthal null variation. A measure of the degree of spectral aliasing is defined as the spectral aliasing ratio, which is calculated as the ratio of the total signal bandwidth to the PRF. If the value of the spectral aliasing ratio is greater than 1, the spectrum is aliased; when the value is equal to 1, the spectrum is just not aliased (e.g., the aliasing threshold in Fig. 5.23). Figure 5.23 shows two possible scenarios. 1. The spectral aliasing ratio decreases as the slope of the walk correction function increases, but the azimuthal null tends to be monotonically increasing. Therefore, the optimal point in this case is the mixing threshold. 2. As the slope of the walk correction function increases, the spectral mixing ratio gradually decreases, and the azimuthal cavitation shows a parabolic trend. Therefore, the optimal point in this case is the lowest point of the parabola. Combining the two cases, if the large oblique spectrum is completely corrected to the positive lateral spectrum by walking correction (as the positive lateral point in Fig. 5.23), the azimuthal null variation cannot be minimized. Therefore, the traditional walk correction method for linear trajectories with large strabismus must take into account its effect on the azimuthal cavitation.

In order to minimize the azimuthal cavitation before the imaging process, the optimal walk correction function is derived as described below.

The optimal linear range walk correction (LRWC) orientation preprocessing schematic is shown in Fig. 5.24, where the three targets A, B and C are located in the same range cell and the other two targets D and E have the same range cell

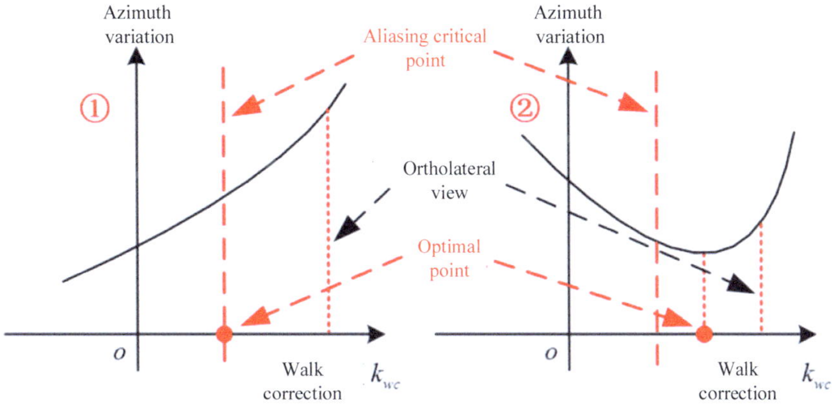

Fig. 5.23 Interaction mechanism between walk correction and azimuth variability

migration (RCM) curves as the RCM curves of target B. The other two targets D and E have the same curvature as the RCM curve of target B. The RCM curves of the three targets A, B and C located in the same range cell have different curvatures (Fig. 5.24a), and the RCM curves of the three targets do not exactly coincide when transformed from the azimuthal time domain to the Doppler domain, and there is azimuthal null variation (Fig. 5.24b). When the optimal LRWC factor is introduced (Fig. 5.24c), the two targets D and E, which have the same bend as the RCM curve of target B, are moved to the range cell where target B is located, and the RCM curves of the three targets D, B and E, which are located in the same range cell, have the same bend (Fig. 5.24d and e). It should be noted that for the sake of convenience, the bends of the RCM curves of targets D, B and E are assumed to be identical, but in reality, there may be some differences in the bends of the three RCM curves, but there must be a minimum difference, which can be minimized by introducing an optimal LRWC factor. Also note that the introduction of the optimal LRWC factor changes the form of the target RCM curve (a significant amount of linearity will appear), which is not given in the figure here.

The LRWC factor can be expressed as $h_{LRWC}(t_a) = k_f t_a$, where k_f is the slope of the LRWC factor. With the introduction of the LRWC factor, the slope course can be expressed as

$$\begin{aligned} R_{LRWC}(t_a; R_0, X) &= R(t_a; R_0, X) + h_{LRWC}(t_a) \\ &= \left(R_0 + k_f t_c\right) + \left[k_1(R_0, X) + k_f\right](t_a - t_c) \\ &\quad + k_2(R_0, X)(t_a - t_c)^2 + k_3(R_0, X)(t_a - t_c)^3 \\ &\quad + k_4(R_0, X)(t_a - t_c)^4 \end{aligned} \tag{5.91}$$

Suppose the initial range unit of target D is R_{0D} and the initial range unit of target B is $R_{0B} = R_0$. After introducing the LRWC factor, the range unit of target D is changed from R_{0D} to R_0, while the range unit of target B remains unchanged, so we

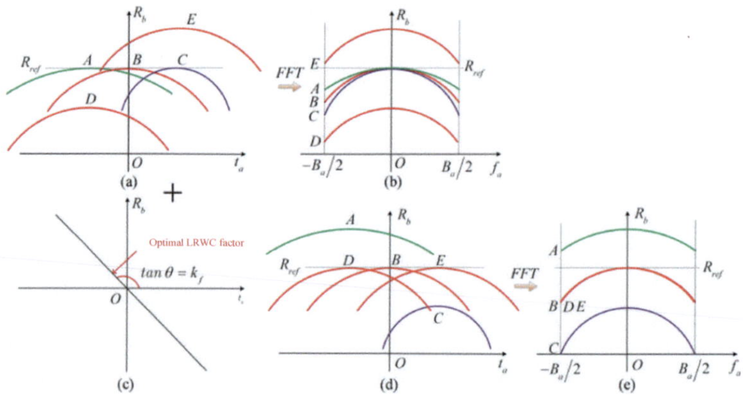

Fig. 5.24 Schematic diagram of optimal LRWC

can obtain

$$R_0 = R_{0D} + k_f t_{cD} \tag{5.92}$$

where $t_{cD} = t_c(X_D)$ and X_D is the azimuthal position of the target D. Therefore, the azimuthal phases of the two-dimensional spectra of target D and B can be expressed as $\phi_a\big(f_a; R_0 - k_f t_{cD}, X_D\big)$ and $\phi_a(f_a; R_0, X_B)$, respectively. In order to minimize the azimuthal cavity, the difference in the azimuthal phase of the two-dimensional spectra of targets D and B needs to be minimized, which can be characterized by the fuzzy function (AF) between targets B and D. The fuzzy function represents the degree of similarity between the two targets, and the smaller the difference in the azimuthal phase between the two targets, the greater the degree of similarity between them.

The orientation signals of targets B and D can be expressed in the Doppler domain as

$$S_{aB}(f_a; R_0, X_B) = \exp(j\phi_a(f_a; R_0, X_B)) \tag{5.93}$$

$$S_{aD}(f_a; R_0, X_D) = \exp\big(j\phi_a\big(f_a; R_0 - k_f t_{cD}, X_D\big)\big) \tag{5.94}$$

Then the fuzzy function between objectives B and D can be expressed as

$$\chi(B, D) = \frac{\int\limits_{f_a} S_{aB}(f_a; R_0, X_B) S_{aD}^*(f_a; R_0, X_D) df_a}{\sqrt{\int\limits_{f_a} |S_{aB}(f_a; R_0, X_B)|^2 df_a} \sqrt{\int\limits_{f_a} |S_{aD}(f_a; R_0, X_D)|^2 df_a}} \tag{5.95}$$

By comparing the above equations and ignoring the magnitude information, the fuzzy function can be simplified as

$$\chi(B,D)=\int_{f_a}\exp\left(j\phi_a(f_a;R_0,X_B)-j\phi_a\left(f_a;R_0-k_ft_{cD},X_D\right)\right)df_a \tag{5.96}$$

The fuzzy function of Eq. (5.96) reflects the correlation coefficient of the position signals of target B and D. The larger the correlation coefficient, the smaller the difference between target B and D, which indicates the smaller the azimuthal null variation. Therefore, when the value of the fuzzy function of Eq. (5.96) is the largest, the azimuthal variation is the smallest.

The Taylor series expansion of the azimuthal signal phase at fa = 0 gives

$$\begin{aligned}\phi_a(f_a;R_B,X)=\phi_a(f_a=0;R_0,X)+f_a\cdot\left.\frac{\partial\phi_a(f_a;R_0,X)}{\partial f_a}\right|_{f_a=0}\\+\frac{f_a^2}{2!}\cdot\left.\frac{\partial^2\phi_a(f_a;R_0,X)}{\partial f_a^2}\right|_{f_a=0}\\+\frac{f_a^3}{3!}\cdot\left.\frac{\partial^3\phi_a(f_a;R_0,X)}{\partial f_a^3}\right|_{f_a=0}+\frac{f_a^4}{4!}\cdot\left.\frac{\partial^4\phi_a(f_a;R_0,X)}{\partial f_a^4}\right|_{f_a=0}\end{aligned} \tag{5.97}$$

Substituting Eq. (5.96) into Eq. (5.97), the fuzzy function can be expressed as

$$\chi(B,D)=\int_{f_a}\exp\left(j2\pi\begin{pmatrix}\Delta\varphi_0(k_f)+\Delta\varphi_1(k_f)f_a+\Delta\varphi_2(k_f)f_a^2\\+\Delta\varphi_3(k_f)f_a^3+\Delta\varphi_4(k_f)f_a^4\end{pmatrix}\right)df_a \tag{5.98}$$

Among them.

$$\Delta\varphi_0(k_f)=\frac{1}{2\pi}\left(\phi_a(f_a=0;R_0,X_B)-\phi_a\left(f_a=0;R_0-k_ft_{cD},X_D\right)\right) \tag{5.99}$$

$$\Delta\varphi_1(k_f)=\frac{1}{2\pi}\left.\left(\frac{\partial\phi_a(f_a;R_0,X_B)}{\partial f_a}-\frac{\partial\phi_a\left(f_a;R_0-k_ft_{cD},X_D\right)}{\partial f_a}\right)\right|_{f_a=0} \tag{5.100}$$

$$\Delta\varphi_2(k_f)=\frac{1}{2\pi\cdot 2!}\left.\left(\frac{\partial^2\phi_a(f_a;R_0,X_B)}{\partial f_a^2}-\frac{\partial^2\phi_a\left(f_a;R_0-k_ft_{cD},X_D\right)}{\partial f_a^2}\right)\right|_{f_a=0} \tag{5.101}$$

$$\Delta\varphi_3(k_f)=\frac{1}{2\pi\cdot 3!}\left.\left(\frac{\partial^3\phi_a(f_a;R_0,X_B)}{\partial f_a^3}-\frac{\partial^3\phi_a\left(f_a;R_0-k_ft_{cD},X_D\right)}{\partial f_a^3}\right)\right|_{f_a=0} \tag{5.102}$$

$$\Delta\varphi_4(k_f)=\frac{1}{2\pi\cdot 4!}\left.\left(\frac{\partial^4\phi_a(f_a;R_0,X_B)}{\partial f_a^4}-\frac{\partial^4\phi_a\left(f_a;R_0-k_ft_{cD},X_D\right)}{\partial f_a^4}\right)\right|_{f_a=0} \tag{5.103}$$

Therefore, minimizing the azimuthal null variation can be modeled as the following optimization problem:

$$k_{f\_opt} = \arg\left(\max_{k_f}|\chi(A, B)|\right) \tag{5.104}$$

where k_{f_opt} is the slope of the optimal LRWC factor. It should be noted that the fuzzy function can be solved by the numerical product method.

The spatial distribution of targets at the simulation points is shown in Fig. 5.25. The targets are distributed in the latitude and longitude planes. In the figure, target 1 is the center of the scene, and target 1 and target 2 are located in the same range unit. The fuzzy function in Eq. (5.98) is solved by numerical calculation. When the LRWC factor is not introduced (the slope of the LRWC factor is 0), target 2 is located in the range cell R_{ref}. When the LRWC factor is introduced, target 2 is no longer located in the range cell R_{ref}. One of the targets 3 (virtual circle) is moved to the range cell Rref. Therefore, for each target 3, we can calculate the fuzzy functions of both target 1 and target 3, as well as the slope of the LRWC factor that moves target 3 to the range cell R_{ref}. The calculation of the fuzzy function with the slope of the LRWC function is shown in Fig. 5.26a, which shows that the AF value is maximum when the slope of the LRWC factor is 5.262.

To further verify the effect of the optimal LRWC orientation preprocessing, the two-dimensional spectrum of the center point target is used to match the filtered image of target 3, and the entropy value is used to quantitatively evaluate the image quality of target 3, which varies with the slope of the LRWC factor as shown in Fig. 5.26b. The results in Fig. 5.26aa and b show that when the slope of the LRWC factor is 5.262, the image quality of target 3 is the highest and the azimuthal null variation is the lowest, and the LRWC factor is the optimal LRWC factor.

The 2D spectrum of target 1 is used to match the filtered imaging process of target 3 at the reference range unit R_{ref}, and the results are shown in Fig. 5.27. Figure 5.27a

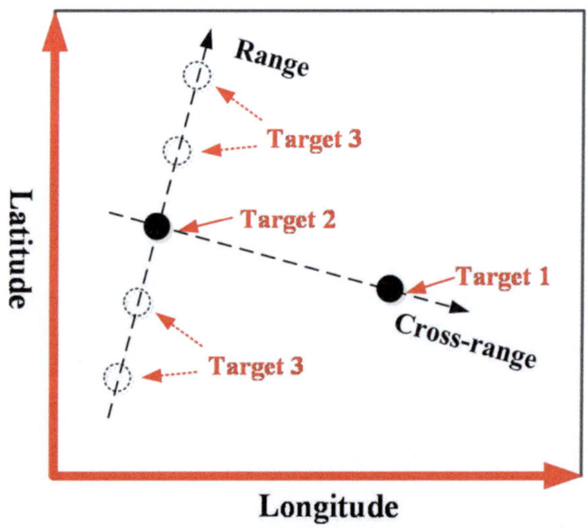

Fig. 5.25 Spatial distribution of simulation point targets

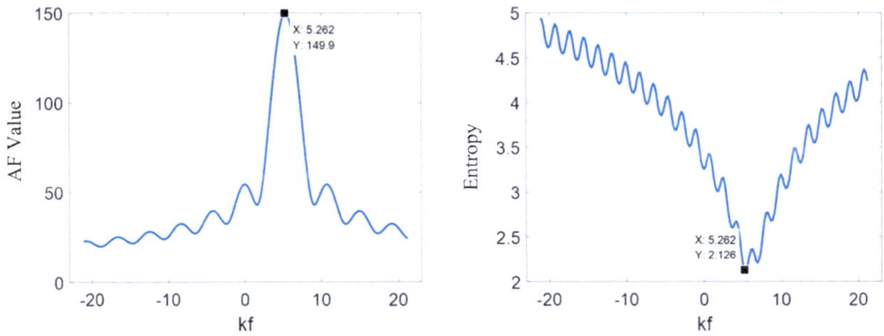

(a) **AF numerical calculation results** (b) **Imaging evaluation results**

Fig. 5.26 AF numerical calculation results and imaging evaluation results

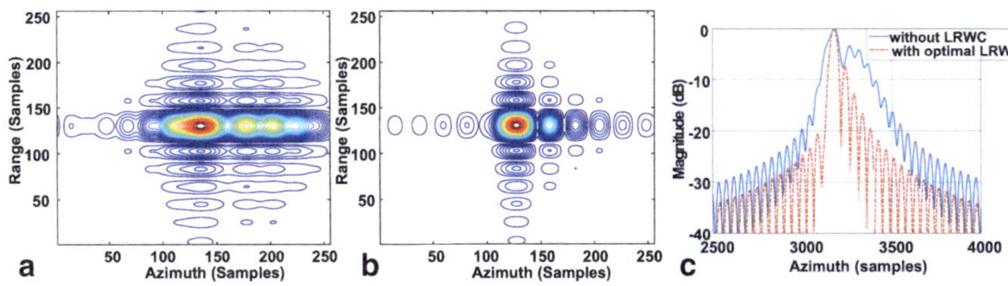

Fig. 5.27 Target 3 matched filtering imaging results **a** Two dimensional contour line map without LRWC **b** Optimal LRWC 2D Contour line map **c** Azimuth profile

shows the 2D contour map of target 3 without LRWC, Fig. 5.27b shows the 2D contour map of target 3 after the optimal LRWC is introduced, and Fig. 5.27c shows the orientation profiles in both cases. The results show that when the optimal LRWC is introduced, the resolution improvement is larger and the side flaps are suppressed.

5.4.2 Wave Number Domain Imaging Method Based on Higher-Order Singular Value Decomposition

For the large two-dimensional null variation problem introduced by large oblique view, a two-dimensional interpolation kernel can be obtained by introducing high-order singular value decomposition (HO-SVD) [14, 15], which can accomplish both range null correction and orientation null correction, and the method takes into account the two-dimensional coupling of range and orientation. The phase of the two-dimensional spectrum is expressed as

$$\begin{aligned}\Phi(f_r, f_a; X, R_0) &= (\Phi(f_r, f_a; X, R_0) - 2\pi f_a t_c) - 2\pi f_a t_c \\ &= \Phi_1(f_r, f_a; X, R_0) - 2\pi f_a t_c\end{aligned} \tag{5.105}$$

For each Doppler frequency f_a, a higher-order singular value decomposition of the phase $\Phi_1(f_r, f_a; X, R_0)$ yields

$$\Phi_1(f_r, f_a; X, R_0) = \Theta_1(R_0)\Pi_1(X)\Delta_1(f_r) + \Theta_2(R_0)\Pi_2(X)\Delta_2(f_r) + \ldots \quad (5.106)$$

Here, only the case where the eigencomponent is 1 is considered, $\Phi_1(f_r, f_a; X, R_0) = \Theta_1(R_0)\Pi_1(X)\Delta_1(f_r)$, if a similar Stolt interpolation operation is performed: $\Delta_1(f_r) \to \Delta_1$, and then the two-dimensional null-variance correction can be completed by IFFT.

According to Sect. 5.4.1 and the summary above, the general steps of the large strabismic nonlinear trajectory imaging algorithm are as follows.

(1) The slope of the optimal walk correction function ˆkwc is first obtained using an optimization method based on the action mechanism modeled in Sect. 5.4.1.
(2) The optimal walk correction function is used to correct the range walk of the radar echoes after the range pulsation to solve the spectral overlap problem and minimize the azimuthal null variation.
(3) Finally, the two-dimensional null correction is performed. Since the beam-domain algorithm can accurately perform the signal null correction using the no-approximation interpolation operation, the beam-domain algorithm is used for the two-dimensional null correction. However, considering the highly nonlinear nature of the interpolation kernel, the signal spectrum is decomposed and analyzed by introducing the higher-order singular value decomposition as a mathematical tool to obtain an accurate 2-D interpolation kernel.

To simplify the higher-order decomposition process, the higher-order singular value decomposition is separated in two dimensions, and the range- and azimuthal-directed signals are analyzed using SVD and verified by imaging, respectively. The first is the SVD analysis of the two-dimensional spectrum.

1. Simulation results and analysis

First, the two-dimensional null-variance properties of the two-dimensional spectrum are analyzed by SVD [16], and considering the need for subsequent imaging algorithms, the quadratic polynomial slope range model is expanded in a Taylor series at the moment when the center of the beam passes the target to obtain

$$\begin{aligned} R(t_a; R_0, X) = R_0 &+ k_1(R_0, X)(t_a - t_c) + k_2(R_0, X)(t_a - t_c)^2 \\ &+ k_3(R_0, X)(t_a - t_c)^3 + k_4(R_0, X)(t_a - t_c)^4 \end{aligned} \quad (5.107)$$

where, t_c denotes the azimuthal moment when the center of the beam passes the target, and still using the level inversion method, the corresponding two-dimensional spectrum of the signal can be expressed as

$$
\begin{aligned}
SS(f_r, f_a; R_0, X) &= \int_{t_a} Ss(f_r, t_a; R_0, X)\exp(-j2\pi f_a t_a)dt_a \\
&\triangleq \exp(j\phi(f_r, f_a; R_0, X)) \\
&\approx \exp(j\phi_r(f_r, f_a; R_0) + j\phi_a(f_a; R_0, X))
\end{aligned}
\tag{5.108}
$$

where $\phi(f_r, f_a; R_0, X)$ denotes the phase of the two-dimensional spectrum; $\phi_r(f_r, f_a; R_0)$ denotes the azimuthal signal independent of the range frequency; $\phi_a(f_a; R_0, X)$ denotes the range signal independent of the azimuthal position, and the expressions of the three can be expressed as

$$
\begin{aligned}
\phi(f_r, f_a; R_0, X) = &-\pi\frac{f_r^2}{\gamma} - \frac{2\pi}{c}R_0(f_r + f_\ominus) - 2\pi f_a t_c \\
&+ \pi\frac{c}{2k_2(R_0, X)(f_c + f_r)}\left(f_a + \frac{(f_c + f_r)}{c}k_1(R_0, X)\right)^2 \\
&+ \pi\frac{k_3(R_0, X)c^2}{4k_2^3(R_0, X)(f_c + f_r)^2}\left(f_a + \frac{(f_c + f_r)}{c}k_1(R_0, X)\right)^3 \\
&+ \pi\frac{\left(9k_3^2(R_0, X) - 4k_2(R_0, X)k_4(R_0, X)\right)c^3}{32k_2^5(R_0, X)(f_c + f_r)^3} \\
&\left(f_a + \frac{(f_c + f_r)}{c}k_1(R_0, X)\right)^4
\end{aligned}
\tag{5.109}
$$

$$
\begin{aligned}
\phi_a(f_a; R_0, X) &= \phi(f_r, f_a; R_0, X)|_{f_r=0} \\
&= -\frac{2\pi}{\lambda}R_0 - 2\pi f_a t_c + \pi\frac{\lambda}{2k_2(R_0, X)}\left(f_a + \frac{k_1(R_0, X)}{\lambda}\right)^2 \\
&+\pi\frac{\lambda^2 k_3(R_0, X)}{4k_2^3(R_0, X)}\left(f_a + \frac{k_1(R_0, X)}{\lambda}\right)^3 \\
&+\pi\frac{\lambda^3\left(9k_3^2(R_0, X) - 4k_2(R_0, X)k_4(R_0, X)\right)}{32f_c k_2^5(R_0, X)}\left(f_a + \frac{k_1(R_0, X)}{\lambda}\right)^4
\end{aligned}
\tag{5.110}
$$

$$
\begin{aligned}
\phi_r(f_r, f_a; R_0) &= \phi(f_r, f_a; R_0, X) - \phi_a(f_a; R_0, X) \\
&\approx \phi_r(f_r, f_a; R_0)|_{f_r=0} + \left.\frac{\partial\phi_r(f_r, f_a; R_0)}{\partial f_r}\right|_{f_r=0}\cdot f_r \\
&+ \frac{1}{2!}\left.\frac{\partial^2\phi_r(f_r, f_a; R_0)}{\partial f_r^2}\right|_{f_r=0}\cdot f_r^2 + \frac{1}{3!}\left.\frac{\partial^3\phi_r(f_r, f_a; R_0)}{\partial f_r^3}\right|_{f_r=0}\cdot f_r^3
\end{aligned}
\tag{5.111}
$$

It should be noted that the derivative does decoupling treatment for orientation and range, so the orientation and range coupling terms can be ignored. The range direction signal expression is a function of both the range frequency and the range position, so the range direction signal is a range null variable. The azimuthal signal expression is a function of both Doppler and azimuthal position, so the azimuthal signal is also azimuthally null.

(1) Range to signal SVD null-variance analysis

The range direction signal after SVD can be expressed as

$$\begin{aligned}\phi_r(f_r,f_a;R_0)=&u_{r1}(R_0,f_a)v_{r1}(f_r,f_a)\\&+u_{r2}(R_0,f_a)v_{r2}(f_r,f_a)+\cdots\end{aligned} \tag{5.112}$$

It should be noted that the range null-variance characteristics in Eq. are for each Doppler frequency fa, so the subsequent SVD null-variance analysis is developed for one Doppler frequency only. In general, as the range resolution increases and the range mapping band expands, the range cavitation increases further [17, 18]. At the same time, the increase in azimuthal resolution (corresponding to the increase in synthetic aperture time) also increases the degree of null variation of the coefficients of each order of the fourth polynomial, thus increasing the range null variation. For a fixed range resolution, the range signal can be accurately represented by one eigencomponent when the range mapping band is narrow, but as the range mapping band widens, the range signal needs to be accurately represented by two or even more than two eigencomponents.

The trend of range signal SVD null-variance analysis with range mapping band is shown in Fig. 5.28, where the horizontal coordinate represents the width of the range mapping band and the vertical coordinate represents the remaining phase error represented by one eigencomponent. When the phase error is less than π/4 of the reference value, the range signal is considered to be accurately represented by one eigencomponent. In the case of only one eigencomponent, only one interpolation operation is needed to correct the range null variation, which is a simple process. Considering that it is difficult to reach 100 km with the continuous wave system SAR satellites in the millimeter band, one "SVD-interpolation" operation can be performed to correct the range variation of the entire range mapping band, and the analysis result will be an important basis for the design of the subsequent imaging algorithm.

(2) Azimuthal signal SVD null-variance analysis

Unlike the range signal, when the orientation signal needs to be represented by two eigencomponents, the focus processing by the orientation subaperture method involves the problem of subaperture splicing, which may make the imaging processing more difficult. The trend of the azimuthal SVD null-variance analysis with the azimuthal scene is shown in Fig. 5.29. When the azimuthal cavitation correction is accurately represented by one feature component, the error exceeds 0.75 rad regardless of how small the azimuthal scene is, and when the azimuthal cavitation correction is represented by two feature components, the azimuthal scene

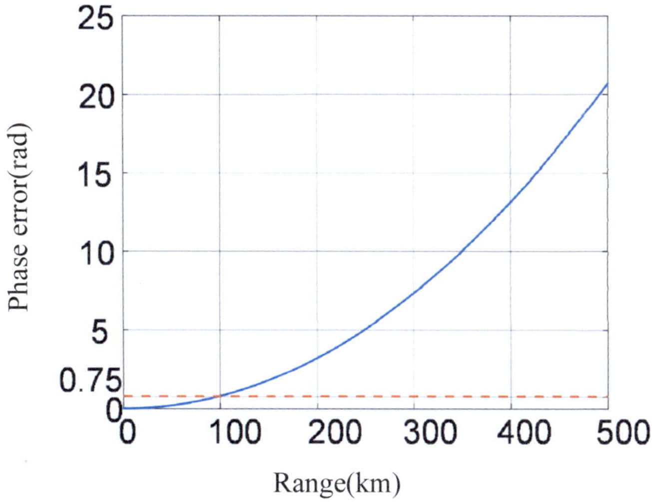

Fig. 5.28 The trend of SVD spatial variation analysis of distance signals with distance mapping bands

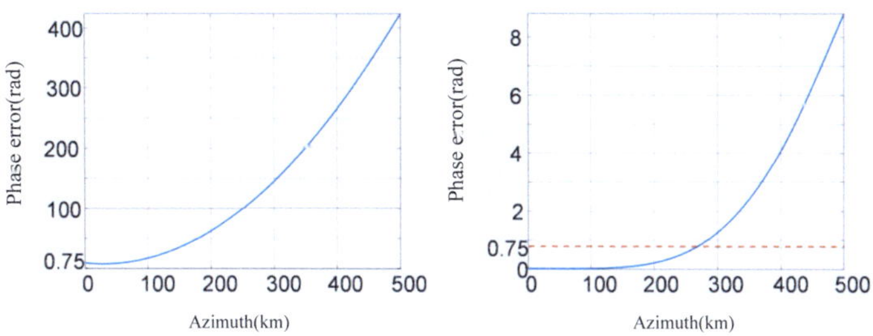

(a) Representation of one feature components **(b) Representation of two feature components**

Fig. 5.29 The trend of SVD spatial variation analysis of azimuth signals with azimuth scenarios

can reach a maximum of about 270 km, which can fully satisfy the SAR signal processing requirements in the on-board millimeter wave regime. Therefore, in order to achieve a larger azimuth scene size, the subsequent imaging algorithm design needs to consider the first two feature components after SVD, and the processing should adopt a full-aperture processing strategy.

2. Two-dimensional SVD-based and optimal LRWC imaging algorithms

(1) Range null correction based on chunked range SVD

After introducing the optimal LRWC factor, the two-dimensional spectrum of the target is changed accordingly, and the two-dimensional spectrum after introducing the optimal LRWC factor can be obtained by simply making $R_0 = R_0 + k_{f\_opt}t_c$, $k_1(R_0, X) = k_1(R_0, X) + k_{f\_opt}$. According to the results of the range signal SVD null-variance analysis, there may be two or more feature components after the range signal SVD, and the recommended method is to divide the whole range mapping band into multiple range blocks. Since one feature component can handle a large range mapping band, only several range blocks are required to meet the accuracy requirements. First, to ensure that the range signal within a range block can be accurately represented by one eigencomponent, the sum of the remaining eigencomponents needs to be less than the reference value π/4, i.e.

$$|\phi_r(f_r, f_a; R_0) - u_{r1}(R_0, f_a)v_{r1}(f_r, f_a)| < \varepsilon = {}^{\pi}/_{4} \tag{5.113}$$

where $\varepsilon = \pi/4$ is the reference value. Therefore, the range signal within a range block can be expressed as

$$\phi_r(f_r, f_a; R_0) = u_{r1}(R_0, f_a)v_{r1}(f_r, f_a) \tag{5.114}$$

where, for each Doppler frequency f_a, $u_{r1}(R_0, f_a)$ is a polynomial function on R_0 and $v_{r1}(f_r, f_a)$ is a polynomial function on f_r. A Taylor series expansion of $u_{r1}(R_0, f_a)$ at $R_0 = 0$ and a Taylor series expansion of $v_{r1}(f_r, f_a)$ at $f_r = 0$ give, respectively

$$\begin{aligned} u_{r1}(R_0, f_a) &= u_{r1}(R_0 = 0, f_a) + \left.\frac{\partial u_{r1}(R_0, f_a)}{\partial R_0}\right|_{R_0=0} \cdot R_0 \\ &\quad + \left.\frac{1}{2!}\frac{\partial^2 u_{r1}(R_0, f_a)}{\partial R_0^2}\right|_{R_0=0} \cdot R_0^2 + \ldots \\ &\triangleq B_0(f_a) + B_1(f_a)R_0 + B_2(f_a)R_0^2 + \ldots \end{aligned} \tag{5.115}$$

$$\begin{aligned} v_{r1}(f_r, f_a) &= v_{r1}(f_r = 0, f_a) + \left.\frac{\partial v_{r1}(f_r, f_a)}{\partial f_r}\right|_{f_r=0} \cdot f_r \\ &\quad + \left.\frac{1}{2!}\frac{\partial^2 v_{r1}(f_r, f_a)}{\partial f_r^2}\right|_{f_r=0} \cdot f_r^2 + \ldots \\ &\triangleq A_0(f_a) + A_1(f_a)f_r + A_2(f_a)f_r^2 + \ldots \end{aligned} \tag{5.116}$$

By performing one interpolation operation on Eq. (5.116), the interpolation kernel can be expressed as

$$A_0(f_a) + A_1(f_a)f_r + A_2(f_a)f_r^2 + \ldots \to \alpha + \beta f_r \tag{5.117}$$

where α and β are determined by the interpolation kernel corresponding to the Doppler center frequency: $v_{r1}(f_r, f_{dc}) \to \alpha + \beta f_r$, where fdc is the Doppler center frequency. After the interpolation operation, the phase of the range signal can be expressed as

$$\phi_r(f_r, f_a; R_0) = \left(B_0(f_a) + B_1(f_a)R_0 + B_2(f_a)R_0^2 + \ldots\right) \cdot (\alpha + \beta f_r) \tag{5.118}$$

At this time, both Doppler and range frequency information exist in Eq. (5.118), and coupled with range position information, RCMC cannot be completed directly by matched filtering, so it is organized into the following expression:

$$\begin{aligned}\phi_r(f_r, f_a; R_0) = &-\alpha \cdot \left(B_0(f_a) + B_1(f_a)R_0 + B_2(f_a)R_0^2 + \ldots\right) \\ &+ \beta \cdot \left(B_0(f_a) + B_1(f_a)R_0 + B_2(f_a)R_0^2 + \ldots\right)f_r\end{aligned} \tag{5.119}$$

In Eq. (5.119), the first term on the right side of the equation is independent of the range frequency and is related to the range position and Doppler, so it can be compensated by matching filtering in the range Doppler domain, and the compensation function can be expressed as

$$H_{res}(R_0, f_a) = \alpha \cdot \left(B_0(f_a) + B_1(f_a)R_0 + B_2(f_a)R_0^2 + \ldots\right) \tag{5.120}$$

The phase of the compensated range signal can be expressed as

$$\phi_r(f_r, f_a; R_0) = \beta \cdot \left(B_0(f_a) + B_1(f_a)R_0 + B_2(f_a)R_0^2 + \ldots\right)f_r \tag{5.121}$$

where, for each range position R_0, ξ can be determined by the interpolation kernel corresponding to the Doppler center frequency: $\beta \cdot \left(B_0(f_{dc}) + B_1(f_{dc})R_0 + B_2(f_{dc})R_0^2 + \ldots\right)f_r \to \xi(R_0)f_r$. The phase of the interpolated range signal can be expressed as

$$\phi_r(f_r, f_a; R_0) = \xi(R_0)f_r \tag{5.122}$$

By performing a range IFFT on Eq. (5.122), the range signal can be focused, and the expression of the signal in the range time domain—azimuthal Doppler domain is

$$sS(t_r, f_a; R_0, X) = \text{sinc}(B_r(t_r - f_1(R_0))) \cdot \exp(j\phi_a(f_a; R_0, X)) \tag{5.123}$$

where $f_1(R_0){=}f_1(\xi(R_0))$ is a function on R0 that determines the range focus position of the target.

(2) Azimuthal null-variation correction based on azimuthal series SVD

After RCMC, the phase $\varphi(f_r, f_a; X, R_0)$ of the two-dimensional spectrum of the signal can be reduced to $\varphi(f_a; X)$, where the Doppler information f_a and the azimuthal position information X are heavily coupled. Singular value decomposition (SVD) is a mathematical tool that can be effectively used to separate the coupling properties of the two. After SVD, the phase of the two-dimensional spectrum of the signal can be expressed as

$$\varphi(f_a; X) = u_1(X)v_1(f_a) + u_2(X)v_2(f_a) + \ldots \tag{5.124}$$

The phase of the two-dimensional spectrum of the signal is separated into the sum of several eigencomponents, each of which is the product of a function on fa and a function on X.

Typically, the first eigencomponent $u_1(X)v_1(f_a)$ in Eq. (5.124) has the largest value, and the values of the subsequent eigencomponents decrease in order. Intuitively, if the phase has only one eigencomponent after SVD and meets the accuracy requirement, $\varphi(f_a; X) = u_1(X)v_1(f_a)$, then only one interpolation operation $v_1(f_a) \to f_a$ is needed to perform uniform orientation focus on all orientation targets. However, considering the long synthetic aperture time of the satellites, the phase of the signal 2D spectrum usually cannot be accurately represented by a single eigencomponent. The eigencomponents of the SVD are shown in Fig. 5.30. The values of the first two eigencomponents are in the order of thousands to tens of thousands of radians (Fig. 5.30a and b), and the sum of the remaining eigencomponents is much smaller than π/4. Therefore, only the first two eigencomponents need to be considered for the subsequent azimuthal null change, and the remaining eigencomponents can be ignored.

After RCMC, the envelopes of all targets have been corrected to straight lines, and the next azimuthal pulse pressure needs to consider the azimuthal null variation characteristics of the targets. Considering that the two-dimensional spectrum of the signal after SVD is represented by two eigencomponents, the azimuthal null correction consists of two parts. (1) The first SVD-interpolation operation corrects the azimuthal null of the first eigencomponent. (2) The second SVD-interpolation operation corrects the azimuthal null of the second eigencomponent. The second SVD-interpolation operation corrects the azimuthal null of the second eigencomponent. Due to the pipelined nature of the two SVD-interpolation operations, this method is called the azimuthal tandem SVD method. It should also be noted that an azimuthal nonlinear variational scale (ANCS) function needs to be introduced to bridge the two SVD-interpolation operations, and the detailed principle and derivation of the azimuthal tandem SVD will be given in the following. The flow chart of the azimuthal null-variance correction algorithm is shown in Fig. 5.31.

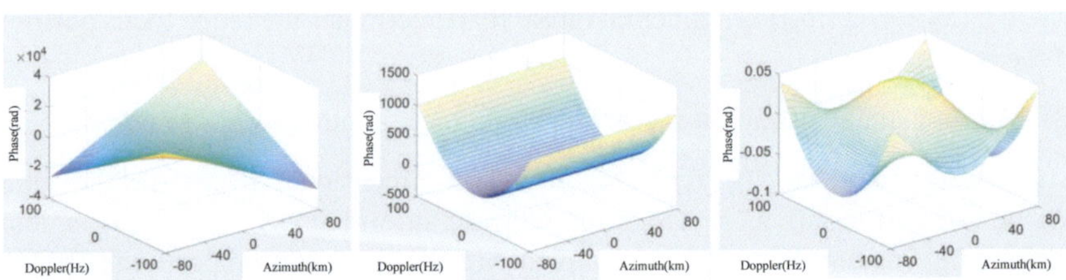

(a) First feature component (b) Second feature component (c) Sum of residual feature components

Fig. 5.30 Characteristic components of the two-dimensional spectrum of the signal after SVD

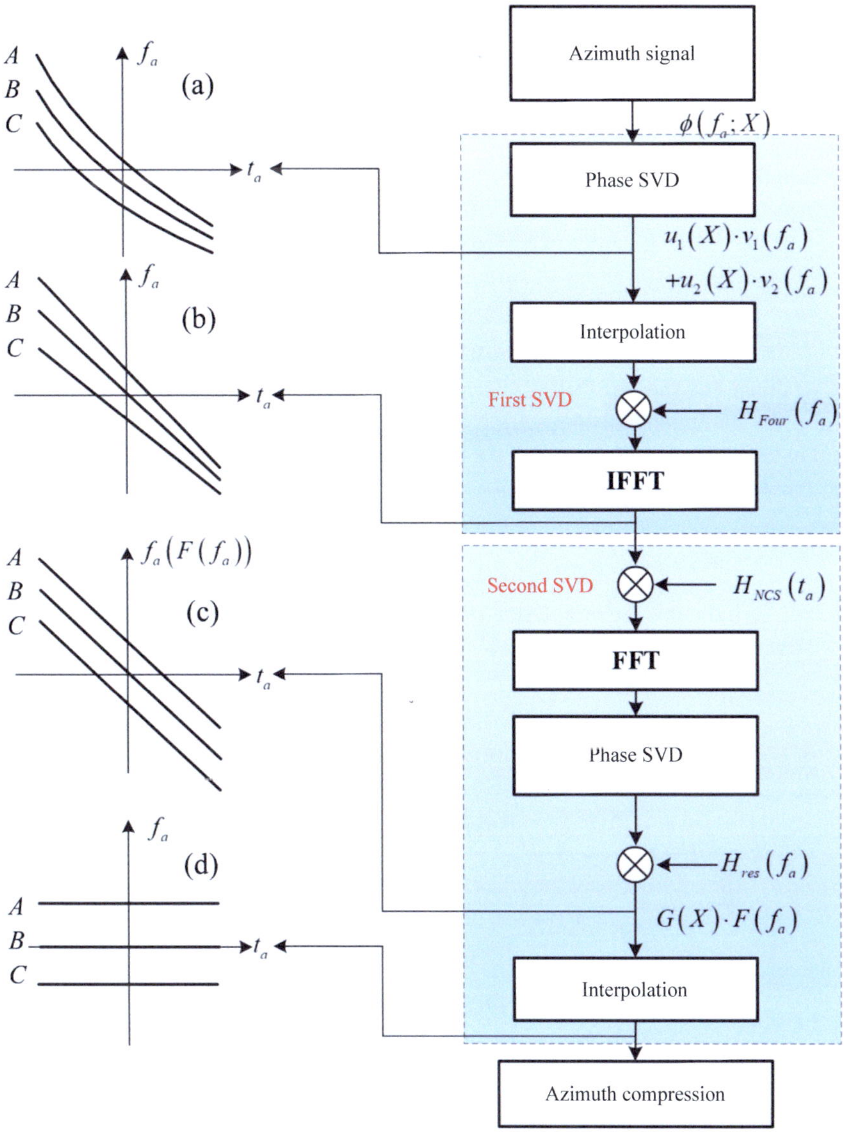

Fig. 5.31 Flow chart of the azimuthal null-variance correction algorithm

The two-dimensional spectrum of the signal after the azimuthal null correction is independent of the range information, and only the first two eigencomponents are considered, the phase of the two-dimensional spectrum after the first SVD can be expressed as

$$\varphi_{fSVD}(f_a; X) = u_1(X)v_1(f_a) + u_2(X)v_2(f_a) \tag{5.125}$$

where the subscript f_{SVD} denotes the first SVD; $u_1(X)$ and $u_2(X)$ are polynomial functions on X, $v_1(f_a)$ and $v_2(f_a)$ are polynomial functions on f_a, and the polynomial coefficients of each order of the four can be solved by the numerical method of polynomial fitting. First, the first interpolation operation $v_1(f_a) \rightarrow v_1$ is implemented, where the square frequency changes from f_a to f'_a and the signal phase can be simplified as

$$\begin{aligned}\varphi_{fSVD}(f_a; X) &= u_1(X) \cdot v_1 + u_2(X) \cdot v_2\left(f_a, f_a'\right) \\ &\approx U_0(X) + U_1(X) f_a + U_2(X) f_a^2 \\ &\quad + U_3(X) f_a^3 + U_4(X) f_a^4\end{aligned} \tag{5.126}$$

At this point, the first "SVD-interpolation" operation has been completed, eliminating the azimuthal null of the tertiary phase term, implying that there is no azimuthal null in the tertiary phase term and the quadratic term, so the two can be compensated by matching filtering, and the matching filtering function can be expressed as

$$H_1(f_a) = \exp\left(-jU_3(X_0) f_a^3\right) \cdot \exp\left(-jU_4(X_0) f_a^4\right) \tag{5.127}$$

where, X_0 indicates the orientation position of the target in the center of the scene.

After compensation, the phase of the signal can be expressed as

$$\varphi_{fSVD}(f_a; X) = U_0(X) + U_1(X) f_a + U_2(X) f_a^2 \tag{5.128}$$

Among them.

$$\begin{aligned}U_1(X) \approx U_1(X)|_{X=X_0} &+ X \cdot \left.\frac{\partial U_1(X)}{\partial X}\right|_{X=X_0} + \frac{X^2}{2!} \cdot \left.\frac{\partial^2 U_1(X)}{\partial X^2}\right|_{X=X_0} \\ &+ \frac{X^3}{3!} \cdot \left.\frac{\partial^3 U_1(X)}{\partial X^3}\right|_{X=X_0} + \frac{X^4}{4!} \cdot \left.\frac{\partial^4 U_1(X)}{\partial X^4}\right|_{X=X_0}\end{aligned} \tag{5.129}$$

$$\begin{aligned}U_2(X) \approx U_1(X)|_{X=X_0} &+ X \cdot \left.\frac{\partial U_2(X)}{\partial X}\right|_{X=X_0} + \frac{X^2}{2!} \cdot \left.\frac{\partial^2 U_2(X)}{\partial X^2}\right|_{X=X_0} \\ &+ \frac{X^3}{3!} \cdot \left.\frac{\partial^3 U_2(X)}{\partial X^3}\right|_{X=X_0} + \frac{X^4}{4!} \cdot \left.\frac{\partial^4 U_2(X)}{\partial X^4}\right|_{X=X_0}\end{aligned} \tag{5.130}$$

It should be noted that the specific expression for the constant term U0 (X) is not given since it does not affect the azimuthal null-variance correction.

Next, a second "SVD-interpolation" operation is performed. Before calibrating the azimuthal null variation of the quadratic phase term, an azimuthal nonlinear variational scale (ANCS) function is introduced for bridging, which takes the form of a cubic polynomial with third-order coefficients defined by C_{ncs}, then the ANCS function can be expressed as

$$H_{NCS}(t_a) = \exp\left(jC_{ncs}t_a^3\right) \tag{5.131}$$

where C_{ncs} is the third-order coefficient of the ANCS function, whose solution procedure will be given in detail below. It should be noted that if the second SVD is performed directly, there are still two eigencomponents of the phase of the 2D spectrum, while the ANCS function is useful in that after introducing it in the azimuthal time domain and transforming the signal to the Doppler domain, the phase of the 2D spectrum of the signal can be precisely represented by one eigencomponent.

With the introduction of the ANCS function, the phase of the signal in the azimuthal time domain can be expressed as

$$\begin{aligned}\varphi_{fSVD}^{ncs}(t_a; X) &= \varphi_{fSVD}(t_a; X) + C_{ncs}t_a^3 \\ &= U_0(X) - \frac{U_1^2(X)}{4U_2(X)} - \frac{\pi U_1(X)}{U_2(X)}t_a - \frac{\pi^2}{U_2(X)}t_a^2 + C_{ncs}t_a^3\end{aligned} \tag{5.132}$$

where the superscript ncs denotes the azimuthal nonlinear variational scale and $\varphi_{fSVD}(t_a; X)$ is the phase form of the signal transformed from the Doppler domain to the azimuthal time domain, transforming the signal from the azimuthal time domain to the Doppler domain yields

$$\begin{aligned}S(f_a; X) &= \int\limits_{t_a} \exp\left(j\varphi_{fSVD}^{ncs}(t_a; X)\right) \cdot e^{-j2\pi f_a t_a} dt_a \\ &= \exp\left(j\varphi_{fSVD}^{ncs}\left(t_a\left(t_a^*\right); X\right)\right) \\ &\triangleq \exp\left(j\varphi_{fSVD}^{ncs}(f_a; X)\right)\end{aligned} \tag{5.133}$$

Among them.

$$t_a^* = \frac{3C\left(\frac{U_1(X)}{2\pi}\right)^2 + 2\pi f_a}{2\left(\frac{3C_{ncs}U_1(X)}{2\pi} + \frac{\pi^2}{U_2(X)}\right)} \triangleq \frac{D(X) + 2\pi f_a}{E(X)} \tag{5.134}$$

The phase of the signal in the Doppler domain after the introduction of the ANCS function can be expressed as

$$\begin{aligned}\varphi_{fSVD}^{ncs}(f_a; X) &= \frac{8\pi^3 C_{ncs}}{E^3(X)} f_a^3 + \left(\frac{12\pi^2 C_{ncs} \cdot D(X)}{E^3(X)} - \frac{2\pi^2}{E(X)} \right) f_a^2 \\ &+ 2\pi \left(\frac{3C_{ncs} \cdot D^2(X)}{E^3(X)} - \frac{D(X)}{E(X)} + \frac{U_1(X)}{2\pi} \right) f_a \\ &+ U_0(X) + \frac{C_{ncs} \cdot D^3(X)}{E^3(X)} + \frac{D^2(X)}{2E(X)} - C_{ncs} \left(\frac{U_1(X)}{2\pi} \right)^3 \\ &\triangleq \varphi_t(X) f_a^3 + \varphi_s(X) f_a^2 + \varphi_f(X) f_a + \varphi_0(X) \end{aligned} \tag{5.135}$$

where $\varphi_t(X)$, $\varphi_s(X)$, $\varphi_f(X)$ and $\varphi_0(X)$ are the coefficients of the cubic phase term, the quadratic phase term, the primary phase term and the constant phase term, respectively, the first three of which can be expressed by the Taylor series as

$$\begin{aligned}\varphi_t(X) &= \frac{8\pi^3 C_{ncs}}{E^3(X)} \\ &\approx a_1(C_{ncs}) + a_2(C_{ncs}) \cdot X + a_3(C_{ncs}) \cdot X^2 \\ &= \varphi_t(X)|_{X=X_0} + \left.\frac{\partial \varphi_t(X)}{\partial X}\right|_{X=X_0} \cdot X + \frac{1}{2} \left.\frac{\partial^2 \varphi_t(X)}{\partial X^2}\right|_{X=X_0} \cdot X^2 \end{aligned} \tag{5.136}$$

$$\begin{aligned}\varphi_s(X) &= \frac{12\pi^2 C_{ncs} \cdot D(X)}{E^3(X)} - \frac{2\pi^2}{E(X)} \\ &\approx b_1(C_{ncs}) + b_2(C_{ncs}) \cdot X + b_3(C_{ncs}) \cdot X^2 \\ &= \varphi_s(X)|_{X=X_0} + \left.\frac{\partial \varphi_s(X)}{\partial X}\right|_{X=X_0} \cdot X + \frac{1}{2} \left.\frac{\partial^2 \varphi_s(X)}{\partial X^2}\right|_{X=X_0} \cdot X^2 \end{aligned} \tag{5.137}$$

$$\begin{aligned}\varphi_f(X) &= 2\pi \left(\frac{3C_{ncs} D^2(X)}{E^3(X)} - \frac{D(X)}{E(X)} + \frac{U_1(X)}{2\pi} \right) \\ &\approx c_1(C_{ncs}) + c_2(C_{ncs}) \cdot X + c_3(C_{ncs}) \cdot X^2 \\ &= \varphi_f(X)\big|_{X=X_0} + \left.\frac{\partial \varphi_f(X)}{\partial X}\right|_{X=X_0} \cdot X + \frac{1}{2} \left.\frac{\partial^2 \varphi_f(X)}{\partial X^2}\right|_{X=X_0} \cdot X^2 \end{aligned} \tag{5.138}$$

It should be noted that the effect of the constant phase term on the follow-up can be neglected. Also, $\varphi_t(X)$, $\varphi_s(X)$, $\varphi_f(X)$ and $\varphi_0(X)$ are related to the coefficients C_{ncs} of the ANCS function, so that their coefficients of each order after the Taylor series expansion are all functions with respect to C_{ncs}. A reasonable choice of C_{ncs} allows the signal to be accurately represented by an eigencomponent after the second SVD, i.e.

$$\begin{aligned}\varphi_{sSVD}^{ncs}(f_a;X) &= F(X)G(f_a) + \varphi_{inv}(f_a) + \varphi_v(f_a;X) \\ &\approx F(X)G(f_a) + \varphi_{inv}(f_a)\end{aligned} \tag{5.139}$$

where the subscript s_{SVD} denotes the second SVD; $F(X)G(f_a)$ is the only eigencomponent after SVD, $F(X)$ as a function of X, and $G(f_a)$ as a function of f_a; $\varphi_{inv}(f_a)$ is the residual phase of azimuthal non-null; $\varphi_v(f_a;X)$ is the residual phase of azimuthal null. In general, this phase is negligible when the value of $\varphi_v(f_a;X)$ is much smaller than the reference value π/4. Next, the coefficient C_{ncs} of the ANCS function is derived.

First combine the linear phase term and the quadratic phase term, the first-order and second-order coefficients of $\varphi_s(X)$ and $\varphi_f(X)$ need to be proportional, i.e.

$$\frac{b_2(C_{ncs})}{c_2(C_{ncs})} = \frac{b_3(C_{ncs})}{c_3(C_{ncs})} \triangleq \zeta(C_{ncs}) \tag{5.140}$$

The coefficients of the ANCS function C_{ncs} can be solved by substituting the expressions of $b_2(C_{ncs})$, $c_2(C_{ncs})$, $b_3(C_{ncs})$ and $c_3(C_{ncs})$ into the above equation. When the linear and quadratic phase terms are combined, the phase of the signal can be expressed as

$$\begin{aligned}\varphi_{sSVD}^{ncs}(f_a;X) = &\underbrace{\varphi_{sf}(X)\left(\zeta'(C_{ncs})f_a^3 + f_a^2 + 1/\zeta(C_{ncs})f_a\right)}_{F(X)G(f_a)} \\ &+ \underbrace{a_1(C_{ncs})f_a^3 + b_1(C_{ncs})f_a^2 + c_1(C_{ncs})f_a}_{\varphi_{inv}(f_a)} \\ &+ \underbrace{\left(a_3(C_{ncs}) - \zeta'(C_{ncs})b_3(C_{ncs})\right)X^2 f_a^3 + \varphi_0(X)}_{\varphi_v(f_a;X)}\end{aligned} \tag{5.141}$$

where $\zeta'(C_{ncs}) = a_2(C_{ncs}) \big/ b_2(C_{ncs})$. Based on the expression of Eq. (5.141), the following expression can be deduced

$$F(X) = \varphi_{sf}(X) = b_2(C_{ncs})X + b_3(C_{ncs})X^2 \tag{5.142}$$

$$G(f_a) = \zeta'(C_{ncs})f_a^3 + f_a^2 + 1/\zeta(C_{ncs})f_a \tag{5.143}$$

$$\phi_{inv}(f_a) = a_1(C_{ncs})f_a^3 + b_1(C_{ncs})f_a^2 + c_1(C_{ncs})f_a \tag{5.144}$$

$$\phi_v(f_a;X) = \left(a_3(C_{ncs}) - \zeta'(C_{ncs})b_3(C_{ncs})\right)X^2 f_a^3 + \varphi_0(X) \tag{5.145}$$

The non-null phase term in Eqs. (5.142)–(5.145) can be compensated by matched filtering, and the compensation function can be expressed as

$$H_{res}(f_a) = \exp\left(-ja_1(C_{ncs})f_a^3\right) \cdot \exp\left(-jb_1(C_{ncs})f_a^2 - jc_1(C_{ncs})f_a\right) \tag{5.146}$$

After compensation, the phase of the signal can be expressed as

$$\varphi_{sSVD}^{ncs}(f_a; X) = \varphi_{sf}(X)\left(\zeta'(C_{ncs})f_a^3 + f_a^2 + 1/\zeta(C_{ncs})f_a\right) \tag{5.147}$$

After the second interpolation operation: (ς ′ (C) fncs3a + f2a + 1/ς (Cncs) fa) → fa, the phase of the two-dimensional spectrum of the signal after the second "SVD-interpolation" operation can be expressed as

$$\varphi_{sSVD}^{ncs}(f_a; X) = \varphi_{sf}(X) \cdot f_a \tag{5.148}$$

At this point, the azimuth IFFT of Eq. (5.148) can be performed to accurately focus the entire azimuth scene target.

(3) Algorithm simulation

The imaging algorithm proposed in this section consists of three parts: (1) orientation pre-processing based on the optimal LRWC, which aims to minimize the azimuth null variation and further expand the azimuth scene size or improve the resolution; (2) range null correction based on the chunked range SVD, which corrects the range null variation of the range signal and completes the accurate RCMC of the large range mapping band; (3) azimuth null correction based on the azimuth tandem SVD, which corrects the severe azimuth null variation and completes the accurate focus of the whole azimuth scene. The flow chart of this algorithm is shown in Fig. 5.32.

The azimuthal tandem SVD method is selected as the reference algorithm here. After imaging the whole scene target with the reference algorithm, the 2D contour and range and orientation profiles of target 1 are shown in Fig. 5.33, and the corresponding imaging results of target 2 are shown in Fig. 5.34. The results show that the reference algorithm has a more serious scattering phenomenon at the point target at the edge of the azimuth scene. The corresponding imaging results are shown in Figs. 5.35 and 5.36. Both the center and edge point targets of the scene are well focused, which verifies the effectiveness of the proposed algorithm in this section.

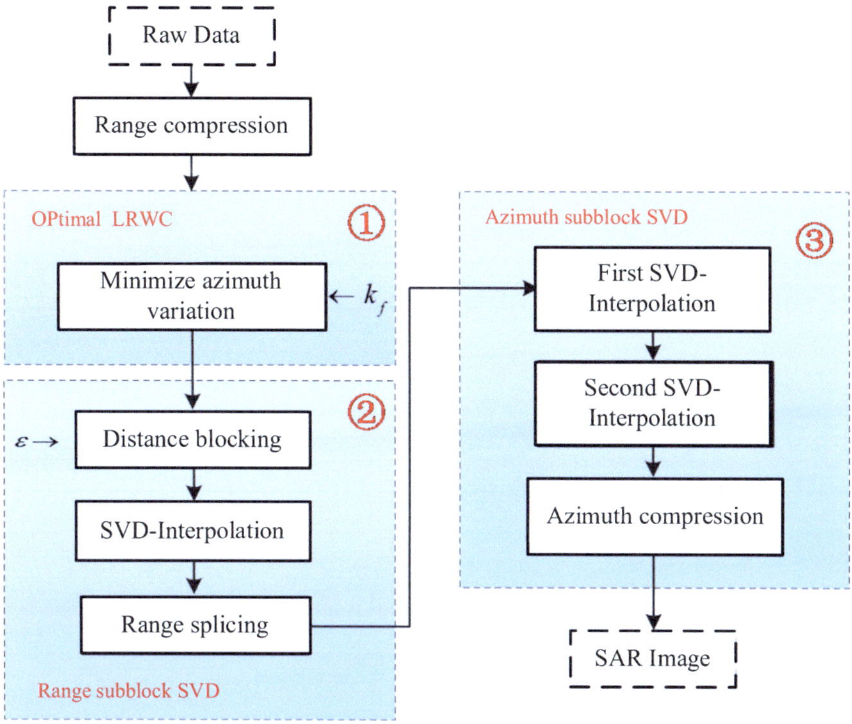

Fig. 5.32 Flow chart of azimuth spatial variation correction algorithm

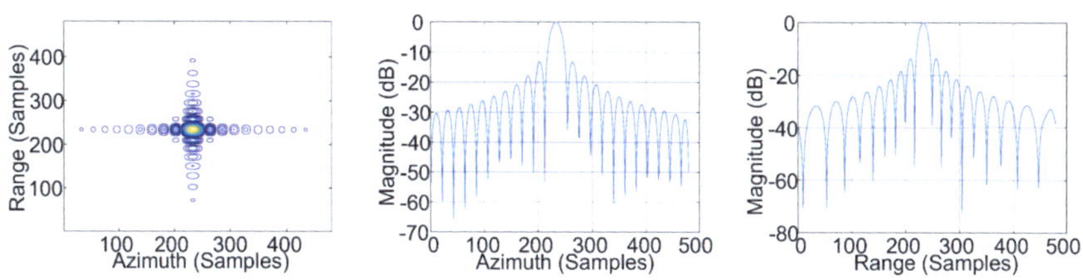

(a) **Two dimensional Contour line** (b) **Azimuth profile** (c) **Range profile**

Fig. 5.33 Reference algorithm PT1 imaging results

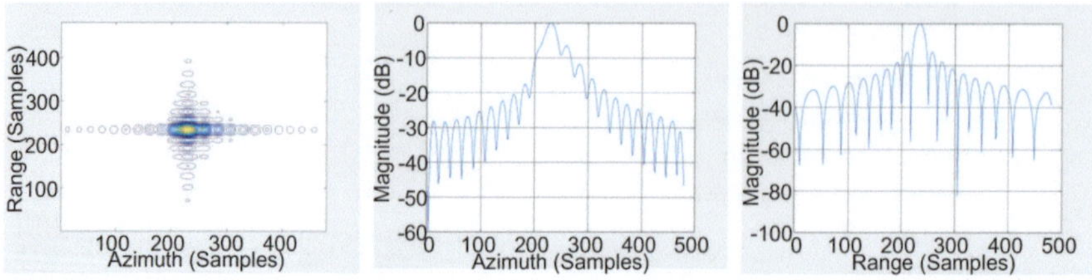

(a) Two dimensional Contour line (b) Azimuth profile (c) Range profile

Fig. 5.34 Reference algorithm PT3 imaging results

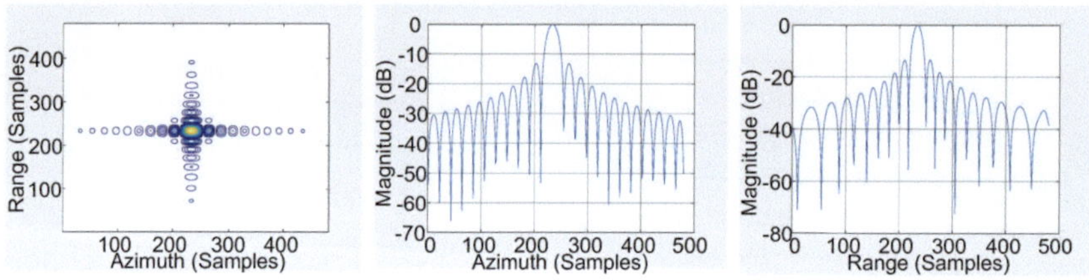

(a) Two dimensional Contour line (b) Azimuth profile (c) Range profile

Fig. 5.35 PT1 imaging results of the proposed algorithm

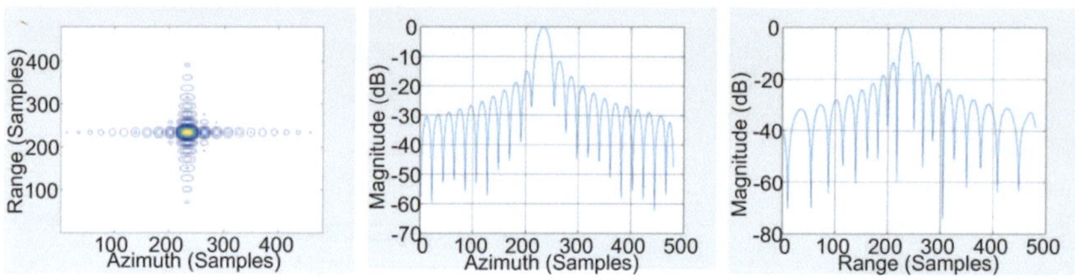

(a) Two dimensional Contour line (b) Azimuth profile (c) Range profile

Fig. 5.36 PT3 imaging results of the proposed algorithm

References

1. Huai Chao, Wang Wenyan. InSAR formation satellite full-zero Doppler attitude guidance study. Shanghai Space, 2014, 31(6): 5.12.
2. He Zhihua, He Feng, Huang Haifeng, et al. Analysis and experimental verification of the effect of distributed SAR phase synchronization error. Journal of Astronautics, 2012, 33(3): 353–357.
3. Li S-Q, Yu W-D. Implementation scheme and experimental verification of distributed satellite SAR phase synchronization. Journal of Electronics and Information, 2012, 34(2): 5.9.
4. Sherwin C W, Ruina J P, Rawcliffe R D. Some early developments in synthetic aperture radar systems. IRE Trans on MIL, 1962, 6(2): 111–115.
5. Jing Guobin. Research on airborne/on-board ultra-high resolution SAR imaging technology. Xi'an: Xi'an University of Electronic Science and Technology, 2018.

6. Li Chunsheng, Yang Wei, Wang Pengbo. A review of satellite-based SAR imaging processing algorithms. Journal of Radar, 2013, 2(1): 111–122.
7. Y. Liang. FM continuous wave SAR signal processing. Xi'an: Xi'an University of Electronic Science and Technology, 2009.
8. Griffiths H D. Synthetic aperture processing for full-deramp radar altimeters. Electronics Letters, 2002, 24(7): 371–3.
9. Liang Y, Guo L, Xing M D, et al. An equivalent positive side-view processing method for strabismic FMCW SAR. Journal of Electronics, 2009, 37(6): 1159–1164.
10. Yamaguchi Y, Mitsumoto M, Sengoku M, et al. Synthetic aperture FM-CW radar applied to the detection of objects buried in snowpack. Geoscience and Remote Sensing IEEE Transactions on, 2002, 32(1): 11–18.
11. Meta A, Hoogeboom P. Development of signal processing algorithms for high resolution airborne millimeter wave FMCW SAR[C]//Radar Conference. 2005 IEEE International, IEEE, 2005.
12. Meta A, de Wit J J M, Hoogeboom P. Development of a high resolution airborne millimeter wave FM-CW SAR[C]//Radar Conference, 2004. eurad. first European. IEEE, 2004.
13. Weib M, Ender J H G. A 3D imaging radar for small unmanned airplanes-ARTINO[C]// Radar Conference, 2005. European, IEEE, 2005.
14. Shumin Geng. Research on key technologies of FM-CW SAR signal processing. Changsha: National University of Defense Technology, 2008.
15. Wang R, Loffeld O, Nies H, et al. Focus FMCW SAR data using the wave number domain algorithm. IEEE Transactions on Geoscience and Remote Sensing, 2010, 48(4): 2109–2118.
16. Dewit J J M, Meta A, Hoogeboom P. Modified range-doppler processing for FM-CW synthetic aperture radar. IEEE Geoscience and Remote Sensing Letters, 2006, 3(1): 83–87.
17. Luo Y, Song H, Gao Y, et al. Modified frequency scaling processing for FMCW SAR[C]/ /Geoscience and Remote Sensing Symposium (IGARSS), 2012 IEEE International, IEEE, 2012.
18. Li Zhenyu, Chen Splash-Lai, Liang Yi, et al. Large strabismus SAR imaging method with Doppler central null correction. Journal of Xi'an University of Electronic Science and Technology, 2016, 43(3): 6.

Chapter 6
Ka-Band Synthetic Aperture Radar in the Pulsed Regime

6.1 Overview

For Ka-band spaceborne SAR, its losses in atmospheric propagation are greater than other low-frequency bands, requiring a higher power aperture product to meet the requirements of imaging quality. To ensure a certain observation width, the antenna size of the traditional single-transmit and single-receive method is limited, and the shorter wavelength of the Ka-band further worsens this problem. The small antenna makes it difficult for Ka-band SAR to consider the requirements of power-aperture product and beam width at the same time in single-transmit and single-receive operation, especially in ultra-high resolution or wide swath imaging.

From the introduction of DBF-SCORE technology in Chap. 3, it can be seen that this technology can synthesize a narrow receiving beam with high gain by using a multi-aperture antenna, and achieve high receiving gain in the entire illuminating area through the scanning of the narrow beam, so as to support the spaceborne SAR system to obtain a larger swath with a smaller transmitting antenna, and ensure the signal-to-noise ratio of the system, optimize the range ambiguity and reduce the influence of rain clutter. The above characteristics of DBF-SCORE technology make it the core technical means to solve the high power-aperture product requirements of spaceborne Ka-band SAR under the pulsed system.

The characteristics of high center frequency and large available bandwidth in the Ka-band can realize spaceborne high-resolution imaging and support more refined ground remote sensing observation. Taking advantage of the short wavelength of Ka-band and sensitivity to interferometric phase, sufficient baseline length can be obtained on a single-satellite platform to achieve high-precision interferometry [1–3].

Therefore, this chapter focuses on the characteristic application directions of spaceborne Ka-band SAR, introduces the spaceborne Ka-band high-resolution imaging method and the Ka-band high-precision interferometry method, and verifies and analyzes the proposed method through the airborne measured data or simulation.

H. Wang, *The Millimeter Wave Synthetic Aperture Radar Technology*,
https://doi.org/10.1007/978-981-97-1044-7_6

6.2 Pulsed Spaceborne SAR High-Resolution Imaging Processing

Under the demand for high-resolution imaging, the influence of non-ideal factors that could be ignored in the past on imaging quality is gradually reflected, such as: orbital nonlinearity, "stop-and-go" model error, two-dimensional frequency domain residual error, etc., this section will analyze these errors and propose corresponding compensation methods. Finally, the echo simulation data is used to quantitatively analyze and prove the algorithm.

6.2.1 Modeling of Azimuth Multichannel Sliding-Spotlight SAR Echoes

The azimuth multi-channel sliding beam-gathering spaceborne SAR mode is an effective means to achieve high resolution in azimuth. As Fig. 6.1, in the azimuth multichannel sliding spotlight mode, the azimuth beam gradually changes from forward strabismus to backward strabismus during one observation period. In-scene targets can achieve longer synthetic aperture time than strip mode, resulting in higher azimuth resolution. N receive channels simultaneously sample the echo signal scattered by the target, and the echo data obtained is N times that of the single-channel sliding beam. Therefore, the use of azimuth multichannel technology can reduce PRF by N times, which can achieve a larger width observation in the distance direction without reducing the azimuth resolution.

The geometric model of the azimuth multichannel sliding beam-based spaceborne SAR is shown in Fig. 6.2, and the azimuth beam rotates clockwise with the movement of the platform during the imaging work, pointing to the virtual rotation point. Among them, Tw represents the irradiation time of this imaging X_s, representing the azimuth scene width $X_f(r)$, the beam footprint length, and the r distance from the zero Doppler moment SAR sensor to the center of the scene. Similarly, azimuth multichannel sliding The geometric model of the azimuth multichannel sliding beam-based spaceborne SAR is shown in Fig. 6.2, and the azimuth beam rotates clockwise

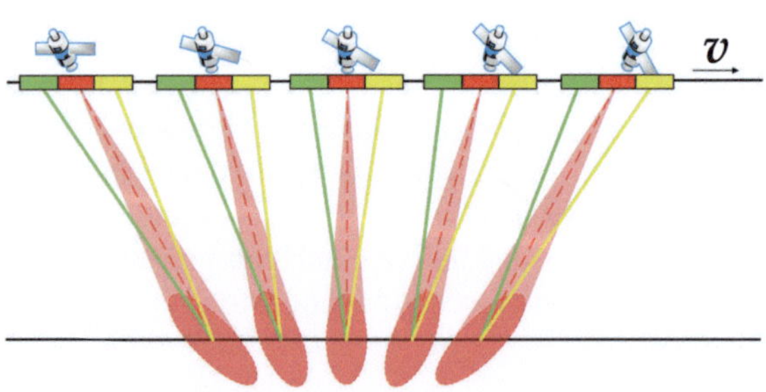

Fig. 6.1 Azimuth multichannel sliding spotlight mode

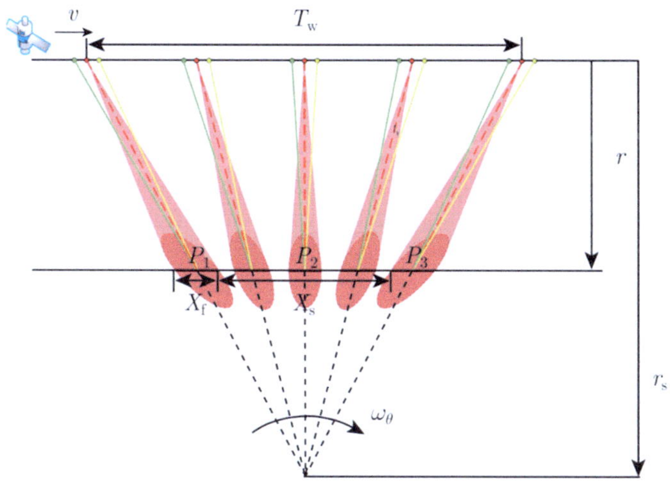

Fig. 6.2 Geometric model of azimuth multichannel sliding spotlight

with the movement of the platform during the imaging work, pointing to the virtual rotation point. Among them, Tw is the irradiation time of this imaging, standing for the azimuth scene width, the beam footprint length, and the distance from the zero Doppler moment SAR sensor to the center of the scene. Similarly, azimuth multichannel sliding spotlight spaceborne SAR has two special parameters: beam scanning angular velocity ω_θ and the distance from the SAR sensor to the virtual rotation point r_s. In azimuth multichannel sliding beam clustering mode, ω_θ both are r_s positive.

Due to beam rotation, the resolution of azimuth multichannel sliding spotlight mode is different from that of azimuth multichannel strip map mode, and it is also necessary to multiply a rotation factor $\alpha(r)$. According to the geometric model of Fig. 6.2 and the r_s positive values, it can be seen that in the azimuth multichannel sliding spotlight spaceborne SAR mode $\alpha(r) < 1$. Based on the $\alpha(r)$ value, the geometric and mathematical models of azimuth multi-channel sliding spotlight mode and azimuth multi-channel TOPS mode can be unified.

6.2.2 Azimuth Multi-Channel Reconstruction Method

From the connection between the single-channel echo signal and the multi-channel echo signal, the data received by each receiving antenna can be equivalent to the result of down sampling of conventional striped SAR data through the transfer function. The transfer functions for each receive channel are as follows:

$$H_i(f) = \exp\left\{-j\frac{d_i}{2v}2\pi f - j\frac{\pi f_r d_k^2}{4v^2}\right\} \cong \exp\left\{-j\frac{d_i}{2v}2\pi f\right\} \tag{6.1}$$

If the echo signal N received by each receiving antenna can be used to eliminate spectral aliasing and reconstruct the signal spectrum before down sampling, the equivalent conversion from the spaceborne SAR echo signal of the high-resolution multi-channel antenna to the traditional strip map SAR echo signal can be realized, which is the principle of the compensation algorithm, as shown in Fig. 6.3.

Let the ideal strip SAR echo signal azimuth spectrum be $S_0(f)$. The system will $S_0(f)$ pass through the system transfer function and $H_i(f)$ downsample, and the signal spectrum of each channel will be obtained $S_i(f)$. If the pulse repetition f_p frequency is, then $S_i(f)$ the frequency range is $\left[-f_p/2, f_p/2\right]$

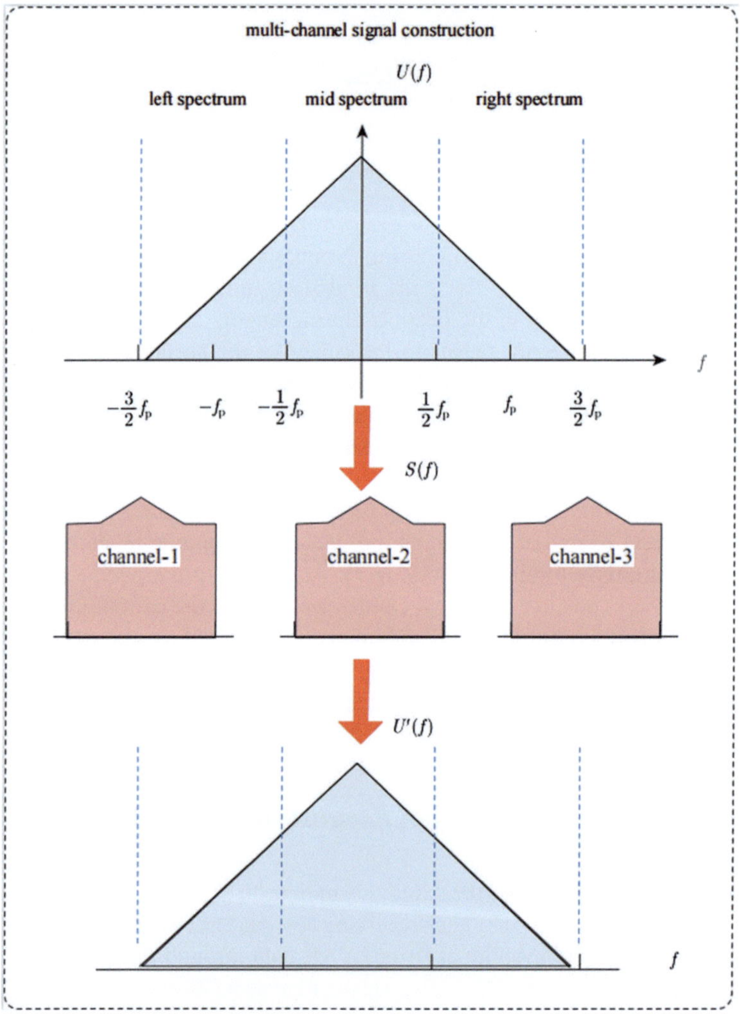

Fig. 6.3 Multi-channel signal generation and compensation

$$T_i(f) = \sum_{k=0}^{N-1} S_0\big(f + k \cdot f_p\big) \cdot H_i\big(f + k \cdot f_p\big), \quad f \in \left[-\frac{N}{2} f_p, -\frac{N}{2} f_p + f_p\right] \tag{6.2}$$

$$\begin{pmatrix} T_0(f) \\ T_1(f) \\ \vdots \\ T_{N-1}(f) \end{pmatrix}^T = \begin{pmatrix} S_0(f) \\ S_0(f + f_p) \\ \vdots \\ S_0(f + (N-1) \cdot f_p) \end{pmatrix}^T \cdot \mathbf{H}(\mathrm{f}) \tag{6.3}$$

Among them,

$$\mathbf{H}(f) = \begin{bmatrix} H_0(f) & H_1(f) & \cdots H_{N-1}(f) \\ H_0(f + f_p) & H_1(f + f_p) & \cdots H_{N-1}(f + f_p) \\ \vdots & \vdots & \vdots \\ H_0(f + (N-1) \cdot f_p) & H_1(f + (N-1) \cdot f_p) & \cdots H_{N-1}(f + (N-1) \cdot f_p) \end{bmatrix} \tag{6.4}$$

The essence of the reconstruction algorithm is to find the filter matrix $\mathbf{G}(f)$ so that

$$\begin{pmatrix} S(f) \\ S(f + f_p) \\ \vdots \\ S(f + (N-1) \cdot f_p) \end{pmatrix}^T = \begin{pmatrix} T_0(f) \\ T_1(f) \\ \vdots \\ T_{N-1}(f) \end{pmatrix}^T \cdot \mathbf{G}(f) = \begin{pmatrix} S_0(f) \\ S_0(f + f_p) \\ \vdots \\ S_0(f + (N-1) \cdot f_p) \end{pmatrix}^T \tag{6.5}$$

This requires $\mathbf{G}(f) \cdot \mathbf{H}(f) = \mathbf{I}$: Therefore, the filter matrix $\mathbf{G}(f)$ is the $\mathbf{H}(f)$ inverse of the matrix.

Since $\mathbf{H}(f)$ it is a matrix function with frequency f as the independent variable. In order not to perform an inverse operation for each frequency point, the decomposition is performed first $\mathbf{H}(f)$. $t_i = d_i/2v$ then the transfer function of each channel $H_i(f)$ can be written as

$$H_i(f) = \exp\{-j2\pi f \cdot t_i\} \tag{6.6}$$

Then, $\mathbf{H}(f) = \mathbf{P} \cdot \mathbf{Q}(f)$ where,

$$\mathbf{P} = \begin{bmatrix} 1 & 1 & \cdots 1 \\ e^{-j2\pi f_p \cdot t_0} & e^{-j2\pi f_p \cdot t_1} & \cdots e^{-j2\pi f_p \cdot t_{N-1}} \\ \vdots & \vdots & \vdots \\ e^{-j2\pi \cdot (N-1) \cdot f_p \cdot t_0} & e^{-j2\pi \cdot (N-1) \cdot f_p \cdot t_1} & \cdots e^{-j2\pi \cdot (N-1) \cdot f_p \cdot t_{N-1}} \end{bmatrix} \tag{6.7}$$

$$\mathbf{Q}(f)=\begin{bmatrix} e^{-j2\pi f\cdot t_0} & 0 & \cdots & 0 \\ 0 & e^{-j2\pi f\cdot t_1} & \cdots & 0 \\ \vdots & \vdots & & \vdots \\ 0 & 0 & \cdots & e^{-j2\pi\cdot f\cdot t_{N-1}} \end{bmatrix} \tag{6.8}$$

P is a constant matrix, $\mathbf{Q}(f)$ a diagonal matrix. Therefore, the $\mathbf{H}(f)$ inverse matrix of is

$$\begin{aligned} \mathbf{G}(f) &= \mathbf{H}^{-1}(f)=\mathbf{Q}^{-1}(f)\cdot\mathbf{P}^{-1} \\ &= \begin{bmatrix} e^{j2\pi f\cdot t_0} & 0 & \cdots & 0 \\ 0 & e^{j2\pi f\cdot t_1} & \cdots & 0 \\ \vdots & \vdots & & \vdots \\ 0 & 0 & \cdots & e^{j2\pi\cdot f\cdot t_{N-1}} \end{bmatrix}\cdot\mathbf{P}^{-1} \end{aligned} \tag{6.9}$$

Order $\mathbf{K}=\mathbf{P}^{-1}$. Yes

$$\mathbf{G}(f)=\begin{bmatrix} e^{j2\pi f\cdot t_0} & 0 & \cdots & 0 \\ 0 & e^{j2\pi f\cdot t_1} & \cdots & 0 \\ \vdots & \vdots & & \vdots \\ 0 & 0 & \cdots & e^{j2\pi\cdot f\cdot t_{N-1}} \end{bmatrix}\cdot\mathbf{K} \tag{6.10}$$

The first term on the right side of the equation can be thought of as a delay filter that compensates for the delay caused by the distance between the phase center of each receiving antenna and the phase center of the transmitting antenna.

The reconstructed spectrum is

$$\begin{pmatrix} S(f) \\ S\big(f+f_p\big) \\ \vdots \\ S\big(f+(N-1)\cdot f_p\big) \end{pmatrix}^T = \begin{pmatrix} T_0(f) \\ T_1(f) \\ \vdots \\ T_{N-1}(f) \end{pmatrix}^T\cdot\mathbf{G}(f) \tag{6.11}$$

Then there are:

$$S\big(f+k\cdot f_p\big)=\sum_{i=0}^{N-1}T_i(f)\cdot e^{j2\pi f\cdot t_i}\cdot\mathbf{K}(i,k),\quad k=0,1,2,\cdots,N-1 \tag{6.12}$$

When the $f\in[-Nf_p/2,-Nf_p+f_p]$ time comes, take all k of them, and calculate the completed spectrum according to the above formula.

6.2.3 Channel Inconsistency Error Correction

There are many types of errors in multi-channel systems, and channel errors are divided into channel error itself, satellite attitude error, satellite speed error and satellite position error according to the source. From the perspective of signal processing, channel error can be divided into amplitude phase inconsistency error, sampling delay difference and channel position inconsistency error, and phase error is further divided into fixed phase error and phase error along distance.

According to the above analysis, when there is channel inconsistency error, the expression of the azimuth multichannel SAR signal in the time domain of the range frequency domain is:

$$\begin{aligned} S_e^{(i)}(f_\tau, t) = \chi_i e^{j\delta_i} W_r(f_r) w_a\left(t - \tfrac{y_p - d_i/2 - \Delta y_i}{v}\right) \\ \exp\left\{-j\tfrac{4\pi}{c}(f_c + f_\tau) R^e\left(t + \tfrac{d_i}{2v} - \tfrac{y_p}{v}; r\right)\right\} \exp(-j2\pi f_\tau \Delta\tau_i) \end{aligned} \tag{6.13}$$

thereinto

$$R^e\left(t + \frac{d_i}{2v} - \frac{y_p}{v}; r\right) = \sqrt{(\Delta x_i - x_p)^2 + \left(vt + \frac{d_i}{2v} + \Delta y_i - y_p\right)^2 + (h + \Delta z_i - z_p)^2} \tag{6.14}$$

χ_i Represents the channel amplitude inconsistency δ_i error, represents the channel phase inconsistency $\Delta\tau_i$ error, represents the channel delay difference Δx_i, Δy_i represents the sum of the phase center measurement error of each channel and the Δz_i phase center position error of each channel caused by the attitude error, because the satellite platform can carry out relatively stable attitude control during an observation, Therefore, this part of the error can be seen as not changing with time. The multichannel spaceborne SAR channel error geometry is shown in Fig. 6.4.

The azimuth multichannel inconsistency error is represented in the two-dimensional frequency domain model as:

$$\begin{aligned} S_e^{(i)}(f_\tau, f) = \chi_i \exp\left\{j\delta_i - j\tfrac{4\pi}{\lambda}\Delta R_{i,1}\right\} \exp\left\{-j2\pi f_\tau\left(\Delta\tau_i + \tfrac{2\Delta R_{i,1}}{c}\right)\right\}\cdot \\ \sum_{k=K_{\min}}^{K_{\max}} U_0\left(f_\tau, f + k\cdot f_{prf}\right)\exp\left\{j2\pi\tfrac{d_i}{2v}\left(f + k\cdot f_{prf}\right)\right\} \end{aligned} \tag{6.15}$$

Since it $\Delta R_{i,1}$ is constant, it can be combined into the phase error and delay difference, so the above equation can be further reduced to

$$\begin{aligned} S_e^{(i)}(f_\tau, f) = \chi_i \exp\{j\delta_i\} \exp\{-j2\pi f_\tau(\Delta\tau_i)\}\cdot \\ \sum_{k=K_{\min}}^{K_{\max}} U_0\left(f_\tau, f + k\cdot f_{prf}\right)\exp\left\{j2\pi\tfrac{d_i}{2v}\left(f + k\cdot f_{prf}\right)\right\} \end{aligned} \tag{6.16}$$

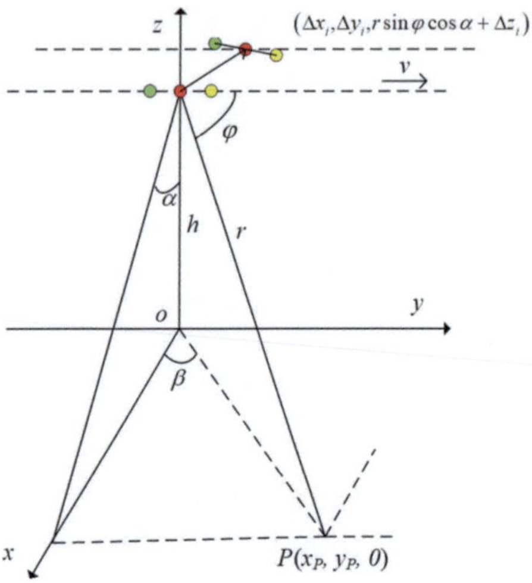

Fig. 6.4 Geometric model of spaceborne SAR channel error in azimuth multi-channel strip map

Through the analysis of the previous section, the phase inconsistency error has a great impact on the high-precision imaging of multi-channel stripe map spaceborne SAR. In order to ensure the image quality, it is necessary to study the phase inconsistency error correction method. This section begins with a brief introduction to several classic correction methods. First, the signal model that considers phase inconsistency error and noise is

$$\mathbf{S}(\tau, f) = \Gamma\mathbf{H}(f)\mathbf{U}_0(\tau, f) + \mathbf{N}(\tau, f) \tag{6.17}$$

(1) Orthogonal Subspace Method (OSM)

The principle of the orthogonal subspace method is based on the spatial spectral characteristics of the signal covariance matrix, which divides the eigenvectors into signal subspace and noise subspace. According to the theory of spectral estimation, the signal subspace and the noise subspace are orthogonal to each other. Based on this, the phase inconsistency error can be estimated.

The covariance matrix of the azimuth multichannel stripe mode spaceborne SAR echo signal with phase inconsistency error is

$$\mathbf{R} = \underset{\tau}{E}\left\{\mathbf{S}(\tau, f)\mathbf{S}(\tau, f)^H\right\} = \Gamma\mathbf{H}(f)\,\underset{\tau}{E}\left\{\mathbf{U}_0(\tau, f)\mathbf{U}_0(\tau, f)^H\right\}\mathbf{H}(f)^{-1}\Gamma^{-1} \tag{6.18}$$

For **R** eigenvalue decomposition, there is

$$R = \sum_{i=1}^{K}\alpha_i g_i g_i^H + \sum_{K}^{N}\alpha_i g_i g_i^H \tag{6.19}$$

thereinto $\alpha_1 > \alpha_2 > \cdots > \alpha_K \gg \alpha_{K+1} = \cdots = \alpha_N = \sigma_n^2$ is a eigenvalue, σ_n^2 indicates noise power, g_i Represents a feature vector. According to matrix theory, the subspace of the space-oriented vector densification with error is consistent with the subspace of the eigenvector that characterizes the signal, and the definition $\mathbf{G} = [g_1, \ldots, g_K]$ Yes:

$$span\left\{\Gamma h_{K_{\min}}(f), \ldots, \Gamma h_{K_{\max}}(f)\right\} = span\{\mathbf{G}\} = span\{g_1, \ldots, g_K\}, \tag{6.20}$$

Then define the noise vector group $\mathbf{V} = [u_{K+1}, \cdots, u_N]$, and define the cost function by using the orthogonality of the signal subspace and the noise subspace

$$J = \sum_k \left\| \mathbf{V}^H \Gamma h_k(f_a) \right\|_2 \tag{6.21}$$

thereinto $\|\bullet\|_2$ Represents the L2 norm, which theoretically exists $\widehat{\Gamma}$ make $J=0$ However, due to noise and systematic errors, there is no analytical solution. Therefore, a feasible solution can be obtained by minimizing the cost function, and the J expand, represented as

$$\begin{aligned} \min_{\widehat{\Gamma}} J &= \min_{\widehat{\Gamma}} \left\{ \sum_k h_k(f)^H \Gamma^H \mathbf{V}\mathbf{V}^H \Gamma h_k(f) \right\} \\ &= \min_x \left\{ \sum_k x^H \mathbf{Q}_k^H(f) \mathbf{V}\mathbf{V}^H \mathbf{Q}_k(f) x \right\} \end{aligned} \tag{6.22}$$

Among them, $\mathbf{Q}_k(f) = \mathrm{diag}\{h_k(f)\}$. Let, $\mathbf{Z}(f) = \sum_k Q_k^H(f)\mathbf{V}\mathbf{V}^H Q_k(f)$ which can be transformed into a quadratic programming problem with constraints, and according to matrix theory, the least squares solution of the problem is expressed as

$$\hat{\mathbf{x}} = \frac{\mathbf{Z}^{-1}(f) w}{w^T \mathbf{Z}^{-1}(f) w} \tag{6.23}$$

Among them, $w = [1, 0, 0, \ldots, 0]^T$.

(2) Signal Space Matching Method (SSCM)

The subspace of the space-oriented vector tensification with the error is consistent with the subspace of the feature vector group Zhang cheng representing the signal, so there is

$$\mathbf{G}\left(\mathbf{G}^H \mathbf{G}\right)^{-1} \mathbf{G}^H = \Gamma \mathbf{H}(f_a) \left(\mathbf{H}^H(f_a) \Gamma^H \Gamma \mathbf{H}(f_a)\right)^{-1} \mathbf{H}^H(f_a) \Gamma^H \tag{6.24}$$

In addition, $\mathbf{B} = \mathbf{G}\mathbf{G}^H$ at the same time $\mathbf{W}(f) = \mathbf{H}(f)\big(\mathbf{H}^H(f)\mathbf{H}(f)\big)^{-1}\mathbf{H}^H(f)$, there is, $\Gamma^H\Gamma = \mathbf{I_N}$, then there is $\mathbf{G}^H\mathbf{G} = \mathbf{I}_K$

$$\mathbf{B} = \Gamma\mathbf{W}(f)\Gamma^H \tag{6.25}$$

Take $\mathbf{B}$ the first column and you can get it

$$B_{i1} - \Gamma_{ii}W(f)_{i1}\Gamma_{11}^* = B_{i1} - W(f)_{i1}e^{j(\delta_i-\delta_1)} = 0 \tag{6.26}$$

Hence there is

$$\delta_i - \delta_1 = \angle\frac{B_{i1}}{W(f_a)_{i1}}, i = 1, 2, \ldots, N \tag{6.27}$$

(3) Adaptive Weighted Least Squares (AWLS)

In order to ensure that the azimuth signal ambiguity meets the system requirement, the equivalent signal bandwidth needs to be less than N·PRF after the azimuth signal reconstruction is completed. Therefore, there are some doppler units that are $f_\Delta + k_0 \cdot f_{prf}$ outside the equivalent bandwidth after reconstruction. Theoretically, the signal energy of this part is approximately zero except for noise. When there is a phase inconsistency error, some of the signal energy leaks out of the effective bandwidth. Therefore, the phase inconsistency error can be estimated by minimizing the signal energy outside the effective bandwidth.

The signal energy at the reconstructed doppler unit $f_\Delta + k_0 \cdot f_{prf}$ is expressed as

$$\left|U_0\big(\tau, f_\Delta + k_0 \cdot f_{prf}\big)\right|^2 = \left|P_{k_0}(f_\Delta)\mathbf{Y}\mathbf{S}(\tau, f_\Delta)\right|^2 \tag{6.28}$$

Rewrite $\mathbf{S}_D(\tau, f_\Delta) = \text{diag}\{\mathbf{S}(\tau, f_\Delta)\}$ and note $\mathbf{Y} = \text{diag}\{y\}$ that the above idea can be expressed as an optimization problem

$$\begin{gathered}\hat{y} = \min_y\left|P_{k_0}(f_\Delta)\mathbf{S}_D(\tau, f_\Delta)y\right|^2 = \min_y y^H\mathbf{A}^H\mathbf{A}y \\ s.t.\ |y_i| = 1,\ i = 1,\ 2,\ \ldots,\ N\end{gathered} \tag{6.29}$$

Thereinto $\mathbf{A} = P_{k_0}(f_\Delta)\mathbf{S}_D(\tau, f_\Delta)$. This equation represents a quadratic transverse module problem that can be solved by the semi-positive definite relaxation method and the eigenvalue relaxation method. The DSO method is similar to the AWLS method, so it will not be repeated.

(4) Spatial correlation coefficient method (SCCC)

The adjacent channel data is correlated processing, ignoring the difference in azimuth envelope, and the signal after azimuth time domain correlation processing has a phase related to the channel equivalent interval

$$\underset{t}{E}\left\{S^{(i-1)}(f_\tau,t)^*S^{(i)}(f_\tau,t)\right\}=\exp\left\{j2\pi f_d\frac{d_i-d_{i-1}}{2v}\right\} \tag{6.30}$$

Considering the phase inconsistency error, the above equation becomes

$$\underset{t}{E}\left\{S^{(i-1)}(f_\tau,t)^*S^{(i)}(f_\tau,t)\right\}=\exp\left\{j2\pi f_d\frac{d_i-d_{i-1}}{2v}\right\}\exp\{j(\delta_i-\delta_{i-1})\} \tag{6.31}$$

Therefore, only the doppler center frequency and phase center distance need to be known to find the channel phase inconsistency error. If the last channel data of a pulse and the data of the adjacent next pulse are also processed, the doppler center frequency can be estimated. However, this method requires more parameter accuracy, and requires good correlation between channels, which is sensitive to scene changes and system parameters.

The rotational invariance of azimuth multichannel spaceborne SAR signals is based on the van der Ment matrix **H**. If there is a phase inconsistency error, the above characteristics no longer hold, and Φ the eigen-root matrix of the matrix is no longer equal to the rotation matrix **D**.

The following is a set of simulation experiments to illustrate the effect of phase inconsistency error on the rotation matrix. Figure 6.5 shows the distribution of eigenvalues in red in the presence of phase inconsistency errors. It can be seen that the eigenvalues **D** of the rotation matrix are equally spaced on the unit circle, and due to the influence of the phase inconsistency error,Φ the eigen-roots of the matrix are randomly distributed near the unit circle. Therefore, this property can be used to Φ estimate the phase inconsistency error by minimizing the difference between the characteristic root matrix and the rotation matrix of the **D** matrix. The optimization problem can be written

$$\hat{\mathbf{x}}=\arg\min_{\delta}\|\mathbf{D}-\lambda(\Phi)\|_2 \tag{6.32}$$

where is the $\|\bullet\|_2$ L2 norm.

In practice, due to crosstalk between individual feature values, iteration using cyclic coordinate gradient descent will converge to erroneous results. To solve this problem, an effective approach is to estimate the phase inconsistency error of adjacent channels using only two adjacent channels at a time

$$\phi_n=e^{j(\delta_n-\delta_{n-1})}d_{\max}=e^{j(\delta_n-\delta_{n-1})} \tag{6.33}$$

where represents ϕ_n the difference between the phase inconsistency error of the nth channel and the nth-1st channel. Therefore,δ_n it can be solved by the following equation

$$\delta_m=\delta_{m-1}+\angle\phi_m \tag{6.34}$$

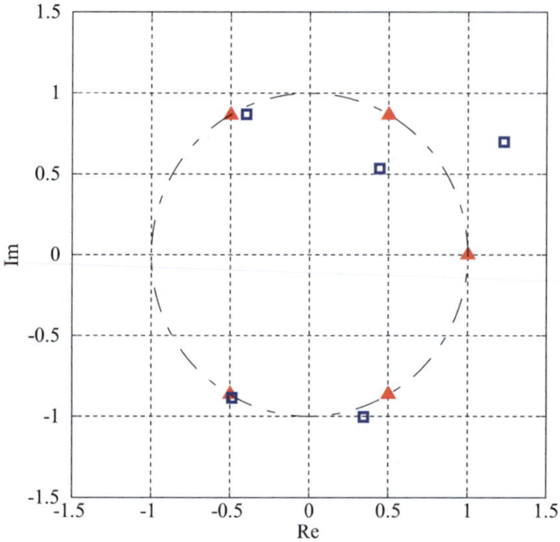

Fig. 6.5 Distribution of rotation matrix **D** and characteristic Φ root of matrix

where,$\delta_0 = 0$ $\angle\cdot$ represents the phase angle operation. Through this rapid estimation strategy, on the one hand, the crosstalk between the errors can be eliminated, and iterative operations can be avoided, which improves the estimation efficiency.

To verify the effectiveness of the phase inconsistency error estimation algorithm presented in this section. The added phase inconsistency error and the estimation result are shown in Table 6.1. The IOS method and AWLS method are used in the simulation to compare with the algorithm proposed in this paper. It is clear from the results that the estimation results of the ESPRIT method proposed in this paper are better than the other two counterpart algorithms.

The following uses a set of measured data to verify the effectiveness of the proposed ESPRIT method. Without correcting the phase error between channels, the signal spectrum reconstruction is carried out by the transfer function method,

Table 6.1 Estimation results of phase inconsistency error

passage	1	2	3	4	5	6
Add error (°)	0	40	−30	18	35	−5
ESPRIT method (°)	0	40.35	−29.58	18.86	35.44	−4.53
Estimation error (°)	0	0.35	0.42	0.86	0.44	0.47
IOS method (°)	0	42.02	−28.39	17.72	33.39	−5.99
Estimation error (°)	0	2.02	1.61	0.28	1.61	0.99
AWLS method (°)	0	44.10	−30.43	10.78	21.50	−10.60
Estimation error (°)	0	4.10	0.43	7.22	13.50	5.60

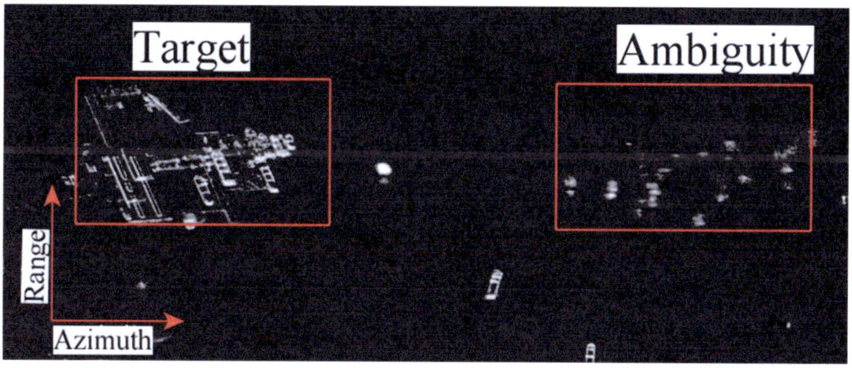

(a) Before the channel inconsistency error is eliminated

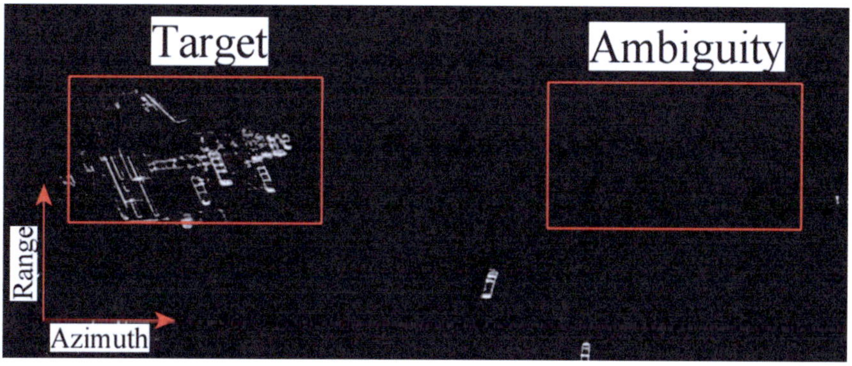

(b) After the channel inconsistency error is eliminated

Fig. 6.6 Before and after comparison of phase inconsistency error compensation between channels

and then the imaging process is performed by CSA, and the results obtained are obviously blurred, as shown in Fig. 6.6a. The ESPRIT method was then used to estimate 0°, 123.2°, 29.2° and 161.0°. Phase error compensation and imaging are then performed, and the resulting image is well-focused and free of false targets, as shown in Fig. 6.6b.

6.2.4 Three-Step High-Resolution Imaging Method

This section studies the problems faced by the imaging processing of azimuth multi-channel sliding spotlight beam spaceborne SAR mode, analyzes the advantages and disadvantages of the range model, studies the data preprocessing methods based on range walking, range compression, improved error compensation and non-uniform derotation, improves the traditional ωk focusing algorithm, proposes a spatial error compensation method for imaging parameters, and gives the block processing process.

The range model is a description of the geometric relationship between the satellite and the observation target, which is directly related to the processing accuracy of the imaging algorithm. With the improvement of resolution and observation swath, higher requirements are put forward for the accuracy of the range model. An excellent range model should provide a concise and efficient description of the satellite-ground relationship under the condition of ensuring a certain accuracy to ensure the accuracy and efficiency of the algorithm implementation. Therefore, this section summarizes the existing range models in spaceborne SAR imaging processing and analyzes the range models suitable for azimuth multi-channel sliding spotlight spaceborne SAR mode.

(1) Strabismus equivalent range model

In the processing of spaceborne SAR data, the strabismus equivalent model is the most widely used. The model assumes that satellites fly in a straight line and have a constant speed of motion

$$R(t; r) = \sqrt{r^2 + (vt)^2 - 2rvt \cos \varphi} \tag{6.35}$$

where the equivalent speed v and equivalent range viewing angle φ It can be calculated by first-order Doppler parameters and second-order Doppler parameters. The advantage of the strabismus equivalent model is that the slant range history is represented by two parameters by using the linear flight assumption, and the main components of each order coefficient in the slant range history are simulated, which is a concise and efficient distance model. Based on the strabismus equivalent model, a series of imaging focusing algorithms are applied to spaceborne SAR processing.

(2) Improved strabismus equivalent distance model

Since the trajectory of the satellite during space flight is not an ideal straight line, there is a certain error in the strabismus equivalent model. Based on this, by adding a time term to compensate for the cubic term error caused by the bending of the satellite flight orbit, an improved equivalent range model can be obtained, and its expression is

$$R(t; r) = \sqrt{r^2 + (vt)^2 - 2rvt \cos \varphi} + \Delta_l t \tag{6.36}$$

thereinto Δ_l is the coefficient of the primary term introduced. At this point, the parameter in the formula (6.36) requires third order doppler to solve. The medium parameter requires third-order doppler to solve. The improved strabismus equivalent distance model, on the basis of inheriting the advantages of the strabismus equivalent distance model, improves the model accuracy by adding a degree of freedom. However, the model only compensates for the linear part of the cubic term error outside the strabismus equivalent distance model with only one linear term, which limits its application in high-resolution imaging processing.

(3) High-order range model

The high-order range model directly uses the high-order doppler parameters to approximate the true range between the satellite and the target, so as to achieve high description accuracy, and its expression is

$$R(t;r) = r - \frac{\lambda f_d}{2}t - \frac{\lambda f_r}{4}t^2 - \frac{\lambda f_{r3}}{12}t^3 - \frac{\lambda f_{r4}}{48}t^4 - \cdots \tag{6.37}$$

Among them, f_{r3} the third-order Doppler parameter, f_{r4} is the fourth-order Doppler parameter. The high-order range model theoretically has high accuracy, but it can be seen that its accuracy is related to the selected order, and it needs to be selected according to the imaging processing requirements. Compared with the strabismus equivalent model, the higher-order range model considers the slant range error above the third order, but ignores the higher-order error above the selected order. From this point of view, the description of the slant range history by the high-order distance model is not efficient and is a clumsy model.

(4) Modified strabismus equivalent range model

The modified strabismus equivalent model still assumes that the satellite is flying in a straight line, but introduces an acceleration variable into the model, which considers the satellite flight process to conform to uniform acceleration of linear motion, and its expression is

$$R(t;r) = \sqrt{r^2 + \left(vt + \frac{a_r}{2}t^2\right)^2 - r\left(2vt + at^2\right)\cos\varphi} \tag{6.38}$$

Where a_r represents the equivalent acceleration variable. The parameters in the modified strabismus equivalent model need to be solved by fourth-order doppler. Compared with the strabismus equivalent distance model, the modified strabismus equivalent model describes the flight trajectory of the satellite more accurately, and realizes the fourth-order approximation of the range history by adding a variable, while retaining the approximation of the higher-order coefficients of the squint equivalent model.

(5) Strabismus equivalent distance model with higher-order compensation

The strabismus equivalent distance model of high-order compensation is also established based on the squint equivalent model, which considers that the satellite trajectory is curved and no longer flies in a straight line, so the high-order range term is added on the basis of the equivalent model to describe the range between the satellite and the target, and its expression is

$$R(t;r) = \sqrt{r^2 + (vt)^2 - 2rvt\cos\varphi} + \frac{\lambda}{12}\Delta f_{r3}t^3 + \frac{\lambda}{48}\Delta f_{r4}t^4 + \frac{\lambda}{240}\Delta f_{r5}t^5 + \cdots \tag{6.39}$$

where Δf_{r3}, Δf_{r4} and Δf_{r5} so represent the compensation coefficients for errors in third, fourth, and fifth order doppler parameters. The parameters can be calculated according to the selected order using the corresponding higher-order doppler parameters. The high-order compensated strabismus equivalent range model improves the accuracy of the strabismus equivalent model, while retaining the basic form of the strabismus equivalent model, which can be well combined with various algorithms derived based on the strabismus equivalent model, which is an efficient and convenient distance model. For most cases, the range model compensation to the fourth order is sufficient to meet the specification requirements.

Based on the above analysis, it can be seen that for the azimuth multichannel sliding spotlight spaceborne SAR, the modified equivalent model and the equivalent model based on high-order compensation can meet the requirements of high-resolution and high-precision imaging processing. Since the equivalent model based on high-order compensation is relatively simple and easier to combine with existing imaging algorithms, it was selected as the range model of azimuth multichannel sliding spotlight spaceborne SAR.

The three-step imaging processing algorithm is proposed based on doppler time–frequency characteristic analysis, and its theoretical basis is: the doppler bandwidth of echo signal in conventional imaging mode is composed of doppler bandwidth corresponding to 3 dB beam width of the antenna beam and doppler frequency shift introduced by azimuth beam rotation in satellite observation time. Based on this conclusion, a three-step imaging processing algorithm is derived.

Before the imaging processing algorithm is proposed, the spatial characteristics of the distance direction of the mixing degree factor are first analyzed. In fact, the angular velocity of the spaceborne SAR imaging mode in azimuth beam scanning ω_φ does not change with the distance gate, but the instantaneous azimuth of the target and satellite changes with the distance gate, so the antenna pattern is corrected to

$$W_a(t) = \sin c^2\left(\frac{L}{\lambda}\cdot\left(\frac{vt}{R_{rs} - R_{rt}(r)} - \omega_\varphi t\right)\right) = \sin c^2\left(\frac{L}{\lambda}\cdot\frac{vt}{R_{rs} - R_{rt}(r)}\cdot\frac{R_{rt}(r)}{R_{rs}}\right) \tag{6.40}$$

where is equal to the sum of the shortest distance from the rotation point to the satellite and the $R_{rt}(r)$ shortest distance from the R_{rs} satellite to the target at different locations $R_{st}(r)$, while still defining the positive direction for upward and negative for vice versa $R_{st}(r) = -r$. It can be seen that the degree factor of blending is a function of the range r, in which case the degree factor is corrected to

$$Y(r) = \frac{R_{rt}(r)}{R_{rs}} = \frac{R_{rs} + R_{st}(r)}{R_{rs}} \tag{6.41}$$

Then modify the azimuth signal expression to

$$S_A(t-t_A;t_A,r)=\mathrm{rect}\left(\frac{Y(r)vt-vt_A}{X_{\Delta\theta}}\right)\mathrm{rect}\left(\frac{t}{T}\right)\mathrm{rect}\left(\frac{vt_A}{X_S}\right)\exp\left\{-j\pi k_r(t-t_A)^2\right\} \tag{6.42}$$

Using the stationary phase principle, the azimuth FFT transformation of the above equation is obtained

$$S_{f_a}(f_a;t_A,r)=\mathrm{rect}\left(-\frac{f_a-k_r\frac{Y(r)-1}{Y(r)}t_A}{B_{\Delta\theta}\frac{1}{Y(r)}}\right)\mathrm{rect}\left(-\frac{f_a-k_rt_A}{k_rT}\right)\mathrm{rect}\left(\frac{vt_A}{X_s}\right)\exp\left\{j\pi\frac{f_a^2}{k_r}-2\pi f_at_A\right\} \tag{6.43}$$

The doppler bandwidth of a target at different distances to a position is a function of the range, i.e

$$B_a=\frac{B_{\Delta\theta}}{|Y(r)|} \tag{6.44}$$

However,$k_\omega T$ the sum $B_{\Delta\theta}$ is not a function of the range, so the correction of the mixing degree factor does not affect the doppler bandwidth of the azimuth full-scene scene B_s, and the following important conclusions can be obtained: the doppler bandwidth of the single-point target at different range changes; Azimuth beam scanning does not affect the full-scene doppler bandwidth, that is, the full-scene doppler bandwidth at different ranges remains unchanged, which is an important prerequisite for the implementation of subsequent imaging algorithms.

Based on the above conclusions, the processing method of the three-step imaging processing algorithm of spaceborne SAR is analyzed.

(1) The first step is De-rotation

When designing the spaceborne SAR system, it is necessary to meet the constraint that the pulse repetition rate f_{prf} is greater than that, but in the sliding beam gathering mode $B_{\Delta\theta}$, the azimuth full-scene doppler bandwidth is usually greater than B_s that, and direct azimuth f_{prf} FFT will cause azimuth spectrum aliasing, so the first step of the three-step imaging processing algorithm is de-rotation operation to compensate for the doppler translation introduced by antenna beam scanning.

Define the de-rotation factor as

$$H_{De-rotation}(t)=\exp\left\{j\pi k_\omega t^2\right\} \tag{6.45}$$

Convolution of the signal $S_A(t-t_A;t_A,r)$ and the de-rotation factor

$$\begin{aligned}S_1(t-t_A;t_A,r)&=S(t-t_A;t_A,r)\otimes_t H_{De-rotation}(t)\\&\approx rect\left[\frac{vt_A}{X_{swath}}\right]\cdot rect\left[-\frac{t}{\frac{\lambda R_{rs}}{Lv}}\right]\cdot rect\left[\frac{t-\frac{R_{rs}}{r}t_A}{T\left(\frac{r-R_{rs}}{r}\right)}\right]\cdot\exp\left\{j\pi k_e(r)(t-t_A)^2\right\}\end{aligned} \tag{6.46}$$

where is defined $k_e(r)$ as

$$k_e(r) = \frac{2v^2}{\lambda(R_{rs} - r)} \tag{6.47}$$

From the spatial geometric relationship model, it is not difficult to obtain the following relationship

$$\left|T\left(\frac{r - R_{rs}}{r}\right)\right| > \left|\frac{\lambda R_{rs}}{Lv}\right| \tag{6.48}$$

It is not difficult to draw the following conclusions through the first step:

(1) After the first step of operation, the targets in different positions in azimuth completely coincide, and the width of the time domain T' is

$$T' = \frac{\lambda |R_{rs}|}{Lv} \tag{6.49}$$

At this point, the azimuth target coincides completely, so the azimuth weighting can be performed after the de-rotation operation.

(2) After the first step of operation, the sampling rate of the azimuth signal has changed, and the sampling rate after the change f'_{prf} :

$$f'_{prf} = \frac{N|k_\omega|}{f_{prf}} \tag{6.50}$$

where N is the number of azimuth FFT points.

(3) In order to avoid azimuth aliasing, it is necessary to properly carry out the zero-padding operation, that is, to satisfy, $f'_{prf} > B_s$therefore:

$$N > f_{prf}\frac{\lambda |R_{rs}|}{Lv} + f_{prf}T = N_0 + N_a \tag{6.51}$$

Among them, it is the N_a number of azimuth points of the original data, and the number N_0 of zero points is added in the azimuth direction, which is not difficult to find that it N_0 has nothing to do with the same distance.

(4) After the zeroing operation, the azimuth time domain width T_1 is

$$T_1 = \frac{f_{prf}}{|k_\omega|} > \frac{B_{\Delta\theta}}{|k_\omega|} = \frac{\lambda |R_{rs}|}{Lv} = T' \tag{6.52}$$

It is not difficult to see that the azimuth time domain width is T_1 greater than the signal width, so the de-rotation operation does not cause time domain aliasing. That is, after the first step of the de-rotation operation, aliasing occurs in both the azimuth time and frequency domains.

(5) In practice, the convolution operation of the first step of de-rotation can be completed by multiplication and FFT operation to improve the processing efficiency.
(6) For strip map mode and scan mode, since the azimuth beam does not rotate, it is k_ω, so it is equivalent to the first step and an impulse response for convolution, and the signal does not change. In practice, data processing in stripe mode and scan mode can be handled without the first step.
(2) The second step is focusing

After the first step of de-rotation, the signal is spread in azimuth, which can be decoupled and focused. At present, the most used focusing algorithm cores include RD algorithm, CS algorithm, $\omega - k$ etc. Considering the accuracy and efficiency of the algorithm, the CS algorithm is undoubtedly the best processing core, which does not require interpolation operations, and only relies on Fourier transform and multiplication operations to complete the focusing of the image, so the CS algorithm processing core is preferred.

The signal is transformed into the range doppler domain, and the linear scaling technique is used to compensate for the range cell migration correction (RCMC), and the scaling factor is

$$\Phi_1(\tau, f_a; r_{ref}) = \exp\left\{-j\pi b_r(f_a; r_{ref})C_s(f_a)(\tau - \tau_{ref})^2\right\} \tag{6.53}$$

thereinto

$$b_r(f_a; r_{ref}) = \frac{bc^2\left[1 - \left(\frac{\lambda f_a}{2v}\right)^2\right]^{\frac{3}{2}}}{c^2\left[1 - \left(\frac{\lambda f_a}{2v}\right)^2\right]^{\frac{3}{2}} + 2br_{ref}\sin\varphi_{ref}\lambda\left(\frac{\lambda f_a}{2v}\right)^2} \tag{6.54}$$

$$C_s(f_a) = \frac{\sin\varphi_{ref}}{\sqrt{1 - \left(\frac{\lambda f_a}{2v}\right)^2}} - 1 \tag{6.55}$$

$$\tau_{ref} = \frac{2}{c}r_{ref}[1 + C_s(f_a)] \tag{6.56}$$

Thereinto r_{ref} is reference distance,b indicates the frequency modulation slope of the signal, φ_{ref} represents the equivalent angle of view at the reference range. The converted signal is transformed into a two-dimensional frequency domain by range to Fourier transform, multiplied by the range compensation factor $\Phi_2(f_\tau, f_a)$ Complete consistent range migration compensation and range focusing processing:

$$\Phi_2(f_\tau, f_a) = \exp\left\{-j\pi \frac{f_\tau^2}{b_r(f_a; r_{ref})[1 + C_s(f_a)]}\right\} \exp\left\{j\pi \frac{4}{c} f_\tau r_{ref} C_s(f_a)\right\} \tag{6.57}$$

According to the characteristics of the 0.1 m sliding spotlight mode, combined with the previous analysis results, the following operations are performed for orbit bending, "stop-and-go" model error and two-dimensional spectrum error:

(1) Track bending compensation

Most of the existing imaging algorithms are derived based on the approximate distance model, but with the increase of resolution and observation angle, the satellite flight trajectory will be bent, so this approximate distance model has errors, which directly affects the imaging focus quality. Therefore, the numerical calculation method is proposed to obtain the orbital residual high-order error term, and compensate for the orbital bending phenomenon in the range doppler domain.

(2) "Stop–go" approximate model error compensation

The current imaging focusing algorithm mostly adopts the "stop-walk" model, but as the resolution increases, this approximation processing will lead to target defocusing, so it is necessary to compensate for the error of the "stop-walk" approximation model. Considering the need for accurate focusing in the whole scene, this approximation error is compensated in the two-dimensional frequency domain.

(3) High-order spectrum error spectrum compensation

In the two-dimensional spectrum of the echo signal, due to the presence of a high-order residual doppler term, a higher-order residual phase error will be introduced, which will affect the effect of azimuth/range decoupling. The higher-order residual doppler term will introduce the residual phase error $4\pi\Delta r/\lambda$ as follows, and it is derived that the expression of the higher-order doppler residual phase error in the two-dimensional frequency domain is:

$$\begin{aligned}
\Phi_3(f_\tau, f_a; r) &= \exp\left\{j\frac{4\pi}{3} \cdot \Delta f_3 \cdot t_k^3\right\} \\
\Phi_4(f_\tau, f_a; r) &= \exp\left\{j\pi \cdot \Delta f_4 \cdot t_k^4\right\} \\
\Phi_5(f_\tau, f_a; r) &= \exp\left\{j\frac{4\pi}{5} \cdot \Delta f_5 \cdot t_k^5\right\} \\
\Phi_6(f_\tau, f_a; r) &= \exp\left\{j\frac{2\pi}{3} \cdot \Delta f_6 \cdot t_k^6\right\}
\end{aligned} \tag{6.58}$$

Among them, $t_k = r\left(\frac{\cos\varphi}{v} - \frac{f_a \sin\varphi}{v\sqrt{4v^2\left(\frac{1}{\lambda} + \frac{f_\tau}{c}\right)^2 - f_a^2}}\right)$.

After the range compensation is completed, the signal is inverse Fourier transformed, and the signal is transformed into a range doppler domain multiplied by the azimuth compensation factor $\Phi_3(\tau, f_a)$ to complete the compensation of the residual phase introduced by the azimuth hyperbolic phase and linear scaling:

$$\begin{aligned}\Phi_3(\tau, f_a) = &\exp\left\{-j\pi\tfrac{4r}{\lambda}\left[1-\sin\varphi\sqrt{1-\left(\tfrac{\lambda f_a}{2v}\right)^2}\right]\right\}\cdot\exp\{j\pi\tfrac{2r}{v}f_a\cos\varphi\}\\ &\cdot\exp\left\{j\pi\tfrac{4}{c^2}b_r(f_a;r_{ref})[C_s(f_a)+1]C_s(f_a)\left(r\tfrac{\sin\varphi}{\sin\varphi_{ref}}-r_{ref}\right)^2\right\}\end{aligned} \tag{6.59}$$

(3) Step 3: De-ramp

After completing range compression, range migration correction, secondary distance compression and azimuth hyperbolic phase compensation, de-rotation residual phase compensation and azimuth scaling operations are still performed in the range doppler domain. Define the third step de-ramp factor as:

$$H_{Scaling}(f_a) = \exp\left\{j\pi\frac{f_a^2}{k_\omega}\right\}\cdot\exp\left\{-j\pi\frac{f_a^2}{k_e}\right\} \tag{6.60}$$

Among them, the first item completes the de-rotation residual phase compensation, and the second completes the azimuth towards dechirp,

$$k_e = \frac{2v^2}{\lambda(R_{rs}-r_0)} \tag{6.61}$$

After the first step of de-rotation, the signal coincides in the azimuth time domain, and the time–frequency characteristics of the signal currently are like the data in the scan mode. Firstly, the expression of the azimuth signal in the frequency domain after de-rotation is derived, and the signal is transformed into the frequency domain by using the principle of stationary phase.

$$\begin{aligned}S'_{f_a}(f_a;t_A,r) = &\operatorname{rect}\left(-\tfrac{f_a+k_e(r)t_A}{B_{\Delta\theta}\frac{1}{Y(r)}}\right)\cdot\operatorname{rect}\left(-\tfrac{f_a-k_rt_A}{k_rT}\right)\cdot\operatorname{rect}\left(\tfrac{vt_A}{X_s}\right)\\ &\cdot\exp\left\{j\pi\tfrac{f_a^2}{-k_e(r)}-2\pi f_at_A\right\}\end{aligned} \tag{6.62}$$

Because the $k_e(r) = -k_r(Y(r)-1)/Y(r)$ modulation frequency of the secondary phase of the signal changes from k_r to $-k_e$ and coincides in the time domain. In the second step of focusing algorithm processing, including distance migration, secondary compression, etc. are all based on reference k_r, so the residual phase $\Delta\Phi_1$ in frequency domain expression is:

$$\Delta\Phi_1 = -j\pi f_a^2\left(\frac{1}{k_e(r)}+\frac{1}{k_r}\right) = -j\pi\frac{f_a^2}{k_\omega} \tag{6.63}$$

The first compensates for residual phase $\Delta\Phi_1$. The second item completes the signal scaling, in which case the signal expression is:

$$S''_{fa}(f_a;t_A,r)=\mathrm{rect}\left(-\frac{f_a+k_e(r)t_A}{B_{\Delta\theta}\frac{1}{Y(r)}}\right)\cdot\mathrm{rect}\left(-\frac{f_a-k_rt_A}{k_rT}\right)\cdot\mathrm{rect}\left(\frac{vt_A}{X_s}\right)\cdot\exp\left\{j\pi\frac{f_a^2}{-k_e}-2\pi f_at_A\right\}\tag{6.64}$$

Using the stationary phase principle to transform the above equation into the time domain, we can obtain:

$$S_2(t-t_A;t_A,r)=\mathrm{rect}\left[\frac{vt_A}{X_{swath}}\right]\cdot\mathrm{rect}\left[-\frac{t-\frac{r_0-r}{R_{rs}-r}t_A}{\frac{\lambda R_{rs}}{Lv}}\right]\cdot\mathrm{rect}\left[\frac{t-\frac{R_{rs}+r-r_0}{r}t_A}{T\left(\frac{r_0-R_{rs}}{r}\right)}\right]\cdot\exp\{j\pi k_e(t-t_A)^2\}\tag{6.65}$$

Comparing the denominators of the second and third terms of the above equation,

$$\left|\frac{\lambda R_{rs}}{Lv}\right|=\left|\frac{\lambda r}{Lv}\frac{R_{rs}}{r}\right|=\left|\frac{\lambda r}{Lv}\frac{R_{rs}}{r_0-R_{rs}}\frac{r_0-R_{rs}}{r}\right|=\frac{X_{\Delta\theta}}{YT}\left|\frac{r_0-R_{rs}}{r}\right|\tag{6.66}$$

Because $X_{\Delta\theta}<YT$,the second term in the equation is decisive, the time domain broadening is:

$$\Delta T=\frac{X_s}{v}\cdot\left|\frac{r_0-r}{R_{rs}-r}\right|=\frac{X_s}{v}\cdot\left|\frac{\Delta r}{R_{rt}+\Delta r}\right|=\left(Y(r)T-\frac{X_{\Delta\theta}}{v}\right)\cdot\left|\frac{\Delta r}{R_{rt}+\Delta r}\right|\tag{6.67}$$

The following further proves that the broadening that occurs by first analyzing the accumulated complete target is not difficult to obtain by the graphical relationship, and the doppler instantaneous frequency of the target N and N' are equal, so the following relationship can be obtained:

$$k_e(r)\left(\frac{T'}{2}-t_A\right)=k_e(t-t_A)\tag{6.68}$$

So it's not hard to get:

$$\Delta T'=\left[2t_A\left(1-\frac{k_e(r)}{k_e}\right)+\frac{k_e(r)}{k_e}T'\right]-T'=(2t_A-T')\cdot\left(1-\frac{k_e(r)}{k_e}\right)\tag{6.69}$$

Considering that the accumulation of incomplete points may also affect the image, further analysis of the time domain broadening caused by the accumulation of incomplete points, doppler instantaneous frequency for the same target P and P' are equal, then:

$$k_e(r)\left(\frac{T'}{2}-t_D\right)=k_e(t-t_D) \tag{6.70}$$

Therefore, it is not difficult to obtain that the broadening caused by the accumulation of incomplete points is:

$$\Delta T=\left(2t_D-T'\right)\cdot\left(1-\frac{k_e(r)}{k_e}\right) \tag{6.71}$$

In fact, the points of incomplete accumulation need to be discarded after the final processing, so by $t_D=t_A+T'/2$,it is guaranteed that the points of complete accumulation are not affected, which can be rewritten as:

$$\Delta T=2t_A\cdot\left(1-\frac{k_e(r)}{k_e}\right) \tag{6.72}$$

It is not difficult to obtain from the schematic diagram of spatial geometric relations $2t_A=X_s/v$

$$\begin{aligned} t_A &= \frac{1}{2}\left(B_s-\frac{B_{\Delta\theta}}{Y(r)}\right)\frac{1}{k_e(r)}=\frac{1}{2}\left(B_{\Delta\theta}+k_\omega T-\frac{B_{\Delta\theta}}{Y(r)}\right)\frac{1}{k_e(r)} \\ &= \frac{1}{2}\left(\frac{k_\omega T}{k_e(r)}+\frac{B_{\Delta\theta}}{k_e(r)}\left(1-\frac{1}{Y(r)}\right)\right)=\frac{1}{2}\left(Y(r)T-\frac{X_{\Delta\theta}}{v}\right)=\frac{1}{2}\frac{X_s}{v} \end{aligned} \tag{6.73}$$

In summary, as shown in Fig. 6.7, the time domain broadening caused by dechirp is:

$$\Delta T=\frac{X_s}{v}\left(1-\frac{k_e(r)}{k_e}\right)=\frac{X_s}{v}\left(\left|\frac{\Delta r}{R_{rt}+\Delta r}\right|\right) \tag{6.74}$$

Compensation quadratic term:

$$H_{Deramp}(t)=\exp\left\{-j\pi k_e t^2\right\} \tag{6.75}$$

Finally, the FFT transformation can complete the final focusing process, and the target t_A is in the azimuth position $k_e t_A$. At this time, the frequency-domain sampling interval is k_ω/f_{prf}, then the position (dimensionless) of the target offset in the image is $k_e t_A/(k_\omega/f_{prf})$, which defines the equivalent sampling rate at this time as f''_{prf}

$$\frac{k_e t_A}{f''_{prf}(k_\omega/f_{prf})}=t_A \tag{6.76}$$

Further simplification yields:

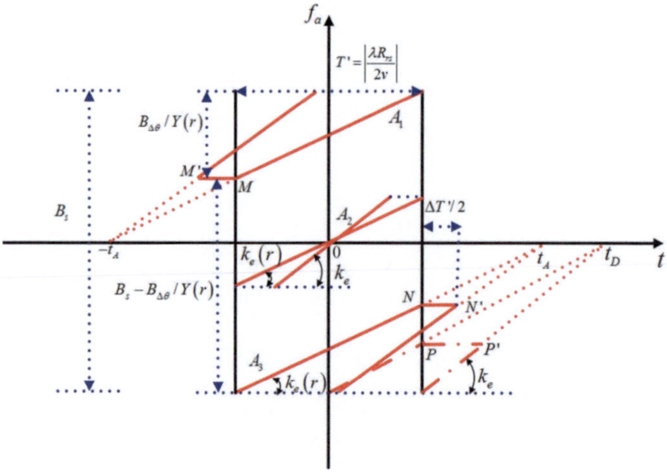

Fig. 6.7 Schematic diagram of directional expansion to the time domain of deramp operation

$$f''_{prf} = \frac{k_e}{(k_\omega/f_{prf})} = \frac{f_{prf}}{Y} \tag{6.77}$$

Therefore, the azimuth output time range is:

$$T'' = Y\frac{N}{f_{prf}} > Y\frac{N_a}{f_{prf}} = YT > X_s \tag{6.78}$$

Therefore, no aliasing will occur in the final image domain.

From the previous analysis, with the improvement of resolution, the synthesis aperture time becomes longer, and the impact of errors caused by nonlinearity of flight trajectory and errors caused by ignoring high-order terms of the two-dimensional spectrum on imaging quality will be highlighted. Therefore, it needs to be compensated during the three-step imaging process, and the corrected three-step imaging processing algorithm is shown in Fig. 6.8.

Figure 6.9 and Table 6.2 show the imaging processing results of point targets, and it can be seen from the simulation results that the processing algorithm can complete the precise focusing processing of ground targets.

Through the 0.1 m simulation study, the effectiveness of the imaging algorithm is verified, which lays a foundation for the subsequent error analysis research and geometric correction algorithm research.

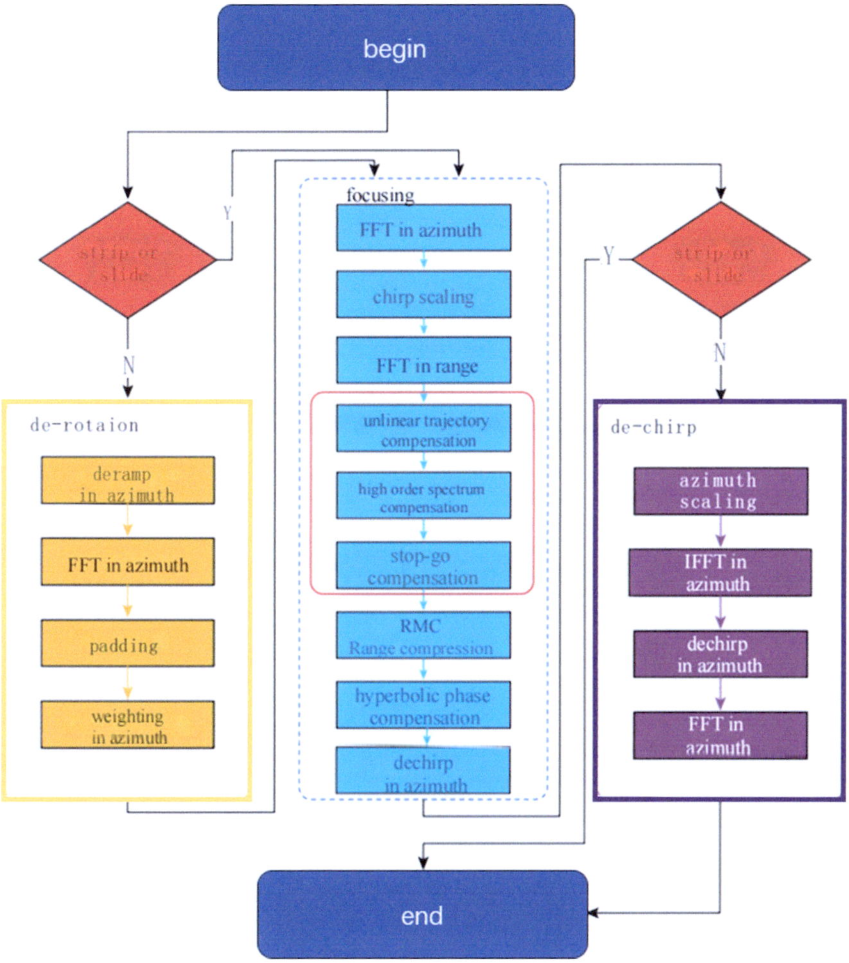

Fig. 6.8 Three-step imaging processing algorithm

6.3 Spaceborne Ka-Band SAR Interorbital Interferometry

Limited by antenna volume and baseline length, the spaceborne radar interferometry system that has been applied on a large scale requires repeat passes or distributed companion to obtain effective interferometric information. Millimeter wave SAR greatly reduces the required antenna aperture at the same beam width, so that interferometry and even multi-mode, multi-baseline InSAR can be achieved by a single satellite [4–6]. Under the general trend of expanding the spectrum of space-based SAR, millimeter-wave-based InSAR will be practiced on a large scale, which will meet the shortcomings of existing low-band interferometry technologies in marine and glacier dynamic monitoring. This section will process interferometry signals for

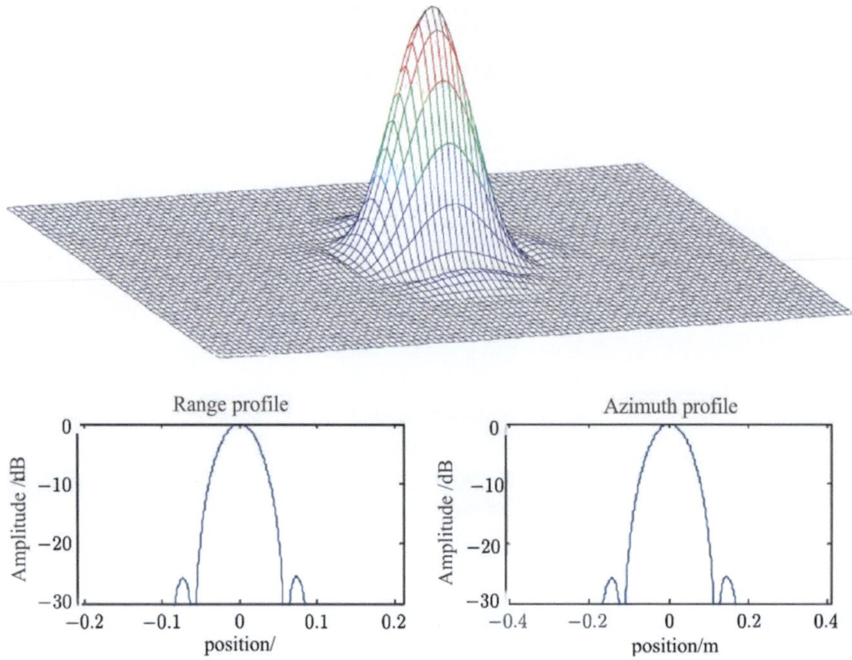

Fig. 6.9 Scene imaging results

Table 6.2 Image quality evaluation results

	Azimuth			Range		
Target number	Resolution (m)	Peak sidelobe ratio (dB)	Integral sidelobe ratio (dB)	Resolution (m)	Peak sidelobe ratio (dB)	Integral sidelobe ratio (dB)
Target 1	0.0891	−25.436	−20.459	0.0454	−25.601	−19.938
Target 2	0.0888	−25.721	−20.451	0.0452	−25.382	−19.763
Target 3	0.0902	−26.119	−21.451	0.0455	−25.670	−20.115
Target 4	0.0913	−26.499	−22.139	0.0456	−25.448	−20.127
Target 5	0.0890	−25.635	−20.394	0.0454	−25.589	−19.929

airborne millimeter wave multi-channel measured data, and establish a full-process millimeter wave dual-mode interferometric information extraction process. And the performance are quantitatively analyzed.

6.3.1 Principle of InSAR

The final appeal of SAR interferometry is absolute elevation acquisition accuracy, and in the actual technical route, due to the actual transfer relationship of the target elevation value, it has undergone the progressive process of terrain geometry, slant range, transmission delay, phase, phase difference, range difference, viewing angle, and measurement height, which will inevitably be affected by noise, decoherence, model accuracy and algorithm efficiency [7]. Among all parameters, the core one is the interferometry baseline, and in a limit situation, the single-transmit and single-receive SAR system also has a baseline of zero length, which can obtain two sets of completely consistent signals; Under the other extremely assumption, the InSAR baseline tends to be infinite, and even if all noise and processing errors can be avoided, all spatial coherence between signals will be lost, and the interferometry phase cannot be extracted [8, 9]. A schematic diagram of InSAR expressed by the range geometry is shown in Fig. 6.10.

In Fig. 6.10, B the baseline length is the baseline; λ is the wavelength; H is the platform height; r_1 and r_2 the range of primary and auxiliary data acquisition, respectively; θ is the angle of incidence; Δr the difference between the main and auxiliary ranges, because r_1 it is much larger Δr, is considered r_1 and r_2 parallel; h is the target height. Then it is easy to derive the formula according to the geometric relationship and the echo delay-phase relationship of electromagnetic wave transmission:

$$\Lambda r = r_2 - r_1 \tag{6.79}$$

$$\Delta r = \frac{\lambda}{2\pi}\Phi \tag{6.80}$$

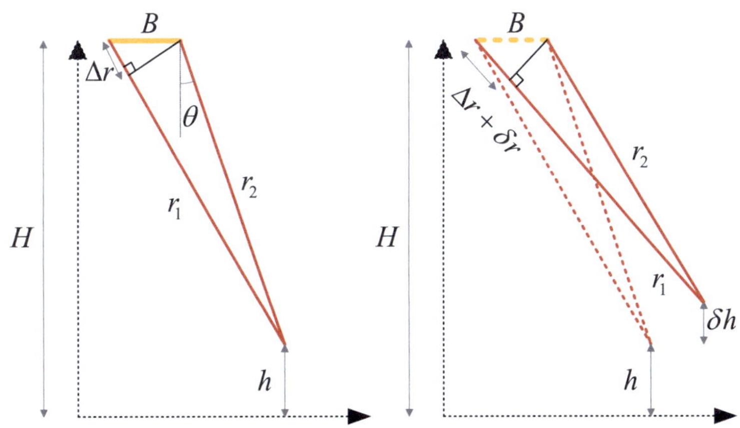

Fig. 6.10 InSAR geometry

thereinto, Φ Primary and auxiliary signal phase difference. Further, to obtain the target elevation information, the formula transforms the amount of angle transformed by the phase information:

$$\theta = acr\sin\left(\frac{\lambda}{2\pi}\frac{\Phi}{B}\right) \tag{6.81}$$

And under this determined geometry, the target elevation value is extracted from the target height-platform height difference:

$$h = H - r\cos(\theta) \tag{6.82}$$

Random noise in interferometry introduces phase error into the measurement result $\delta\Phi$, which is converted into an elevation error amount:

$$\delta h = \lambda r\tan(\theta)/2\pi B \cdot \delta\Phi \tag{6.83}$$

Under the small angle of incidence, it $x = r\tan(\theta)$ represents the ground range, that is, the error amount increases linearly along the width of the width, and it is easy to obtain the relationship between the band, the baseline length and the error transfer coefficient. However, baseline growth also brings further spatial decoherence, and the wavelength-to-baseline ratio shown by JPL in the SRTM program 9×10^{-4} is an appropriate value for analyzing the optimal baseline length [10–12]. For the optimal baseline argument without expanding the analysis here, Fig. 6.11 shows the optimal baseline values for different bands.

Under the same phase difference, according to the λ/B transfer coefficient, in order to obtain the same elevation error at the same ground distance, the Ka-band (wavelength 0.86 cm) baseline only needs one-sixth of the C-band (wavelength 5.6 cm), which greatly improves the possibility of lightweight and multifunctional InSAR.

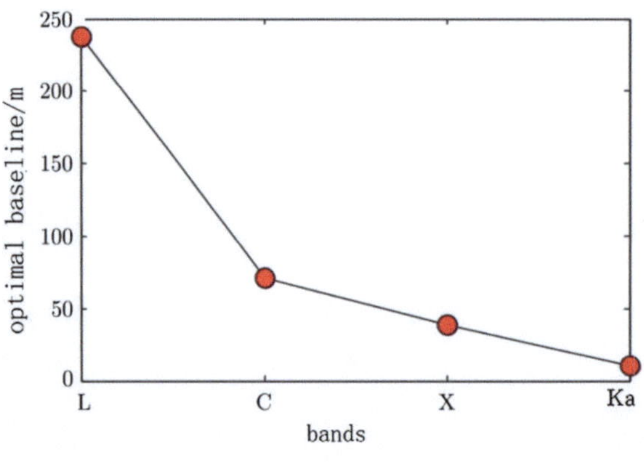

Fig. 6.11 Optimal baselines for different bands

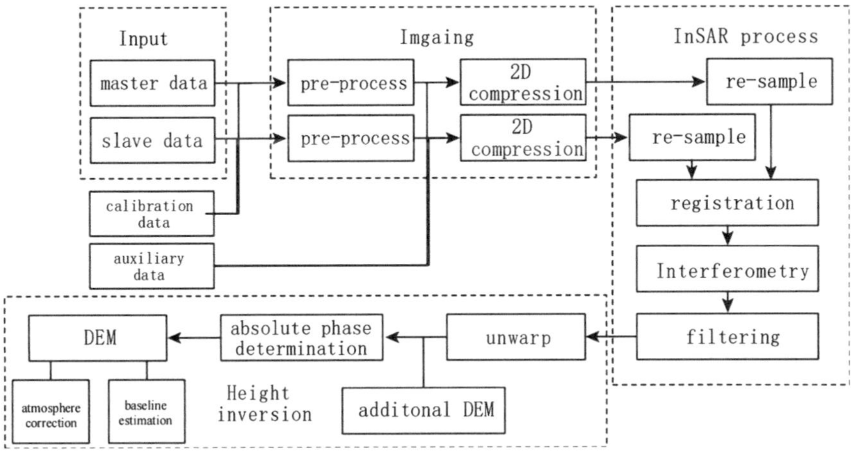

Fig. 6.12 Flow chart of spaceborne interferometric SAR data processing

Generally speaking, from the data acquisition system to the output of InSAR-based DEM products, the main processes that need to be experienced are: raw data acquisition, data unpacking and decoding, imaging, InSAR information extraction, multi-look processing, absolute phase determination based on coarse DEM, geometric correction and geo-calibration, etc., and calibration, atmospheric error correction, baseline measurement and correction are required at the same time [13–16]. Figure 6.12 shows the flowchart.

In this regard, it is necessary to study the techniques of signal processing: phase unwrap, phase filtering, and SAR image registration.

6.3.2 Interferometric Signal Processing Method

1. **Image registration**

Complex image registration is one of the most important steps in the InSAR process, which is used to calibrate the relative displacement between pairs of complex images, and sometimes perform uneven regional local calibration to cope with the effects of viewing angle, shadow overlap, and perspective shrinkage. Common registration strategies are similarity measures based on phase values and similarity measures based on intensity values. The maximum spectrum method is a classical SAR image registration method, and its basic idea is to use InSAR image quality as a registration measure.

Due to the short baseline of the airborne system, there is only a subpixel mismatch between the images, and the registration based on the maximum spectrum measure is performed after 16× interpolation, and the amplitude model of the two SAR images is set as the formula

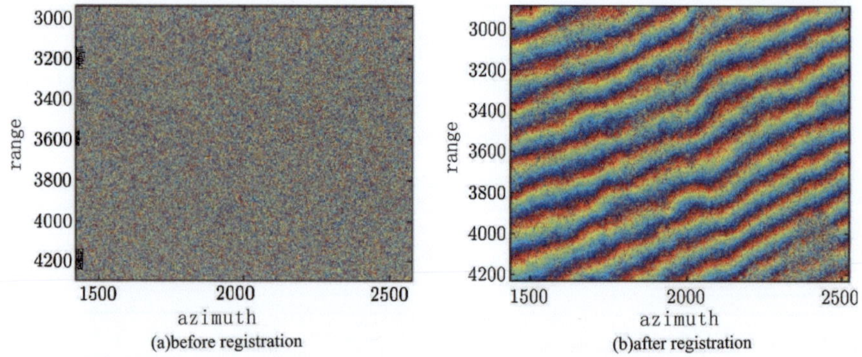

Fig. 6.13 Complex image registration

$$s_1 = |u_1|e^{j\phi_1} \tag{6.84}$$

$$s_2 = |u_2|e^{j\phi_2} \tag{6.85}$$

Then the conjugate InSAR data is determined by the registration measure window

$$s = s_1 s_2^* e^{j(\phi_1 - \phi_2)} \tag{6.86}$$

The transformation to the two-dimensional frequency domain is carried out to obtain the spatial frequency at the offset value f_x and f_y corresponding to the two dimensions, and the time domain shift can be completed by phase complex multiplication. It can be found that there is no obvious coherent information about the pre-registration interferometry phase as shown in Fig. 6.13. After registration, as shown in the figure on the right, interferometry fringes appear in the image with a two-dimensional image offset of (2.125, −8.5).

2. **Phase filtering**

After the registration of the image, the InSAR information can be extracted, and the original interferometry phase can be obtained by conjugate phase finding:

$$\phi = \arg\left[s_1(x, y)s_2^*(x, y)\right] \tag{6.87}$$

Among them, the left figure of Fig. 6.14 is the initial phase of calculation, but because the receiving antenna has a spatial position difference, the target oblique distance information obtained is not only a function of the elevation value of the target point target, but also a function of the range between the radar platform and the target. Even on horizontal ground, there will be range changes in the interferometric phase, which is manifested as a close and distant fringe. Before 2D unwrap, the InSAR phase of the reference ground plane needs to be removed, which is called deleveling, in order to reduce the phase gradient and facilitate phase unwrapping. It

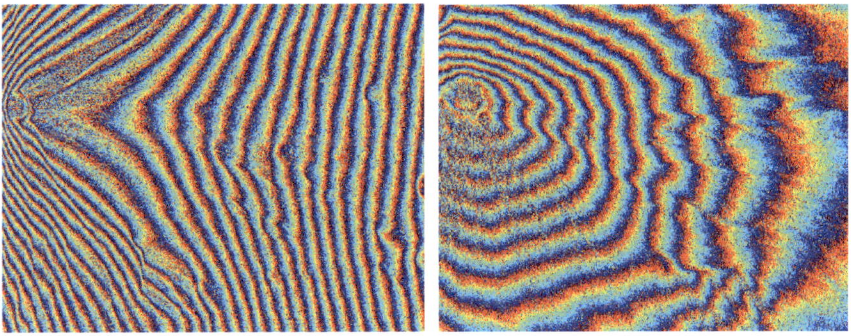

(a) Before flat earth removal (b) After flat earth removal

Fig. 6.14 Flat earth effect

can be approximated that the cosine of the lower viewing angle $\cos\theta = H/r_1$ exists between the height and range, so the angle $\theta = \cos^{-1}(H/r_1)$ and the flat ground at the antenna distance also r_1 have the interferometry phase:

$$\phi_g = \frac{2\pi}{\lambda}(r_1 - r_2) = \frac{2\pi}{\lambda}\left(\sqrt{r_1^2 + B^2 + 2Br_1\cos(\alpha+\theta)} - r_1\right) \tag{6.88}$$

In this way, for each pixel, the InSAR phase of the horizontal ground corresponding to its range is subtracted from the original phase, and the principal value is taken, the interferometry phase after the flat phase is obtained. The right figure in Fig. 6.14 shows the distribution and trend of interferometry fringes after the flat phase is removed, which can roughly represent the topography of the peak.

Due to the influence of system noise, baseline decorrelation, antenna pattern inconsistency and other factors, the two images are not completely coherent, and the phase of the image will be accompanied by randomly distributed components (noise), which will bring serious errors to the ground elevation estimation, and will also bring a large number of residual points, making the subsequent phase unwrapping process more difficult [17]. As shown in Fig. 6.15, there are many phase transitions and glitches before phase filtering, which greatly reduces the phase unwrapping accuracy.

Therefore, in many cases, the interferometric phase must also be denoised during the generation of the interferometric phase. There are two main noise reduction processing methods adopted now: one is the use of multi-look image for noise reduction, which uses multiple image pairs obtained by SAR, which can effectively improve the signal-to-noise ratio of the InSAR phase; The second is filtering. It uses the phase value of a pixel neighborhood pixel to phase smooth the pixel, including mean filtering and median filtering, because the phase is modulated by the circumferential phase, in the process of filtering must pay attention to the edge of the fringe cannot be smoothed. These two types of methods are usually performed after the flat-earth removal. The rotary mean filtering method can effectively avoid phase transitions. Selecting a rectangle with a filter window of 6*6, the phase before and

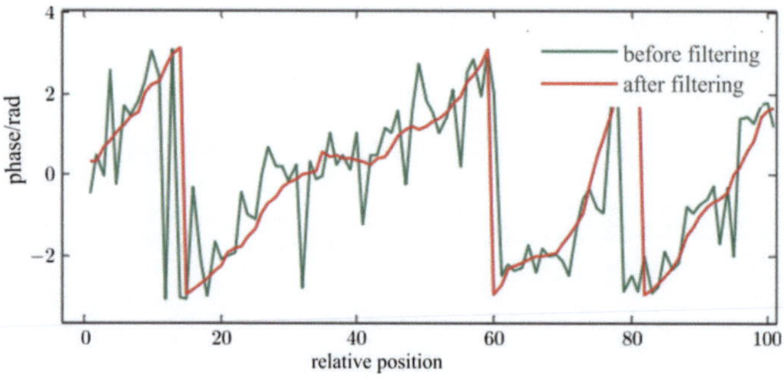

Fig. 6.15 Schematic diagram of phase filtering

after filtering is shown in Fig. 6.16, the phase smoothness after filtering is greatly increased, the noise is reduced, and it is closer to the ideal noise-free situation than before filtering.

3. **Phase unentanglement**

As shown in Fig. 6.17, the results of absolute phase and winding within the circumferential phase, respectively. Phase unwrapping is generally divided into one-dimensional unwrapping and two-dimensional unwrapping, and the one-dimensional scene is relatively simple and has only one unwinding path. In the two-dimensional case, it is more complicated, and there may be multiple paths between two sides of the two-dimensional space.

Especially in the presence of strong noise, the unwrapping performance is severely limited by the choice of unwinding path. In general, there is no absolutely valid path mechanism. The least squares method of unwrapping is equal weight analysis, and the basic idea is to solve for the smallest deviation between the true phase gradient and the wrapping phase gradient [18]. It is considered that the phase of the data points under the independent grid is wound by 2π, and the principal value phase is expressed as the formula

$$\begin{aligned} &\psi_{ij} = \phi_{ij} + 2\pi k, k \text{ integer} \\ &-\pi < \psi_{ij} \le \pi, i = 0 \ldots M-1, j = 0 \ldots N-1 \end{aligned} \tag{6.89}$$

where is the ϕ_{ij} real phase and the ψ_{ij} winding phase. Defining the winding operator W implements the procedure $W\{\Phi_{ij}\} = \psi_{ij}$. The phase gradient is calculated in two dimensions respectively to obtain the formula

$$\begin{aligned} \Delta_{ij}^{x} &= W_{ij}\{\psi_{i+1,j} - \psi_{i,j}\} \\ \Delta_{ij}^{y} &= W_{ij}\{\psi_{i,j+1} - \psi_{i,j}\} \end{aligned} \tag{6.90}$$

The least squares solution minimizes Φ_{ij} the following evaluation function

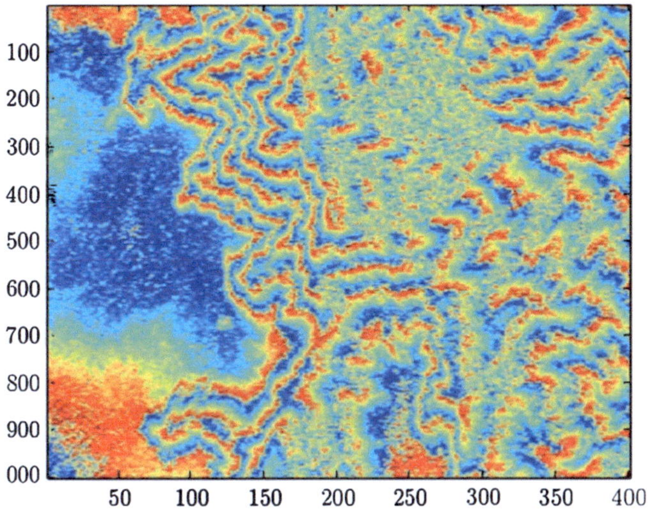

(a)interferometric phase before filtering

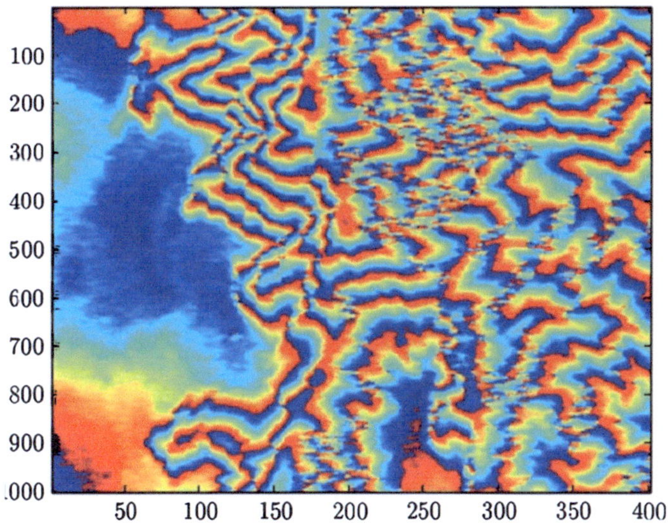

(b)interferometric phase after rotating mean Filtering

Fig. 6.16 Phase filtering

$$\sum_{i=0}^{M-2}\sum_{j=0}^{N-1}\left(\phi_{\mathrm{i+1,j}}-\phi_{\mathrm{i,j}}-\Delta_{\mathrm{ij}}^{x}\right)^2+\sum_{i=0}^{M-1}\sum_{j=0}^{N-2}\left(\phi_{\mathrm{i,j+1}}-\phi_{\mathrm{i,j}}-\Delta_{\mathrm{ij}}^{y}\right)^2 \tag{6.91}$$

Obtained after simplification

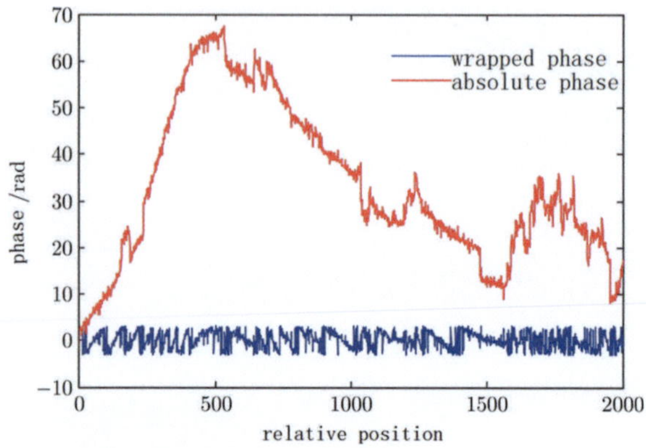

Fig. 6.17 Phase unwrapping example

$$\left(\Phi_{\mathrm{i+1,j}} + \Phi_{\mathrm{i-1,j}} + \Phi_{\mathrm{i,j}} + \Phi_{\mathrm{i,j-1}} - 4 \cdot \Phi_{\mathrm{i,j}}\right) = \rho_{\mathrm{i,j}} \tag{6.92}$$

where is represented $\rho_{\mathrm{i,j}}$ as a formula

$$\rho_{\mathrm{i,j}} = \left(\Delta_{\mathrm{ij}}^{x} - \Delta_{\mathrm{i-1,j}}^{x}\right) + \left(\Delta_{\mathrm{ij}}^{y} - \Delta_{\mathrm{i,j-1}}^{y}\right) \tag{6.93}$$

It is easy to find that the above equation is the Poisson equation

$$\frac{\partial^2}{\partial x^2}\phi(x, y) + \frac{\partial^2}{\partial y^2}\phi(x, y) = \rho(x, y)$$

discretized form. Apply a fast DCT transform to express the desired solution

$$\hat{\phi}_{i,j} = \frac{\hat{\rho}_{i,j}}{2\left(\cos\frac{\pi i}{M} + \cos\frac{\pi i}{N} - 2\right)} \tag{6.94}$$

Solving the above equation yields the phase of the unwrapping. Using the 3D model shown on the left side of Fig. 6.18 as the terrain input, add white noise to the acquired interferometry phase to achieve a signal-to-noise ratio of 5dB.

The 2D and 3D plots of the unwrapping phase are shown in Fig. 6.19.

6.3.3 *Ka-Band SAR Experiment and Data Processing*

The analysis data comes from the millimeter wave DBF-SAR system, and the antenna configuration is shown in Fig. 6.20.

Under the compact integrated antenna design, the total front area is less than 0.5 m^2. Due to the short baseline, it is easy to know that the antenna phase center

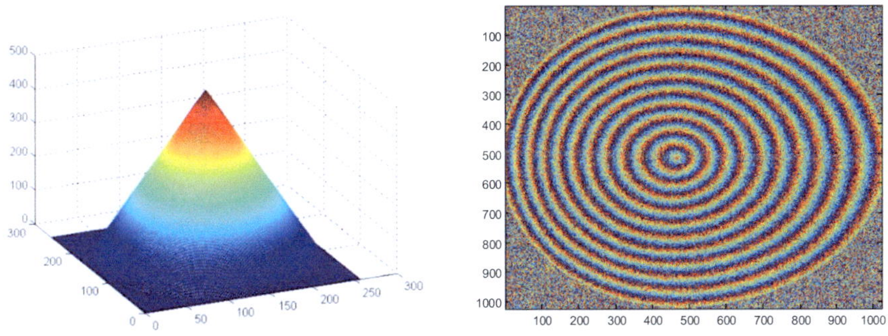

(a) Input terrain (b) wrapped phase with noise

Fig. 6.18 Input phase

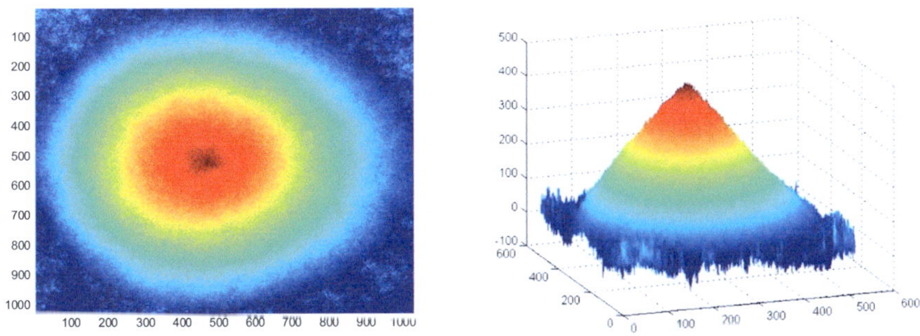

(a) 2-D unwrapped phase (b) 3-D unwrapped phase

Fig. 6.19 Unwrapping results

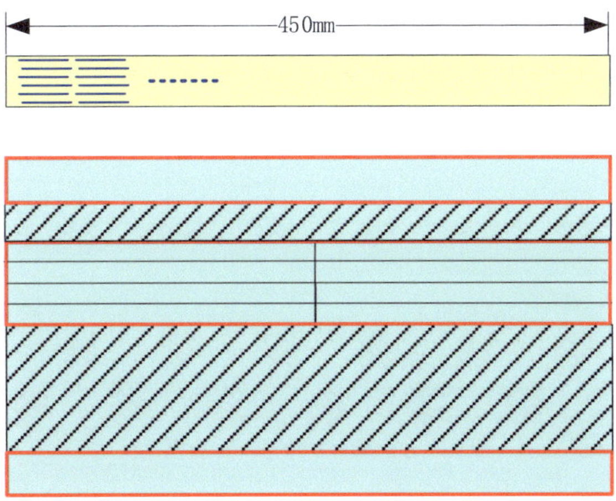

Fig. 6.20 DBF-SAR antenna array

difference along the cross-track is less than one resolution unit, and only subpixel registration needs are required, and coarse registration is not required. As shown in Fig. 6.21, the cross-sectional area of the backscattered after normalization of the water surface decays rapidly at about 10°. Therefore, in order to achieve high echo power on the water surface, a small incidence angle is required [19, 20]. After power calculation, 8° is selected as the appropriate angle of incidence, and the incidence range is approximately 6° to 10° NRCS (Normalized Radar Cross Section) better than 7 dB.

As shown in Fig. 6.22, the decomposition of the airborne platform error is better than 1.1 m in full width with margin.

Airborne Interferometric information extraction requires preprocessing, complex image subpixel registration, interferometry extraction, interferometry phase filtering,

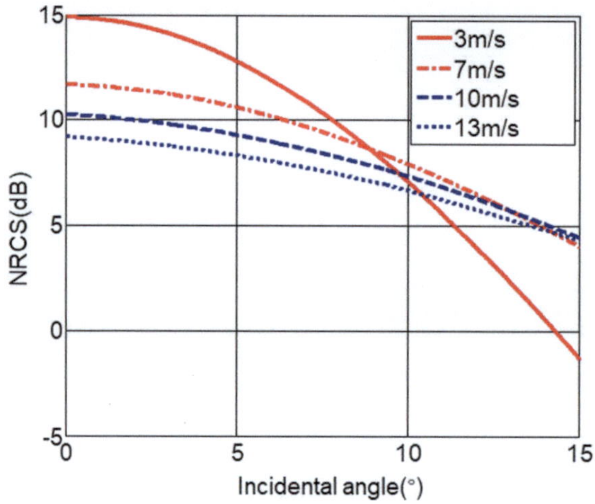

Fig. 6.21 RCS at different angles of incidence

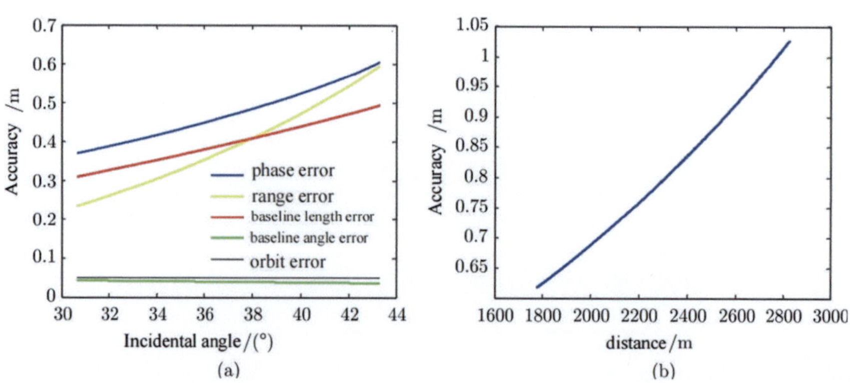

Fig. 6.22 Error budget and full performance

and phase unwrapping. The inter-channel signal does not have complete coherence due to beam pointing differences, component differences, processing errors and other factors, the first thing to note is the need to use a consistent imaging algorithm and the same Doppler estimation parameters, otherwise the envelope and phase at the corresponding pixels of the image do not have corresponding meaning, greatly reducing coherence. SAR is a two-dimensional band-limited signal that is not coherent in the frequency domain spectrum and can be eliminated by prefiltering. As shown in Fig. 6.23, the rectangular filter in the frequency domain can effectively eliminate out-of-band noise and non-overlapping spectrum.

The extracted InSAR phase after registration is shown in Fig. 6.24, which corresponds to the short baseline and long baseline, respectively.

It is easy to see that the flat-earth phase changes more at the long baseline. Due to the existence of the flat-earth phase, the treatment is carried out and the phase is unentangled to achieve terrain recovery. The above operation was performed on the SAR data of a mountainous area, and the image of interferometry filtering and

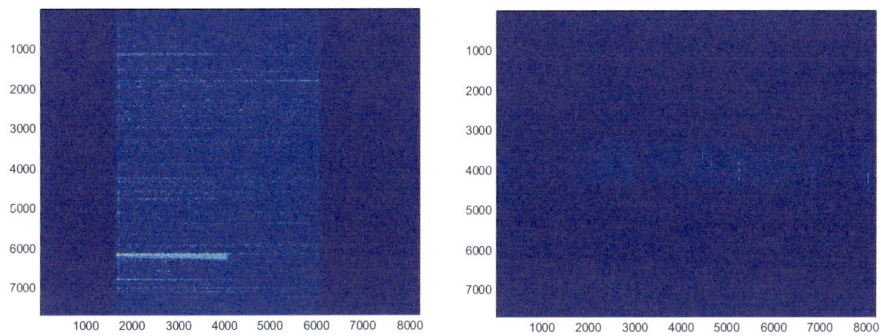

(a) frequency domain in range (b) frequency domain in azimuth

Fig. 6.23 Schematic diagram of prefiltering

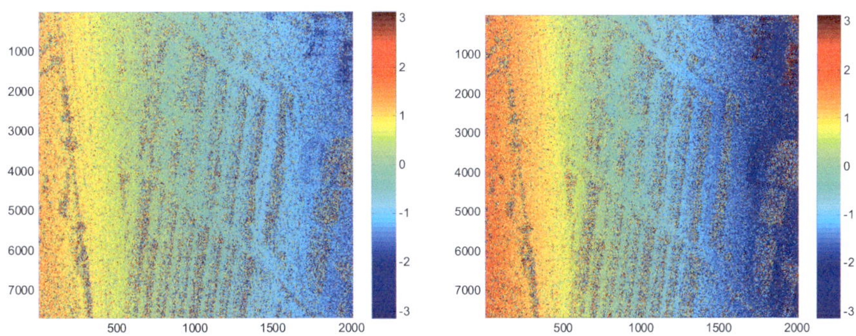

(a) Short baseline phase (b) Long baseline phase

Fig. 6.24 Schematic diagram of long and short baseline interferometry phase

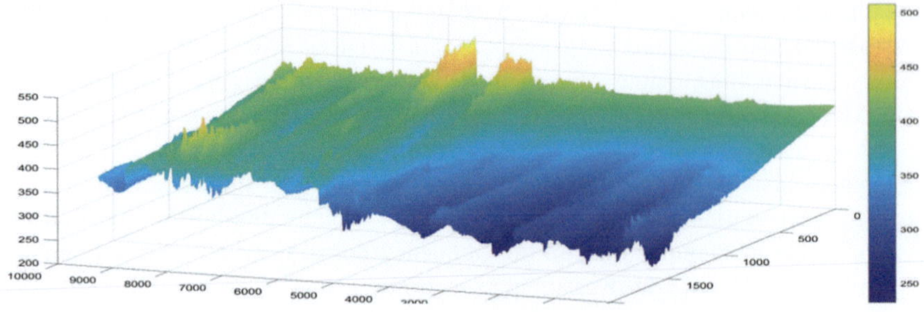

Fig. 6.25 Extraction results of interferometric information

unwinding was obtained as follows. The results of altimetry in Fig. 6.25 are verified by multiple validation points, and the accuracy is better than 0.8 m.

6.4 Spaceborne Ka-Band SAR Along Track Interferometry

Unlike the spatial baseline required for cross track, along track interferometry requires the observed geometry to be as consistent as possible. The SAR imaging principle shows that the high resolution comes from Doppler modulation, and the directly measured baseband Doppler comes from the stable relative motion trend between the ground stationary target and the linear motion platform, and when the target is accompanied by radial velocity during the imaging process, it is bound to attach a phase modulation different from the stationary target to the phase term.

According to public literature, SRTM, TerraSAR-X, Gaofen-3, etc. have completed the orbital ATI using different baseline configurations. Due to the design relationship of approximate binding of baseline and wavelength, airborne ATI can be easily realized in the decimeter level baseline in the millimeter band, and the speed information of the scene can be obtained. It is necessary to combine Ka DBF-SAR to complete theoretical analysis and measured data processing, and verify the equivalence of the spaceborne millimeter wave multi-receiver interferometry system.

6.4.1 Principle of Along Track Interferometry

Under the condition of stripmap, the clutter component of the scene echo meets the zero doppler condition. Set the observation scene as two antennas probing and sampling the scene successively in the same motion mode along the track, and their phase centers have a baseline length B_x; The platform flight altitude is H_p and speed V_p. At this time, according to the specific transmission and reception mode, there are three differences in ATI configuration: (1) standard ATI, single antenna transmission, double antenna reception; (2) Ping-pong type ATI, dual antennas transmit and receive respectively; (3) Dual baseline mode ATI, where the two antennas receive adjacent pulses transmitted by the other antenna.

The Ka-band has high phase sensitivity at the same baseline, or only a short baseline is required to achieve the same phase sensitivity.

6.4.2 DBF-SCORE ATI Processing Method

According to the data, the backscattering of the water surface above 10° is extremely weak, and the normalized radar cross-sectional area will be more than 20 smaller than at the incidence angle of 5°. In this regard, the small incidence angle is used to observe the scene and the echo gain is improved by DBF technology. By following the track baseline, the ATI information with time difference is extracted. A DBF-SCORE-based interferometry method along the trajectory is proposed to extract the velocity information of the target region by DBF synthesis, imaging, registration, phase extraction and time-domain intensity masking processing, and performs 10 in azimuth and 2 in range multi-look processing. The processing flow is summarized in Fig. 6.26.

At present, the existing analysis of ATI performance mainly realizes error decomposition for two sets of single-channel data, and analyzes the influence of each error component on radial velocity measurement. In order to solve the ATI for high-resolution and wide-width applications, the millimeter-wave multi-channel ATI system under the advantages of lightweight and integrated transceiver antennas in the millimeter-wave band is analyzed, and the influence of inter-channel phase error and arrival angle estimation error on InSAR coherency is evaluated, so as to provide a quantitative reference for subsequent multi-channel ATI data analysis. According to Fig. 6.27, the geometric relationship of obtaining ATI information is expressed.

Assuming that the radar platform has flight altitude H and speed, V_p is equipped with two sets of sequential distribution receiving antennas, and has a long phase center distance along the trajectory, and B has multiple receiving channels in the distance upward. Set the coordinate system $x - y - z$ to correspond to the orbit, cross-track, and altitude direction. The target is located at the angle of incidence θ, oblique distance,r_0 and has a three-way velocity, $\left(v_x, v_y, v_z\right)$ which can be obtained as the radial velocity of the target

$$v_r = -v_z \cos\theta + v_y \sin\theta \tag{6.95}$$

Ideally, the interferometry phase of the target due to velocity is

$$\phi_v = \frac{2\pi}{\lambda}\frac{B_x}{V_p} v_r \tag{6.96}$$

ATI fuzzy velocity is the target velocity that brings the 2π rad phase

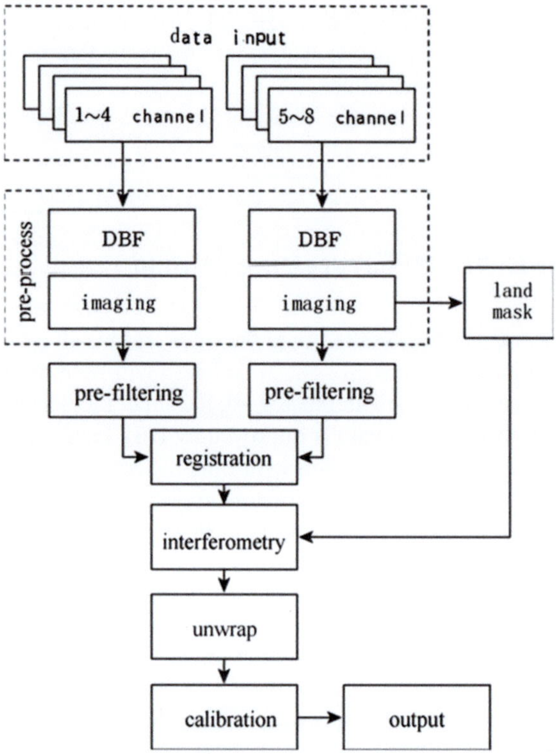

Fig. 6.26 Completion of DBF-SCORE ATI

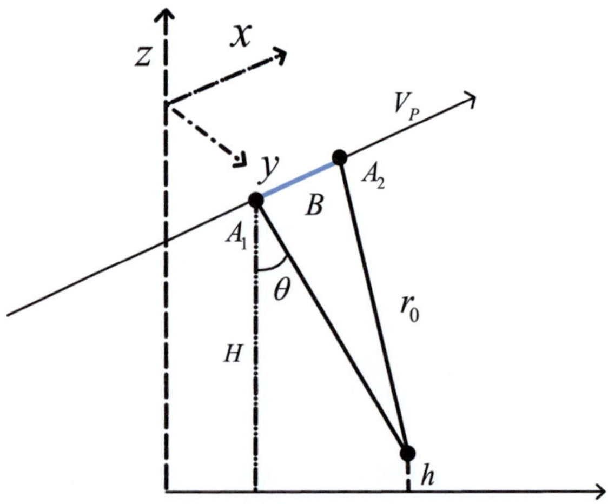

Fig. 6.27 Geometric diagram of obtaining ATI information

$$v_{\mathrm{amb}} = \frac{\lambda}{2B_{\mathrm{x}}} V_{\mathrm{p}} \tag{6.97}$$

The theoretical speed range at which the system can be observed without blurring is $-v_{\mathrm{amb}}/2 < v_{\mathrm{r}} < v_{\mathrm{amb}}/2$.

The main limiting factors of interferometric performance along the track are described below, and the influence of limiting factors on DBF-SAR performance is quantitatively analyzed. The interferometric process is preceded by separate imaging to obtain two focused images.

(1) Along-track speed

If the velocity component v_{x} is so large that it does not match the azimuth reference equation ϕ_{a}, the target is defocused in azimuth. Use a second-order approximation,ϕ_{a} for

$$\phi_{\mathrm{a}} = -\frac{2\pi}{\lambda}\frac{x^2}{2r_0} = -\frac{2\pi}{\lambda}\frac{V_{\mathrm{p}}^2 t_{\mathrm{a}}^2}{2r_0} \tag{6.98}$$

where is the t_{a} azimuth time, ignoring the radial velocity component, and the phase history of the target is the formula

$$\phi_{\mathrm{t}} = -\frac{2\pi}{\lambda}\frac{\left(V_{\mathrm{p}} - v_{\mathrm{x}}\right)^2 t_{\mathrm{a}}^2}{2r_0} \tag{6.99}$$

The residual phase at the end of the synthetic aperture is

$$\phi_{\mathrm{r}} = -\frac{2\pi}{\lambda r_0}\left[V_{\mathrm{p}}^2 - \left(V_{\mathrm{p}} - v_{\mathrm{x}}\right)^2\right]\left(\frac{T_{\mathrm{a}}}{2}\right)^2 \tag{6.100}$$

where is the T_{a} synthetic aperture time corresponding to L_{a} the azimuth antenna length $\lambda r_0/L_{\mathrm{a}}V_{\mathrm{p}}$. The requirement ϕ_{r} is less than $\pi/2$, and assumes $V_{\mathrm{p}} \gg v_{\mathrm{x}}$ that the constraint on the speed along the track is $|v_{\mathrm{x}}| \le L_{\mathrm{a}}^2 V_{\mathrm{p}}/2\lambda r_0$.

(2) Cross-track speed

The vertical trajectory speed of the target v_r also causes poor image focusing, and additional linear terms are added to the phase history, which in turn changes the Doppler center frequency and causes the separation of the moving target's azimuth spectrum from the clutter. The Q_{Dop} offset formula of the target azimuth spectrum is denoted by

$$Q_{\mathrm{Dop}} = \left|\frac{2v_{\mathrm{r}}/\lambda}{2V_{\mathrm{p}}/L}\right| = \left|\frac{v_{\mathrm{r}}L}{V_{\mathrm{p}}\lambda}\right| \tag{6.101}$$

Only part of the target bandwidth is processed, so the azimuth resolution is $1 - Q_{\mathrm{Dop}}$ reduced by a factor and the signal-to-noise ratio of a single pixel of the target is reduced by a factor $1/(1 - Q_{\mathrm{Dop}})(1 - Q_{\mathrm{Dop}})^2$.

(3) Range migration

Since the additional range migration due to radial velocity also causes the moving target to defocus, when the azimuth spectrum of the target is aliased relative to the PRF, the uncompensated distance migration between adjacent pulses is expressed as a formula

$$\Delta\rho_{\mathrm{n}} = n_{\mathrm{Dop}}\frac{\lambda}{2} \tag{6.102}$$

where n_{Dop} is an integer multiple of the target doppler offset relative to the processor Doppler offset relative to the PRF. The total extra distance migration over synthetic aperture time is

$$\rho_{\mathrm{n}} = \Delta\rho_{\mathrm{n}} T_{\mathrm{int}}\mathrm{PRF} = n_{\mathrm{Dop}}\frac{\lambda^2 r_0 PRF}{2L_{\mathrm{a}}V_{\mathrm{p}}} \tag{6.103}$$

The radial velocimetry accuracy can be derived from the error sub-contribution by differentiation, and its sum of squares can characterize the overall interferometric velocimetry capability. The DBF-ATI system defines the axis heading under the coordinate $x - y - z$, A_1 and A_2 expresses two sets of multi-channel receiving antennas along the orbit and has a baseline B along the course, which is the sum of the orbital height H and a the radius of the earth, and the viewing angle θ of the target point. Speed measurement accuracy

$$\sigma_{V_{\mathrm{r}}} = \sqrt{\begin{array}{l}\left(\frac{\partial V_{\mathrm{r}}}{\partial V_{\mathrm{P}}}\right)^2\sigma_V^2 + \left(\frac{\partial V_{\mathrm{r}}}{\partial B}\right)^2\sigma_B^2 + \left(\frac{\partial V_{\mathrm{r}}}{\partial \Phi_{12}}\right)^2\sigma_{\Phi_{12}}^2\\ +\left(\frac{\partial V_{\mathrm{r}}}{\partial B_{\mathrm{y}}}\right)^2\sigma_{B_{\mathrm{y}}}^2 + \left(\frac{\partial V_{\mathrm{r}}}{\partial B_{\mathrm{z}}}\right)^2\sigma_{B_{\mathrm{z}}}^2 + \left(\frac{\partial V_{\mathrm{r}}}{\partial h}\right)^2\sigma_{\mathrm{h}}^2\\ +\left(\frac{\partial V_{\mathrm{r}}}{\partial a}\right)^2\sigma_{\mathrm{a}}^2 + \left(\frac{\partial V_{\mathrm{r}}}{\partial R}\right)^2\sigma_{\mathrm{r}_0}^2\end{array}} \tag{6.104}$$

Among them, the error source means the measurement uncertainty of aircraft speed, baseline length along the track, interferometry phase, baseline y-axis, baseline z-axis, observation platform height, orbital altitude and observation range, and the error contribution factors are respectively $\sigma_{V_{\mathrm{P}}}, \sigma_B, \sigma_{\Phi_{12}}, \sigma_{\mathrm{B_y}}, \sigma_{\mathrm{B_z}}, \sigma_{\mathrm{h}}, \sigma_{\mathrm{a}}, \sigma_{\mathrm{r_0}}$

$$\frac{\partial V_{\mathrm{r}}}{\partial V_{\mathrm{P}}} = \frac{1}{B}\left(\frac{\lambda}{2\pi}\Phi_{12} - B_{\mathrm{z}}\cos\theta - B_{\mathrm{y}}\sin\theta\right) \tag{6.105}$$

$$\frac{\partial V_{\mathrm{r}}}{\partial S_{\mathrm{V}}} = -\frac{2v_{\mathrm{amb}}}{\lambda B}\left(\frac{\lambda}{2\pi}\Phi_{12} - B_{\mathrm{z}}\cos\theta - B_{\mathrm{y}}\sin\theta\right) \tag{6.106}$$

$$\frac{\partial V_{\mathrm{r}}}{\partial \Phi_{12}} = \frac{v_{\mathrm{amb}}}{\pi} \tag{6.107}$$

$$\frac{\partial V_{\mathrm{r}}}{\partial B_{\mathrm{y}}} = \frac{2v_{\mathrm{amb}}}{\lambda}\sin\theta \tag{6.108}$$

$$\frac{\partial V_{\mathrm{r}}}{\partial B_{\mathrm{z}}} = \frac{2v_{\mathrm{amb}}}{\lambda}\cos\theta \tag{6.109}$$

$$\frac{\partial V_{\mathrm{r}}}{\partial h} = -\frac{2v_{\mathrm{amb}}r_{\mathrm{e}}}{\lambda a r_0}\left(B_{\mathrm{z}} + \frac{1}{\tan\theta}B_{\mathrm{y}}\right) \tag{6.110}$$

$$\frac{\partial V_{\mathrm{r}}}{\partial a} = -\frac{2v_{\mathrm{amb}}}{\lambda}\left(\frac{1}{r_0} - \frac{\cos\theta}{a}\right)\left(B_{\mathrm{z}} + \frac{1}{\tan\theta}B_{\mathrm{y}}\right) \tag{6.111}$$

$$\frac{\partial V_{\mathrm{r}}}{\partial r_0} = -\frac{2v_{\mathrm{amb}}}{\lambda}\left(\frac{1}{a} - \frac{\cos\theta}{R}\right)\left(B_{\mathrm{z}} + \frac{1}{\tan\theta}B_{\mathrm{y}}\right) \tag{6.112}$$

In the polyop number is N_{L}, the interferometry phase can be expressed as

$$\hat{\phi} = \arctan\left[\frac{\mathrm{Im}\left\{\sum_{n=1}^{N_L} s_1^{(n)} s_2^{*(n)}\right\}}{\mathrm{Re}\left\{\sum_{n=1}^{N_L} s_1^{(n)} s_2^{*(n)}\right\}}\right] \tag{6.113}$$

thereinto s_1 and s_2 Don't indicate A_1 and A_2 After multichannel synthesis, complex signal with standard deviation of interferometry phase is

$$\sigma_{\Phi_{12}} = \frac{1}{\sqrt{2N_{\mathrm{L}}}}\frac{\sqrt{1-\gamma^2}}{\gamma} \tag{6.114}$$

The standard deviation of the interferometry phase is a function of the coherence coefficient γ. The total coherence coefficient can be expressed as the product of the coefficients

$$\gamma = \gamma_{\mathrm{tgt}} \cdot \gamma_{\mathrm{dop}} \cdot \gamma_{\mathrm{pro}} \cdot \gamma_{\mathrm{asr}} \tag{6.115}$$

thereinto, $\gamma_{\mathrm{tgt}} \cdot \gamma_{\mathrm{dop}} \cdot \gamma_{\mathrm{pro}} \cdot \gamma_{\mathrm{asr}}$. The target signal-to-noise ratio coherence coefficient, spatial coherence coefficient, Doppler coherence coefficient, processing coherence coefficient, and fuzzy coherence coefficient are respectively. Among them, the target signal-to-noise is greatly affected by the distance to the multi-channel, and the three factors mainly include channel fixed phase error, spatial decoherence, beam pointing error and wave arrival angle error to reduce the signal-to-noise ratio, affect the quality of the synthesized signal, and then reduce the ATI effect.

Consider the same scene SAR track image group, Eq. (6.120) and formula (6.121) corresponds to the resolution unit image value a_1 and a_2 respectively

$$a_1 = s_1 + n_1 \tag{6.116}$$

$$a_2 = s_2 + n_2 \tag{6.117}$$

thereinto, s_1 and s_2 is the signal of interest,n_1 and n_2 Corresponds to the noise component. Has a phase difference ϕ_v

$$s_2 = s_1 \exp(-j\phi_v) \tag{6.118}$$

Under ideal conditions, the two sets of sequential receive channels are perfectly identical. n_1 and n_2 are independent noise. Then the coherence coefficient for this resolving unit is

$$\gamma_{tgt} = \frac{\langle a_1 a_2^* \rangle}{\sqrt{\langle a_1 a_1^* \rangle \langle a_2 a_2^* \rangle}} \tag{6.119}$$

Get after removing the non-related crossover term

$$\gamma_{tgt} = \frac{\langle s_1 s_2^* \rangle}{\sqrt{\left(\langle s_1 s_1^* \rangle + \langle n_1 n_1^* \rangle\right)\left(\langle s_2 s_2^* \rangle + \langle n_2 n_2^* \rangle\right)}} \tag{6.120}$$

Simplify to

$$\gamma_{tgt} = \frac{S \exp(j\phi_v)}{S + N} = \frac{\exp(j\phi_v)}{1 + 1/\mathrm{SNR}} \approx \frac{1}{1 + 1/\mathrm{SNR}} \tag{6.121}$$

thereinto, $N = \langle n_1 n_1^* \rangle = \langle n_2 n_2^* \rangle$; $S = \langle s_1 s_1^* \rangle = \langle s_2 s_2^* \rangle$; $S \exp(j\phi_v) = \langle s_1 s_2^* \rangle$. Thermal noise is the main element of incoherent noise. In practical applications, multi-channel synthesis has multiple error forms and inherent decoherence effects, mainly fixed phase error, wave arrival angle estimation error, and decoherence caused by the inherent spatial distribution of channels. Defines the radar cross-sectional area of the observed target in the radar echo as σ_{tgt}, the equivalent backscatter coefficient of the radar system is σ_0. Then we get the equation that takes the logarithm (6.126), the signal-to-noise ratio of the target along the viewing angle

$$G_{\mathrm{tgt}} = \sigma_{\mathrm{tgt}} - \sigma_0 \tag{6.122}$$

(1) Channel fixed phase error

Let the first channel signal echo n signal, after fast time compensation, is no longer modulated by the echo viewing angle, such as

$$s_n(t) = s_{\mathrm{ref}}(t)\exp(j\phi_{\mathrm{n}}) = rect\left(\frac{t-t_0}{T_{\mathrm{s}}}\right)\exp\left(\mathrm{j}\pi K_{\mathrm{r}}(t-t_0)^2\right)\exp\left(-\mathrm{j}\frac{4\pi r_0}{\lambda}\right)\exp(j\phi_{\mathrm{n}}) \tag{6.123}$$

thereinto $s_{\mathrm{ref}}(t)$ is the reference channel signal, ϕ_{n} is the residual fixed phase error of each channel relative to the reference channel, K_{r}, t_0 is the oblique distance r_0, then sum to get the DBF synthesis signal, ie

$$s_{\mathrm{dbf}}(t) = \sum_{n=1}^{N} s_{\mathrm{n}}(t) \tag{6.124}$$

It can be found that when the signal of each channel of the DBF receiving system has a fixed phase error, it will not directly affect the energy accumulation caused by single-channel pulse compression, but after multi-channel synthesis, it will form a cancellation in each corresponding distance unit. Define the fixed phase gain loss as

$$G_{\mathrm{p}} = 10\log 10\left(\frac{S_{\mathrm{dbf}}}{N S_{\mathrm{ref}}}\right) \tag{6.125}$$

thereinto $S_{\mathrm{dbf}} = \langle s_{\mathrm{dbf}} s_{\mathrm{dbf}}^{*} \rangle$; $S_{\mathrm{ref}} = \langle s_{\mathrm{ref}} s_{\mathrm{ref}}^{*} \rangle$.

(2) Wave arrival angle estimation error

In fact, when the ground has high undulation, the wave arrival angle estimation obtained by the ideal flat earth geometry has a non-negligible error, and the direction pointed by the SCORE method is defined as,$\theta_{\mathrm{s}}(t)$ and the target echo direction h of the true elevation is θ_0, then the angle estimation error is the formula

$$\Delta\theta = \theta_0 - \theta_{\mathrm{s}}(t_0) \tag{6.126}$$

thereinto t_0 Due to the true two-way delay of the echo, the echo is not received by the maximum gain synthesized by the multichannel according to the ideal state, but because the gain loss is caused by the terrain.G_0 is the equivalent pattern of the receiving antenna, and the loss amount is

$$G_{\mathrm{l}} = \frac{G_0(\theta_{\mathrm{s}} + \Delta\theta)}{G_0(\theta_{\mathrm{s}})} \tag{6.127}$$

(3) Space decoherence

Under ideal tracks, the influence of spatial decoherence under the interferometric system along the track can usually be ignored, but for multi-channel systems, due to the difference in incidence angle brought by distance to multi-channel, the inherent signal difference of radar echo will be caused and the coherence on the spatial scale will be reduced. Referring to the interferometric spatial decoherence coefficient, the spatial baseline gain loss under multichannel is defined as

$$G_{\text{spatial}} = 10 \log 10\left(1 - \frac{2 \cos\theta|\delta\theta|\rho_y}{\lambda}\right) \tag{6.128}$$

thereinto ρ_y For distance resolution, the maximum difference in viewing angle between channels $\delta\theta$ Approximately $(N-1)d/r_0$.

Then the signal-to-noise ratio of the DBF-SAR image is

$$\text{SNR} = G_{\text{tgt}} + G_p + G_l + G_{\text{spatial}} \tag{6.129}$$

Using the system parameters in Table 6.3, the influence of multi-channel synthesis on ATI speed measurement performance was analyzed.

After applying multi-channel synthesis, the coherence coefficient of interferometry image can be effectively improved, and through the monovision simulation results of Fig. 6.28, it can be obtained that the coherence coefficient of the single-channel data of the radar system will be lower than 0.5 when the target normalized RCS is less than −20dB, and the corresponding value of the four-channel system can reach 0.67. When the target scattering cross-sectional area −10dB is greater, the multi-channel coherence coefficient improvement limit is less than 0.1.

By substituting the error source configuration in Table 6.4, the ideal speed measurement performance of multiple channels is simulated. Under the current error distribution, it can be found that the interferometry phase error is the main error amount, and the error source contribution at the ground distance of 700km of a single channel mainly includes phase error contribution of 0.88m/s, Z-axis baseline error contribution of 0.05 m/s, and Y-axis baseline error contribution of 0.05 m/s.

Table 6.3 Input of simulation parameters

System parameters	Parameter value	remark
Carrier	35.75 GHz	
Track height	500 km	
Bandwidth	100 MHz	Chirp
Transmit antenna height	0.25 m	
Receive antenna height	0.25 m	
The number of receive channels	N = 1, 2, 4, 8, 16	
Transceiver antenna length	5 m	
X-axis baseline length	10 m	
Y-axis baseline length	0 m	
Z-axis baseline length	0 m	
Single-channel NESZ	−30 dB	Typical value
RASR	−40 dB	Typical value
AASR	−25 dB	Typical value
Grid	10 m*10 m	

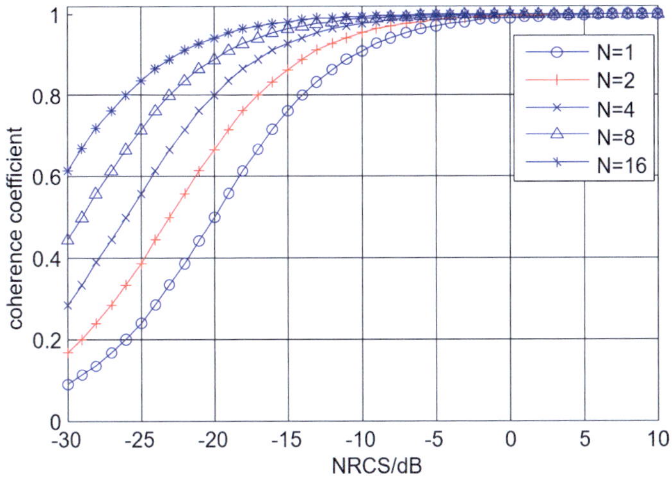

Fig. 6.28 Channel synthetic coherence coefficient

Table 6.4 Multi-source error input

System parameters	Parameter value	Remark
X-axis baseline error	0.01 m	In the direction of the track
Y-axis baseline error	0.1 mm	X–Z surface normal direction
Z-axis baseline error	0.1 mm	Geocentric direction
Platform height error	5 m	
Land surface elevation error	5 m	
Platform speed error	3 m	
Slant error	10 m	

The proximal accuracy of a single channel in the width is greater than 1.4 m/s in the weak scattering scenario, and the proximal accuracy after 16.channel synthesis is 1.24 m/s. In the strong scattering scenario, the multi-channel speed measurement accuracy is concentrated at about 1.25 m/s. The ATI speed measurement accuracy after multi-channel synthesis can be obtained by testing with different ground elevations and the maximum phase error between channels of 15°, and considering spatial decoherence. When the ground distance is small, the elevation error will seriously affect the speed measurement accuracy, and the actual system performance may not be as good as that of the single channel.

It is easy to conclude that DBF-SAR can improve the interferometric performance of the system in the face of a weak scattering environment. In terms of specific implementation, real-time or post-processing of the phase error of multi-channel radar system can effectively improve the system ATI performance in most cases, and for complex terrain environments, it is necessary to perform wave reach angle

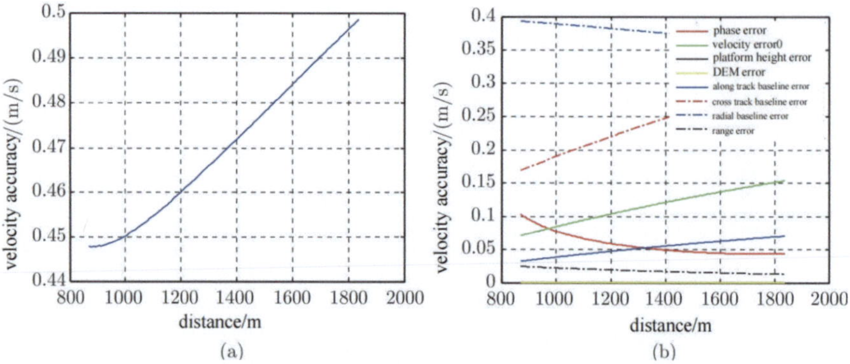

Fig. 6.29 Speed measurement accuracy (left) and decomposition (right)

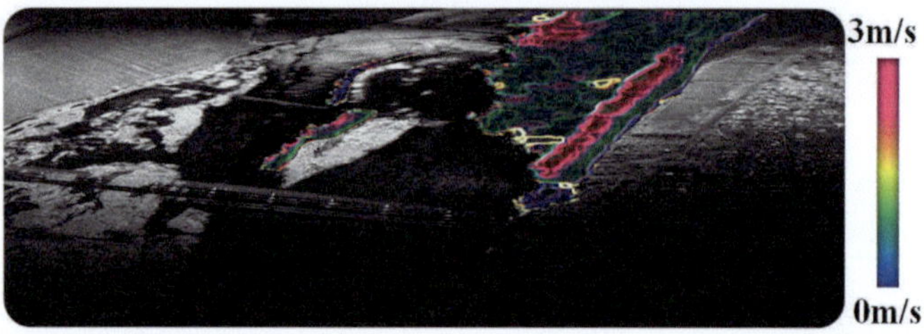

Fig. 6.30 Extraction results of ATI information

estimation based on adaptive or external DEM data to complete high-accuracy fast time weight correction.

6.4.3 ATI Test and Data Processing

Airborne tests were conducted to validate the DBF-SCORE-ATI method. With the four-channel synthesis technique, a larger theoretical echo gain can be obtained, which eliminates the need for a lower viewing angle of 10° or less as in the intersection. As shown in Fig. 6.29, the theoretical speed measurement accuracy of the analytical prototype is better than 0.5 m/s under the condition of retaining margin.

Figure 6.30 is obtained through measured data processing, and it can be found that the velocity extraction conforms to the basic landform of the scene, showing a more obvious degree of differentiation at the riverbank, and due to the shallow water flow, the tidal flats and shoals do not have obvious speed information. The velocity of the water along the river bank is higher and the speed of the center of the river is lower. After multi-point detection by flow meter, the comprehensive speed measurement accuracy is better than 0.48 m/s, which meets the design requirements.

References

1. Bechor NB, Zebker HA. Measuring two-dimensional movements using a single InSAR pair[J]. Geophysical research letters. 2006,33(16):325–333.
2. Hooper A, Zebker HA. Phase unwrapping in three dimensions with application to InSAR time series[J]. JOSA. 2007,24(9):2737–2747.
3. Hooper A. A multi-temporal InSAR method incorporating both persistent scatterer and small baseline approaches[J]. Geophysical Research Letters. 2008,35(16):21–28.
4. G. Connan, H. D. Griffiths, P. V. Brennan, D. et al. Experimental imaging of internal waves by a mm-wave radar [C]. OCEANS '98 Conference Proceedings, Nice, 1998, 2: 619–623.
5. G. Connan, H. D. Griffiths, P. V. Brennan, et al. W-Band radar measurements of laboratory water waves [C]. OCEANS '99 MTS/IEEE, Seattle, WA, 1999, 3: 1333–1337.
6. G. Connan, R. Garello, H. D. Griffiths, et al. Millimeter-Wave radar back-scattering from water waves [C]. The Record of the IEEE 2000 International Radar Conference, Alexandria, VA, 2000: 347–351.
7. Deledalle C-A, Denis L, Tupin F. NL-InSAR: Nonlocal interferogram estimation[J]. IEEE Transactions on Geoscience. 2010,49(4):1441–1452.
8. Parizzi A, Brcic R. Adaptive InSAR stack multilooking exploiting amplitude statistics: A comparison between different techniques and practical results[J]. IEEE Geoscience Remote Sensing Letters. 2010,8(3):441–445.
9. Perissin D, Wang T. Time-series InSAR applications over urban areas in China. IEEE Journal of Selected Topics in Applied Earth Observations Remote Sensing of Environment[J]. 2010,4(1):92–100.
10. Zebker HA, Hensley S, Shanker P, et al. Geodetically accurate inSAR data processor[J]. IEEE Transactions on Geoscience Remote Sensing. 2010,48(12):4309–4321.
11. Liu X, Gao W, Deng Y. Synthesis technique of array-fed shaped-reflector antenna for DBF-SAR application. IEEE Antennas Wireless Propagation Letters [J]. 2011,11:30–33.
12. Sandwell D, Mellors R, Tong X, et al. Gmtsar: An insar processing system based on generic mapping tools[J]. IEEE Transactions on Geoscience Remote Sensing. 2010,48(12):4426–4443.
13. Bordoni F, Younis M, Krieger G, et al. Performance investigation on the high-resolution wide-swath SAR system operating in multisubpulse mode [C]. 2012 IEEE International Geoscience and Remote Sensing Symposium, 2012.
14. Rosen PA, Gurrola E, Sacco GF, et al. The InSAR scientific computing environment [C]. EUSAR 2012; 9th European Conference on Synthetic Aperture Radar; 2012.
15. Adamiuk G, Schaefer C, Fischer C, Heer C, editors. SAR Architectures based on DBF for C- and X-band applications[C]. EUSAR 2014; 10th European Conference on Synthetic Aperture Radar; 2014: VDE.
16. Villano M, Krieger G, Moreira A. A novel processing strategy for staggered SAR[J]. IEEE Geoscience Remote Sensing Letters. 2014,11(11):1891–1895.
17. Bekaert D, Walters R, Wright T, et al. Statistical comparison of InSAR tropospheric correction techniques[C]. Remote Sensing of Environment. 2015,170:40–47.
18. Liu B, He Y. Improved DBF algorithm for multichannel high-resolution wide-swath SAR[C]. IEEE Transactions on Geoscience Remote Sensing of Environment. 2015,54(2):1209–1225.
19. K. S. Kulpa, A. Wojtkiewicz, M. Nalecz, et al. The simple method for analysis of nonlinear frequency distortions in FMCW radar [C]. 13th International Conference on Microwaves, Radar and Wireless Communications, Wroclaw, 2000, 1: 235–238.
20. Patyuchenko A, Younis M, Krieger G, et al. Design and Optimization of the Reflector Antenna for an X/Ka-Band Spaceborne DBF SAR System [C]. European Conference on Antennas and Propagation (EuCAP), 2015.

Chapter 7
Millimeter-Wave Synthetic Aperture Radar Image Application Techniques

7.1 Overview

SAR is a technical means of acquiring the backscattered signals and images of objects on the ground through remote sensing. SAR technology serves various fields such as land, environmental protection, marine, agriculture and forestry, transportation, disaster monitoring, and national defense. In order to further extract information and generate data products that are directly relevant to users, SAR image processing is an essential step in realizing the practical application of SAR data products.

Compared to traditional frequency bands, millimeter-wave SAR images have the advantages of strong ability to depict target contours and optical-like image quality, which provide unique advantages in target detection and recognition. Additionally, the strong backscattering characteristics of objects in the millimeter-wave frequency band make the shadows of targets in millimeter-wave SAR images more complete and clear. By incorporating target shadow information, the descriptive capability of target features can be further enhanced. Furthermore, the shorter wavelength of millimeter waves enables shorter baseline lengths for achieving the same interferometric measurement accuracy, making high-precision interferometric measurements achievable with a single satellite. Moreover, the weaker penetration capability of millimeter waves allows for more accurate reflection of surface heights for water bodies and ice and snow, providing more precise measurement results.

Taking into account the aforementioned characteristics of millimeter-wave SAR images, this chapter focuses on introducing corresponding processing methods. Firstly, a SAR image speckle suppression method based on trainable nonlinear diffusion models is introduced to improve SAR image quality. Based on this, detection and recognition methods for ship targets are presented, along with building target detection and moving target detection methods that utilize target shadow information in millimeter-wave SAR images. Finally, for comprehensive ocean observation, a method for retrieving sea surface heights and ocean surface currents based on single-satellite millimeter-wave interferometric SAR is introduced.

H. Wang, *The Millimeter Wave Synthetic Aperture Radar Technology*,
https://doi.org/10.1007/978-981-97-1044-7_7

7.2 Coherent Speckle Denoising Based on Trainable Nonlinear Diffusion Model

Coherent speckle denoising in synthetic aperture radar (SAR) images has always been a key technology in SAR image processing. Experts at globally have conducted extensive research on coherent speckle noise suppression, which can be mainly categorized into four types of methods: spatial domain algorithms, transform domain algorithms (such as wavelet domain denoising algorithms), non-local filtering, and variational algorithms.

In traditional spatial domain filtering algorithms, stationary filtering algorithms are mainly based on the noise model of coherent speckle. Examples of such filters include Lee, Frost, and Kuan filters, all of which are linear filters developed based on the linear minimum mean-square error (LMMSE) criterion. Building upon stationary filtering algorithms, researchers proposed non-stationary filtering algorithms based on the assumption of locally non-stationary coherent speckle noise within the filtering window. Examples of these algorithms include MAP Gaussian filtering and Gamma MAP filtering.

There are two main types of coherent speckle denoising algorithms based on wavelet transform domain. The first type is based on logarithmic transformation in the wavelet transform domain, and on this basis, these algorithms are optimized according to the Bayesian criterion. The main drawback of wavelet domain denoising algorithms using logarithmic transformation is the deviation of the denoised image's mean value. To address this issue, in recent years, many researchers have proposed a new generation of wavelets with good two-dimensional singularity characterization and shift-invariant properties for coherent speckle denoising. These methods typically preserve image details well but often suffer from difficulties in threshold selection and the presence of significant pseudo-Gibbs artifacts.

In recent years, non-local means (NLM) filtering has been widely used in coherent speckle denoising in SAR. Algorithms such as Bayesian NLM filtering, probabilistic patch-based (PPB) filtering, and Block-Matching 3D (BM3D) have been applied to coherent speckle denoising and achieved good results. However, in these algorithms, the selection or grouping of similar blocks may be influenced by noise, leading to the introduction of some artifacts (ghost artifacts), which is particularly evident in relatively smooth regions. Additionally, these algorithms are computationally intensive.

Variational models have been increasingly applied in the field of coherent speckle denoising. Since total variation (TV) can effectively approximate piecewise constant functions, TV-based coherent speckle denoising methods tend to introduce staircase effects to some extent. To address this limitation, total generalized variation (TGV) has been introduced to image processing. Theoretical analysis and experimental results have shown that TGV-based coherent speckle denoising methods can effectively avoid staircase effects.

Currently, the most effective coherent speckle denoising algorithms are mainly based on the non-local means (NLM) algorithm. Although these algorithms achieve good results, their computational efficiency is poor. To address this problem, a simple and efficient denoising method has been proposed, which not only exhibits good denoising performance but also requires less computation compared to current mainstream coherent speckle denoising methods. This method utilizes the recently proposed image restoration framework based on trainable nonlinear diffusion, which can be understood as a recursive convolutional neural network with trainable activation functions.

In this chapter, we first provide a brief introduction to coherent speckle noise and the trainable nonlinear diffusion process. Then, we propose a coherent speckle denoising method for SAR intensity images based on the trainable nonlinear diffusion model. During the diffusion process, the linear filter, influence function, and other relevant parameters used in the model are trained based on the statistical characteristics of coherent speckle noise in the intensity image, making it suitable for coherent speckle denoising applications. Experimental results demonstrate that the performance of this algorithm is comparable or even superior to that of current mainstream algorithms, while maintaining high computational efficiency.

7.2.1 Trainable Nonlinear Diffusion Process

Firstly, a brief introduction is provided for coherent speckle noise and the trainable nonlinear diffusion process.

Let Y be the observed SAR intensity image. Then Y can be expressed as the product of the true image intensity u and the coherent speckle noise

$$\eta : Y = u\eta \tag{7.1}$$

Note that the coherent speckle noise η in the SAR intensity image follows the Gamma distribution given by:

$$p(\eta) = \frac{1}{\Gamma(L)} L^L \eta^{L-1} e^{-L\eta} H(\eta). \tag{7.2}$$

where H represents the Heaviside function and Γ represents the traditional Gamma function.

A simplified and efficient image restoration framework based on nonlinear diffusion is proposed. This framework incorporates parameterized filters and nonlinear diffusion functions into the traditional nonlinear diffusion model, resulting in a time-variant nonlinear diffusion model, as follows:

$$u_0 = f, \quad u_{t+1} = u_t - \sum_{i=1}^{N_k} k_i^t * \varphi_i^t (k_i * u_t - \psi(u_t)), \quad f, t = 0 \cdots T - 1 \tag{7.3}$$

where u_0. represents the initial state of the diffusion process, k_i^t. represents the nonlinear filter, $k_i * u$ represents the 2D convolution of the image u and the filter kernel, φ_i^t. represents the nonlinear diffusion function, ψ. represents the data fidelity term, and N_k represents the number of filters. Assuming the data fidelity term of the model is $\psi(u_t, f) = \nabla_u D^t(u, f)$, the parameters in each step of the nonlinear diffusion are different and need to be learned through a certain training framework. The typical problem formulation solved by the FoE model is as follows:

$$E(u, f) = \sum_{i=1}^{N_k} \sum_{p=1}^{N} \rho_i^t\left((k_i^t * u)_p\right) + D(u, f) \tag{7.4}$$

where $\rho_{i\,t=0}^{t\,t=T-1}$ represents the penalty function and $D(u, f)$. denotes the fidelity term. The above equation is rewritten as a gradient descent process composed of several iterative steps, optimizing the parameters in each iteration.

The parameter training process of the nonlinear diffusion is described as follows. Given a training dataset consisting of S sets of training samples, where u represents degraded images (such as noisy images, low-resolution images, and blurred images) and u_{gt} corresponds to the ground truth images. A training framework employs a certain loss function (L2 norm is commonly used in this chapter) and trains a model parameter set for each diffusion step. The model parameter set for each diffusion step includes the linear filter parameters, the nonlinear diffusion function, and the data fidelity term. The training framework can be formulated as the following optimization problem:

$$\begin{cases} \min\limits_{\Theta} L(\Theta) = \sum\limits_{s=1}^{S} \ell\left(u_T^s, u_{gt}^s\right) = \sum\limits_{s=1}^{S} \frac{1}{2}\left\|u_T^s - u_{gt}^s\right\|_2^2 \\ s.t. \begin{cases} u_0^s = f^s \\ u_{t+1}^s = u_t^s - \sum\limits_{i=1}^{N_k} \sum\limits_{i=1}^{N_k} \overline{k}_i^t * \varphi_i^t\left(k_i^t * u_t^s\right) - \psi\left(u_t^s, f^s\right), t = 0 \cdots T-1 \end{cases} \end{cases} \tag{7.5}$$

This optimization problem can be solved using the L-BFGS algorithm or stochastic gradient descent. For simplicity, we only use the Joint training framework to simultaneously train the model parameters of the diffusion process. The corresponding parameter gradient $\frac{\partial u_t}{\partial \Theta_t}$ can be represented as:

$$\frac{\partial \ell\left(u_T, u_{gt}\right)}{\partial \Theta_t} = \frac{\partial u_t}{\partial \Theta_t} \cdot \frac{\partial u_{t+1}}{\partial u_t} \cdots \frac{\partial \ell\left(u_T, u_{gt}\right)}{\partial u_T} \tag{7.6}$$

It is worth noting that this model is closely related to convolutional neural networks. Each iteration of the nonlinear diffusion process involves convolution operations with a set of linear filters, making the diffusion process a special type of convolutional network. Specifically, the nonlinear diffusion process can be regarded as a feedback convolutional neural network. Due to this feedback operation, the diffusion process should be classified as a Recurrent network with trainable activation functions.

For addressing coherent speckle noise, we adopt $D(u) = \langle u - Y \log(u), 1 \rangle$ as the data fidelity term. However, if we take $\psi(u, f) = \nabla_u D(u, f) = 1 - \frac{f}{u}$., the following problem arises: during algorithm iterations, when u is very close to zero, the value of zz $\psi(u, f)$. can become very large, leading to highly unstable gradient descent. Therefore, directly computing the gradient descent $1 - \frac{f}{u}$ is not feasible. To address this problem, we employ the Ipiano algorithm for optimization, which can avoid the form $1 - \frac{f}{u}$. The problem formulated by the Ipiano algorithm is:

$$\arg\min_u F(u) + G(u) \tag{7.7}$$

where $F(u)$ is a smooth function (possibly non-convex) and $G(u)$. is a convex function (possibly nonsmooth). The update rule of the Ipiano algorithm is as follows:

$$u^{n+1} = \mathcal{I} + \tau \partial G^{-1} u^n - \tau \nabla F(u^n) \tag{7.8}$$

where τ is the iteration step size. $\mathcal{I} + \tau \partial G^{-1} u$. can be obtained by solving the following optimization problem:

$$\mathcal{I} + \tau \partial G^{-1} u = \arg\min_u \left| u - u^n \right|^2 + \tau G(u) \tag{7.9}$$

7.2.2 Optimization-Based Nonlinear Diffusion Process for Coherent Speckle Denoising

Based on the principle of trainable nonlinear diffusion process discussed in the previous section, an optimization-based nonlinear diffusion process is proposed for coherent speckle denoising. Firstly, the optimization-based nonlinear diffusion process can be seen as optimizing the parameters in a few iterative steps during the solution. Therefore, we combine the data fidelity term $D(u) = \langle u - Y \log(u), 1 \rangle$. and construct the following FoE optimization model:

$$\arg\min_{u>0} E(x) = \sum_{i=1}^{N_k} \sum_{p=1}^{N} \rho_i \left((k_i * u)_p \right) + \lambda \langle u - f \log(u), 1 \rangle \tag{7.10}$$

We can obtain $F(u) = \sum_{i=1}^{N_k} \sum_{p=1}^{N} \rho_i((k_i * u)p)$ and $G(u) = \lambda \langle u - f \log(u), 1 \rangle$. By calculation, we can easily obtain:

$$\nabla F(u) = \sum i = 1^{N_k} K_i^T \varphi_i(K_i u) \tag{7.11}$$

where $K_i \in \mathbb{R}^{N \times N}$. is a highly sparse matrix representing the 2D convolution of the image u and the filter kernel k_i, i.e., $K_i u \Leftrightarrow k_i * u$. We have:

$$\nabla F(u) = \sum_{i=1}^{N_k} k_i^T \varphi_i(k_i u) \tag{7.12}$$

where k_i. is obtained by rotating the 2D filter k_i by 180°, i.e., $K_i^T u \Leftrightarrow ki * u$. The other subproblem, $\mathcal{I} + \tau \partial G^{-1} u.$, takes the following form:

$$(I + \tau \partial G)^{-1}(\tilde{u}) = \arg\min_u \frac{\|u - \tilde{u}\|_2^2}{2} + \tau\lambda \langle u - f \log u, 1\rangle \tag{7.13}$$

The solution to the above equation can be obtained by directly taking the derivative as:

$$\hat{u} = (I + \tau \partial G)^{-1}(\tilde{u}) = \frac{\tilde{u} - \tau\lambda + \sqrt{(\tilde{u} - \tau\lambda)^2 + 4\tau\lambda f}}{2} \tag{7.14}$$

where $\hat{u}$. is always positive when $f > 0$. This update criterion ensures that $u > 0$ during the diffusion process, which guarantees the meaningfulness of $\log(u)$. Therefore, the diffusion process for coherent speckle denoising can be expressed as:

$$u_{t+1} = \frac{\tilde{u}_{t+1} - \lambda^{t+1} + \sqrt{\left(\tilde{u}_{t+1} - \lambda^{t+1}\right)^2 + 4\lambda^{t+1} f}}{2} \tag{7.15}$$

where $\tilde{u}_{t+1} = u_t - \sum_{i=1}^{N_k} \overline{k}_i^{t+1} * \varphi_i^{t+1}\left(k_i^{t+1} * u_t\right)$, and we set $\tau = 1$.

The processing steps of this algorithm can be summarized as shown in Fig. 7.1, where the training parameters for the t-th iteration are denoted as $\Theta_t = \Gamma^t, \Omega^t$, $\Gamma^t = \left(\varphi_i^t, k_i^t\right)_{i=1}^{N_k}$, and $\Omega^t = \lambda^t$. The specific steps are described as follows:

Input the noisy image $u_0 = f$ and the trained parameters $\{\Theta_t\}_{t=1}^T$.

For t = 1 to T - 1

Step 1: Calculate $ut + 1$. using $u_{t+1} = u_t - \sum_{i=1}^{N_k} k_i^{t+1} * \varphi_i^{t+1}(k_i * u_t - \psi(u_t))$

Step 2: Calculate using Eq. (7.15) End Output the denoised image: $u_{\text{output}} = u_T \; u_{t+1}$

Next, we introduce the joint training strategy for the diffusion process.

For the diffusion function in coherent speckle denoising, we need to calculate the gradient of the loss function \ell(u_T^s,u_{gt}^s) with respect to the parameter set $\Theta_t = \left(\lambda^t, \varphi_i^t, k_i^t\right)$, where s denotes the sample index. According to $\frac{\partial \ell(u_T, u_{gt})}{\partial \Theta_t} = \frac{\partial u_t}{\partial \Theta_t} \cdot \frac{\partial u_{t+1}}{\partial u_t} \cdots \frac{\partial \ell(u_T, u_{gt})}{\partial u_T}$, we need to calculate the individual components of $\frac{\partial \ell(u_T, u_{gt})}{\partial \Theta_t}$,

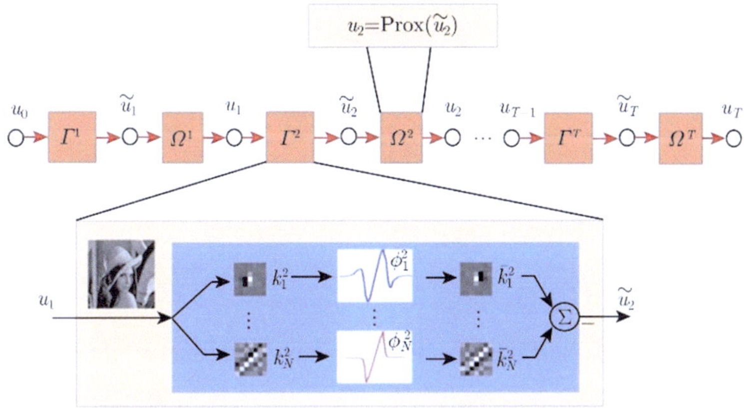

Fig. 7.1 Algorithm flow chart

i.e., $\frac{\partial u_t}{\partial \Theta_t}$, $\frac{\partial u_{t+1}}{\partial u_t}$, and $\frac{\partial \ell(u_T, u_{gt})}{\partial u_T}$. Based on the quadratic cost function, $\frac{\partial \ell(u_T, u_{gt})}{\partial u_T}$ can be represented as:

$$\frac{\partial \ell(u_T, u_{gt})}{\partial u_T} = u_T - u_{gt} \tag{7.16}$$

Meanwhile, $\frac{\partial u_{t+1}}{\partial u_t}$ can be calculated as follows:

$$\frac{\partial u_{t+1}}{\partial u_t} = \frac{\partial \tilde{u}_{t+1}}{\partial u_t} \cdot \frac{\partial u_{t+1}}{\partial \tilde{u}_{t+1}} \tag{7.17}$$

We have:

$$\frac{\partial \tilde{u}_{t+1}}{\partial u_t} = I - \sum_{i=1}^{N_k} K_i^{t+1\,T} \cdot \Lambda_i \cdot \left(\overline{K}_i^{t+1}\right)^T \tag{7.18}$$

where $\Lambda_i = diag(\varphi_i^{t'}(z_1), \cdots, \varphi_i^{t'}(z_p))$, and $\varphi_i^{t'}$ is the first derivative of φ_i^t, and $z = k_i^{t+1} * u_t$. In practical applications, we do not need to construct strict matrices K_i and (related to k_i and ki). Instead, we calculate them thro K_i ugh the convolution operation between the image u and the 2D filter k_i, $ki * u$.

Next, we calculate $\frac{\partial u_t}{\partial \Theta_t}$. we have

$$\frac{\partial u_{t+1}}{\partial \tilde{u}_{t+1}} = diag\big(z_1, \cdots z_p\big) \tag{7.19}$$

Since Θ_t contains $(\lambda^t, \varphi_i^t, k_i^t)$, we need to calculate the partial derivatives of u_t with respect to φ_i^t and k_i^t separately. The derivatives of u_t with respect to φ_i^t and k_i^t can be obtained from the following equations:

$$\frac{\partial u_t}{\partial \varphi_i^t} = \frac{\partial \tilde{u}_t}{\partial \varphi_i^t} \cdot \frac{\partial u_t}{\partial \tilde{u}_t}$$
$$\frac{\partial u_t}{\partial k_i^t} = \frac{\partial \tilde{u}_t}{\partial k_i^t} \cdot \frac{\partial u_t}{\partial \tilde{u}_t} \tag{7.20}$$

The gradient of u_t with respect to λ_t is obtained as:

$$\frac{\partial u_t}{\partial \lambda_t} = \left(z_1, \cdots, z_p\right) \tag{7.21}$$

where $z = \frac{1}{2}\left[-1 + \frac{(\lambda^t - \tilde{u}_t) + 2f}{\sqrt{(\tilde{u}_t - \lambda^t)^2 + 4\lambda^t f}}\right]$. In practical applications, to ensure that λ_t is positive during the training process, we set $\lambda = e^{\beta}$. Therefore, in the actual programming process, we use $\frac{\partial u_t}{\partial \beta_t}$ instead of $\frac{\partial u_t}{\partial \lambda_t}$. $\frac{\partial u_t}{\partial \beta_t}$ can be represented as:

$$\frac{\partial u_t}{\partial \beta_t} = \frac{\lambda^t}{2}\left[-1 + \frac{(\lambda^t - \tilde{u}_t) + 2f}{\sqrt{(\tilde{u}_t - \lambda^t)^2 + 4\lambda^t f}}\right] \tag{7.22}$$

7.2.3 Coherent Speckle Denoising Experimental Analysis

To evaluate the performance of the proposed method, we compare it with four representative image denoising algorithms: Enhanced Lee Filter, DPAD, PPBit, and SAR-BM3D. Comparative experiments use commonly used test image sets in image denoising research. To visually compare the advantages of our algorithm, we provide enlarged local images of the same region in the experimental result images of each method. Quantitative analysis is performed using the peak signal-to-noise ratio (PSNR) and the structural similarity index (SSIM) to measure the denoising performance. The formula for calculating the PSNR is:

$$PSNR(dB) = 10\log_{10}\frac{255^2}{\frac{1}{N}\sum_{i=1}^{N}(u_i - \widehat{u_i})^2} \tag{7.23}$$

where N represents the number of pixels in the image, u represents the noise-free image, and $\hat{u}$ represents the denoised image. The formula for calculating the structural similarity index (SSIM) is:

$$SSIM(\hat{u}, u) = \frac{(2\mu_{\hat{u}}\mu_u + C_1)(2\sigma_{u\hat{u}} + C_2)}{(\mu_{\hat{u}}^2 + \mu_u^2 + C_1)(\sigma_{\hat{u}}^2 + \sigma_u^2 + C_2)} \tag{7.24}$$

where μ_u, $\mu_{\hat{u}}$ represent the mean values of images u and $\hat{u}$, σ_u and $\sigma_{\hat{u}}$ represent their respective standard deviations, and $\sigma_{u\hat{u}}$ represents the covariance between μ_u. and

(a) Original image

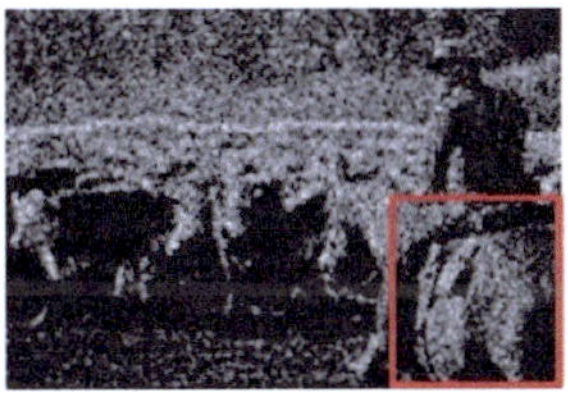

(b) Image after adding noise

(c) Enhanced Lee filtering

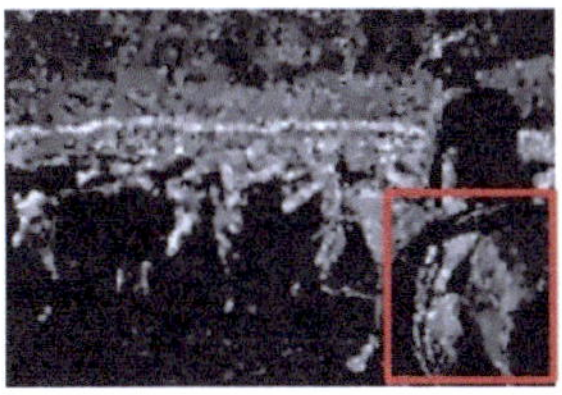

(d) DPAD

(e) PPBit

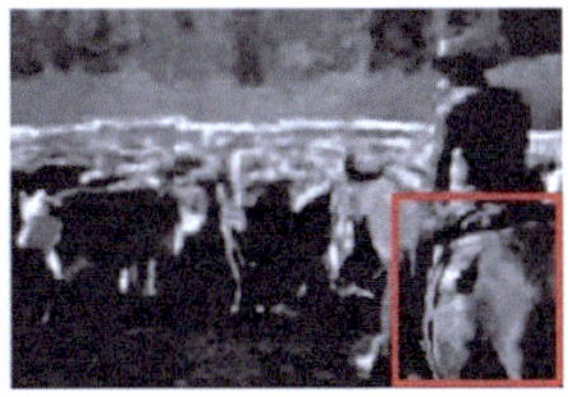

(f) SAR-BM3D

(g) The algorithm proposed in this chapter

Fig. 7.2 Comparison of single-view image denoising results

$\mu_{\hat{u}}$. Additionally, the constants are set as $C_1 = (0.01D)^2$ and $C_2 = (0.03D)^2$, where D represents the dynamic range of the image (255 for 8-bit images). The SSIM value ranges from 0 to 1, with 1 indicating the best reconstruction result and 0 indicating the worst. In the experiments, we selected 8 diffusion steps, a filter size of 7×7, 48 filters for each step, and a training dataset of 400 image samples. First, we processed single-view images. The experimental results in Fig. 7.2 show that compared to the Enhanced Lee Filter, DPAD, and PPBit algorithms, the proposed algorithm in this chapter performs similarly to the SAR-BM3D algorithm, preserving details and small image features. At the same time, compared to the SAR-BM3D algorithm, the proposed algorithm in this chapter obtains clearer edge information. Comparing the results of image denoising and the enlarged local images in Fig. 7.3, it can be observed that in terms of detail preservation and geometric structure preservation, the performance of the proposed algorithm is comparable to SAR-BM3D and superior to the other three algorithms. It can be seen that the speckle interference in the uniform areas of the images after denoising using the SAR-BM3D algorithm is relatively severe. Since the diffusion filters and influence functions used in this chapter's algorithm consider the statistical characteristics of coherent speckle noise during training, it can effectively preserve the structural features of the image while denoising.

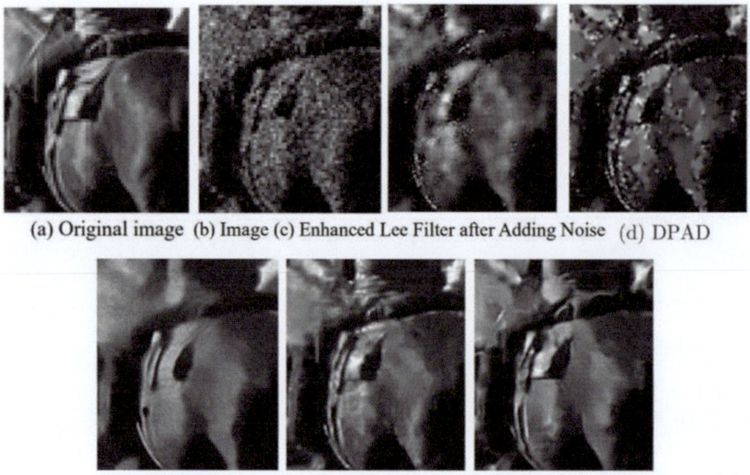

(a) Original image (b) Image (c) Enhanced Lee Filter after Adding Noise (d) DPAD

(e) PPBit (f) SAR-BM3D (g) The algorithm proposed in this chapter

Fig. 7.3 Enlarged local comparison of single-view image denoising results

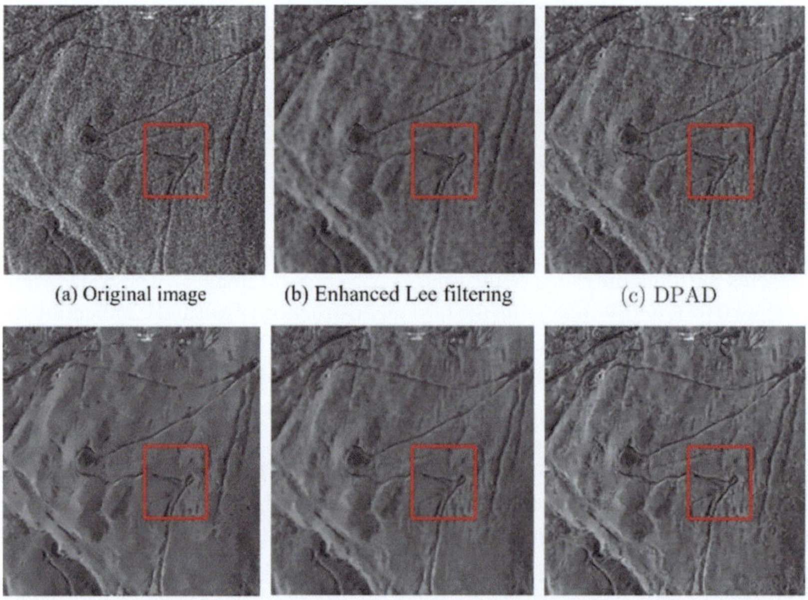

(a) Original image (b) Enhanced Lee filtering (c) DPAD

(d) PPBit (e) SAR-BM3D (f) The algorithm proposed in this chapter

Fig. 7.4 Comparison of denoising results for single-view SAR images

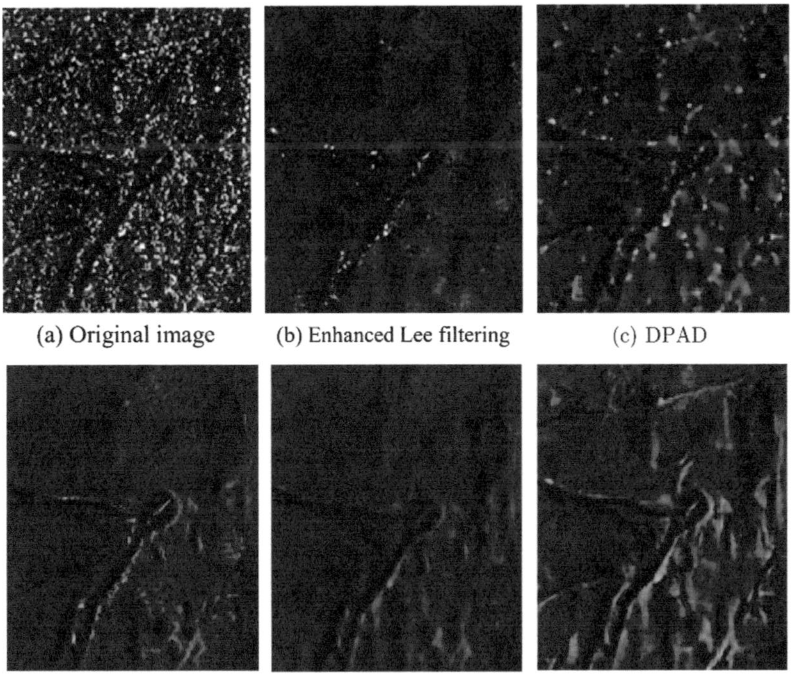

(a) Original image (b) Enhanced Lee filtering (c) DPAD

(d) PPBit (e) SAR-BM3D (f) The algorithm proposed in this chapter

Fig. 7.5 Local zoomed-in comparison of denoising results for single-view SAR images

Through careful visual comparison, we observe that the images processed by the algorithm proposed in this chapter preserve more image details and edge information (Figs. 7.4 and 7.5).

Since the computational complexity of the proposed algorithm mainly lies in the convolution operations of the images, it is highly suitable for parallel implementation. We provide the computation times for denoising two different-sized images using our algorithm, as well as the processing times for the four comparative algorithms. The four comparative algorithms are implemented in MATLAB, but their core parts are compiled using the C language. The algorithm proposed in this chapter is implemented in MATLAB. Table 7.1 also presents the computation times required for implementing our algorithm using GPU parallel processing. Through comparison, it is evident that the proposed algorithm requires less processing time (especially when utilizing GPU parallel computing) compared to the other algorithms.

Table 7.1 Denoising processing times for different-sized images

	Advanced Lee	DPAD	PPBit	SAR-BM3D	Algorithm in this chapter
256 × 256	4.6	20.5	13.2	42.4	3.05 (0.03)
512 × 512	17.8	82.3	48.9	169.1	9.33 (0.09)

7.3 Object Detection and Recognition

7.3.1 *Ship Detection Method Based on Dual-Parameter Global CFAR*

Ship detection in synthetic aperture radar (SAR) images is a research hotspot in marine remote sensing applications. Traditional ship detection algorithms in SAR images suffer from issues such as high computational complexity, inaccurate feature extraction, and high false alarm rates [1–3]. In this chapter, we propose a ship detection algorithm for SAR images based on iterative sidelobe suppression and global constant false alarm rate (CFAR). This algorithm addresses the challenges of high computational complexity and inaccurate feature extraction in traditional SAR ship detection algorithms.

First, the algorithm applies the global CFAR algorithm for preliminary detection. Then, it extracts connected regions and selects the regions of interest (ROIs) based on the number of pixels in each region. Next, it removes the sidelobe interference of the ROIs using an iterative sidelobe suppression method, and extracts their geometric features. Finally, it selects the final detection results based on these geometric features. Experimental results demonstrate that the proposed SAR ship detection algorithm achieves good detection efficiency and effectiveness.

1. **Global CFAR Detection**

CFAR ship detection algorithm has gained wide application due to its simplicity and high efficiency. The basic principle of CFAR detection is to determine an adaptive threshold based on the false alarm probability and the statistical distribution model of clutter in SAR images, and use this threshold to classify pixels as target or background pixels [4, 5].

$$r(x|w_0) < r(x|w_1) \tag{7.25}$$

If the above condition is satisfied, the pixel is classified as a target; otherwise, it is classified as background. The likelihood equivalent form of the above rule is:

$$P(x) = \frac{r(x|w_0)}{r(x|w_1)} > \frac{r(w_0)}{r(w_1)} \tag{7.26}$$

If the above equation is true, classify the pixels as targets, otherwise classify them as backgrounds. For SAR images, assuming that the Prior probability of the ship target and the background clutter is unknown, if the Prior probability of the two is equal, the maximum likelihood criterion for determining the target is:

$$P(x) = \frac{r(x|w_0)}{r(x|w_1)} > 1 \tag{7.27}$$

However, for SAR images, the prior probabilities of background clutter are much higher than the prior probabilities of target pixels, making the assumption of equality in the above case unrealistic. The reasonable decision approach should be to control the probability of the first type of error within a certain range and then find a test that minimizes the probability of the second type of error, which is known as the Neyman-Pearson criterion [6]. When

$$\lambda = \lambda(a) = \frac{r(x|w_1)}{r(x|w_0)} \tag{7.28}$$

and $P_{fa} = \int_a^{\infty} r(x|w_0)\, dx$, the probability of detection error is minimized. The decision criterion becomes:

$$P(x) = \frac{r(t|w_1)}{r(t|w_0)} > \lambda \tag{7.29}$$

where t is obtained from Eq. (7.7):

$$P_{fa} = \int_t^{\infty} r(x|w_0)\, dx \tag{7.30}$$

In Fig. 7.6, Pb (x) is the distribution function of w0, Pt (x) is the distribution function of w1, Pd is the detection accuracy, and Pfa is the false alarm rate. Taking into account both high positive detection rate and low false alarm rate, the optimal threshold T is set to achieve better target detection results.

The traditional general CFAR algorithm process centers around a pixel in the image and sets its background window. Assuming that the clutter background follows a Gaussian distribution, the parameters of the distribution function are estimated by calculating the pixel set within the window [7, 8]. By solving the distribution function and setting the false alarm rate in advance, the CFAR threshold is obtained, and the specific expression is as follows:

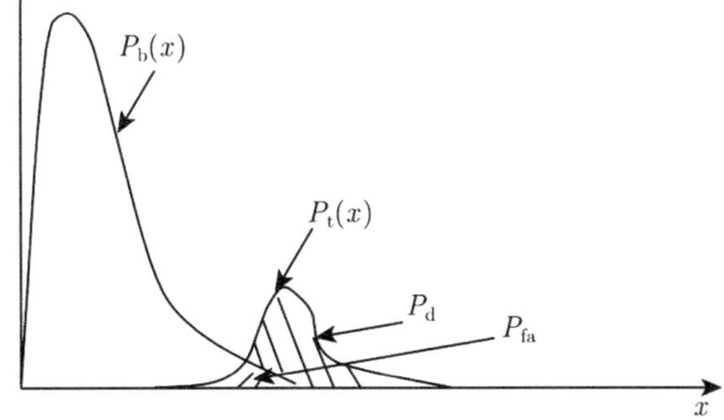

Fig. 7.6 Schematic diagram of detection rate based on threshold

$$1 - P_{fa} = \int_0^T p(x)\,dx \tag{7.31}$$

Compare the threshold with the pixel value of the point, determine whether the point is a target, and output the detection of the pixel point. In the process of obtaining the full image detection results, the background window is the sliding window, and the distribution function is solved. When parameters are set, all pixels repeatedly participate in the calculation, resulting in a large amount of computation and reduced efficiency [9]. However, in actual onboard real-time detection projects, onboard computing storage resources are limited and must be constrained as much as possible. However, traditional CFAR detection algorithms require multiple parameter estimation threshold operations for each pixel. A large amount of redundant calculations have been generated, resulting in high computational overhead. Therefore, this chapter proposes to use the global CFAR detection algorithm to.

Perform preliminary detection, divide the image into blocks, calculate parameters and thresholds once for each block, and only count each pixel once. Higher rate, faster computing speed, and less computational resources are beneficial for adapting to real-time detection on satellites. The basis is that ship pixels in SAR images generally only account for less than 2% of the total number of pixels in the entire image, which affects the background clutter

$$P_{fa} = 1 - \phi((T - \mu)/\sigma) \tag{7.32}$$

The impact can be ignored. In addition, the Gaussian distribution assumption was chosen to replace the commonly used K-distribution and G0 score.

Among them, ϕ Is the cumulative integral function of the Gaussian distribution; μ and σ Represent the mean and standard deviation of clutter, respectively. To ensure similar distribution of ocean clutter in the image, the image size for each detection is set to 3000 × Below 3000. To reduce the missed detection rate under high sea conditions, the threshold is set low, and then constrained and false alarms are removed through feature extraction. Figure 7.7 shows the results after global CFAR detection.

After the preliminary detection using global CFAR and obtaining a binary image, it is necessary to extract target information by connecting pixels based on certain rules. As shown in Fig. 7.8, if pixel o is a target, the eight neighboring pixels (a, b, c, d, e, f, g, h) are the eight neighbors of o, and b, g, d, and e are the four neighbors of o.

In this chapter, a two-pass algorithm is adopted to extract connected regions in the entire image. In the first pass, the image is scanned from the top left to the bottom right, and pixels that are marked as 1 and have not been previously marked are labeled. In the second pass, a depth-first search algorithm is used to merge equivalent labels. During this process, the number of pixels and the coordinate information of each connected region can be extracted. Considering that the area of ship targets is limited, and assuming the resolution is known, the approximate range of the number of pixels representing ship targets in the SAR image can be estimated. Based on

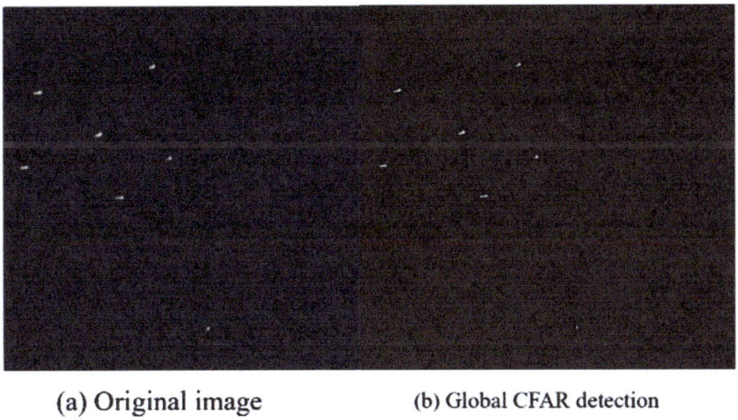

Fig. 7.7 Comparison of global CFAR detection results

Fig. 7.8 Pixel connection relationship

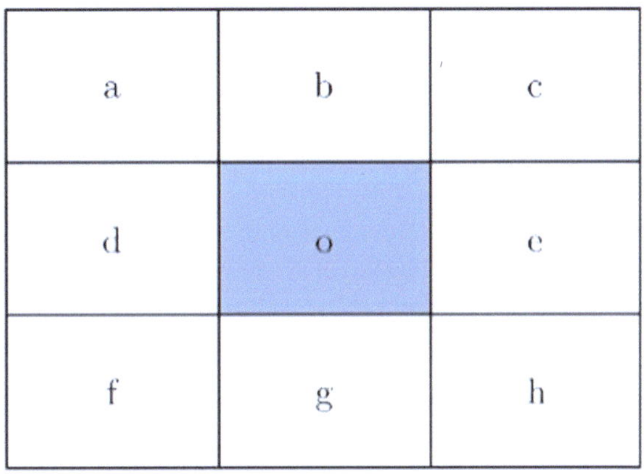

this, preliminary detection results that do not meet the criteria can be eliminated, and ROIs can be selected.

2. **Iterative Sidelobe Correction**

Due to the use of global CFAR detection with low thresholds in the initial detection process, there are high false alarms in the detection results, which may include artificial targets, waves, speckle noise, and other factors. By constraining the number of pixels in connected regions, some false alarms that correspond to small areas, such as waves and speckle noise, smaller than ships, can be eliminated. However, the ability to remove false alarms caused by artificial targets, small rocks, or large waves is limited. Therefore, further feature extraction is required to identify the ROIs and reduce false alarms in the detection results. However, due to the imaging mechanism of SAR, ship targets in SAR images often exhibit cross-shaped sidelobes, as shown in Fig. 7.9, which severely affects the accuracy of feature extraction. To address this

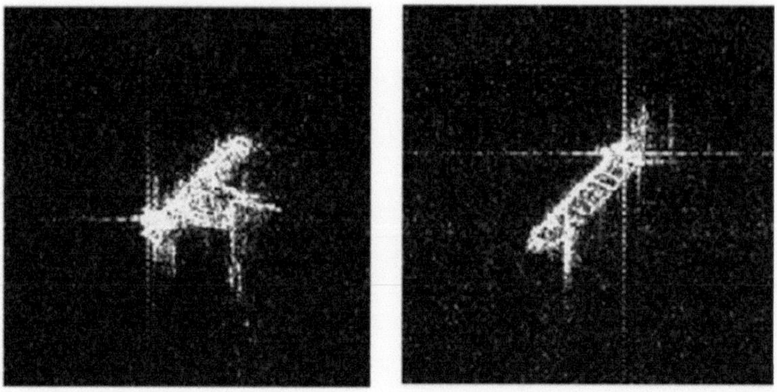

Fig. 7.9 Cross sidelobe phenomenon of ship targets in SAR images

issue, an additional step called iterative sidelobe correction is introduced between the extraction of connected regions and the extraction of geometric features. The purpose of this step is to eliminate the cross-shaped sidelobe effects caused by the SAR imaging mechanism and improve the accuracy of feature extraction for ROIs. To facilitate subsequent operations and reduce computational complexity, each ROI is sliced.

The iterative sidelobe correction technique iteratively performs slice rotation and removal of the sidelobes on both sides of the target. Eventually, the sidelobes on both sides of the ship target are removed, and the ship is rotated to be parallel to the vertical axis. The detailed algorithm process is as follows:

(1) Divide the ship into upper and lower parts and calculate the overall center of gravity of the ship, denoted as P0, and the center of gravity of the upper and lower parts, denoted as P1 and P2, respectively.
(2) Calculate the angle theta between P1, P2, and the vertical axis. Rotate the ship by theta degrees with P0 as the origin.
(3) Iterate the first two steps until theta is smaller than a threshold.
(4) Calculate the number of pixels in each column on both sides of the slice. Set those columns with a count smaller than a threshold T to 0.
(5) Repeat the above four steps 1–3 times until the main axis of the ship is completely parallel to the vertical axis of the image.

After the completion of iterative sidelobe correction, the cross-shaped sidelobes of the ship target are almost completely removed, and the ship is rotated to be parallel to the vertical axis. In the first three steps, the rotation method with continuously adjusted center of gravity is used, which has advantages such as being less affected by sidelobes and higher accuracy compared to traditional methods such as Vusar and Radon transform that calculate the inclination angle before rotation. It can more accurately rotate the ship to be parallel to the vertical axis of the image. The iterative sidelobe correction method combines the tasks of extracting the main axis, removing sidelobes, and rotating the ship to be parallel to the vertical axis in an iterative manner, completing all three tasks simultaneously. Generally, 2–4 iterations are sufficient for the first two steps. Figure 7.10 shows a comparison of the sidelobe correction before and after. It can be observed that this method can effectively remove the sidelobes of the ship target and accurately rotate the main axis of the ship to be parallel to the vertical axis. The minimum bounding rectangle closely aligns with the real ship, demonstrating the feasibility of the proposed algorithm.

3. **Geometric Feature Extraction and Identification**

Geometric features such as length, width, aspect ratio, azimuth, area, perimeter, and other physical quantities provide a direct description of ship targets. In this chapter, three features, namely length, width, and aspect ratio, are used to identify the targets. After the ROI slices undergo iterative sidelobe correction, their geometric features can be extracted through simple traversal and judgment operations, without the need for calculating the minimum bounding rectangle.

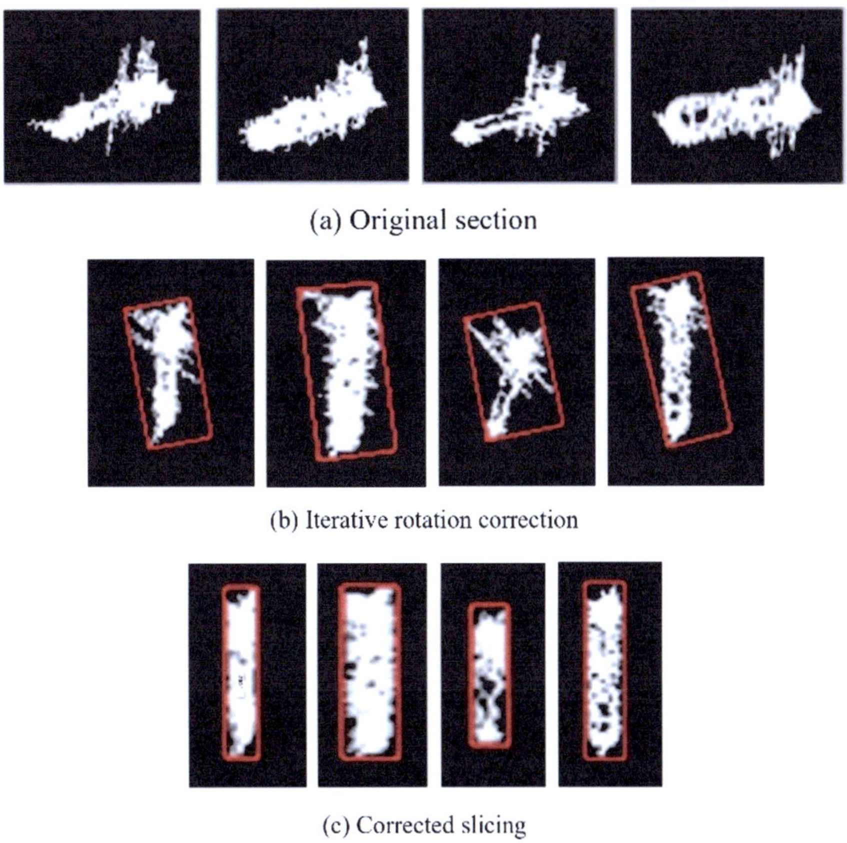

Fig. 7.10 Effect of iterative sidelobe correction

Length feature: Starting from the upper edge of the slice, calculate the number of pixels in each row. If the pixel count is less than 3, continue moving towards the center; otherwise, save the row value as "up" and end the operation. Starting from the lower edge of the slice, calculate the number of pixels in each row. If the pixel count is less than 3, continue moving towards the center; otherwise, save the row value as "down" and end the operation. The length of the target can be represented as:

$$\text{Length} = \text{down} - \text{up} \tag{7.33}$$

Width feature: Starting from the left edge of the slice, calculate the number of pixels in each column. If the pixel count is less than 3, continue moving towards the center; otherwise, save the column value as "left" and end the operation. Starting from the right edge of the slice, calculate the number of pixels in each column. If the pixel count is less than 3, continue moving towards the center; otherwise, save the column value as "right" and end the operation. The width of the target can be represented as:

$$\text{Width} = \text{right} - \text{left} \tag{7.34}$$

Aspect ratio feature: The aspect ratio of a ship is the ratio of its length to width, typically ranging from 5 to 8.

Compared to commonly used methods such as Radon transform to find the longest major axis or calculating the minimum bounding rectangle, the proposed method is less affected by sidelobes, resulting in more accurate and robust results with lower dependency on the CFAR detection algorithm.

Finally, after performing the above steps, the geometric features of each ROI slice in the image are extracted. Based on prior geometric knowledge of ships, feature identification can be applied to filter out most false alarms.

4. **Experimental Results and Analysis**

In this chapter, experiments were conducted using ocean SAR images from GF-3 satellite (single polarization, image resolution of 3 m) to verify the effectiveness of the proposed detection algorithm [6]. The experimental parameters were set as follows: CFAR threshold $T = u + 4.5v$, area constraint for connected regions ranged from 100 to 10,000, the number of iterations in the iterative sidelobe correction was set to 3, the angle constraint in each iteration was set to 6°, and the sidelobe constraint threshold was set to 15.

Experiment 1: A sample image was selected to conduct experiments on iterative sidelobe correction and geometric feature extraction. Some of the experimental results are shown in Fig. 7.11.

Experiment 2: 200 GF-3 images were selected to compare the computation time between the traditional method and the proposed algorithm. Some of the comparison results are shown in Table 7.2.

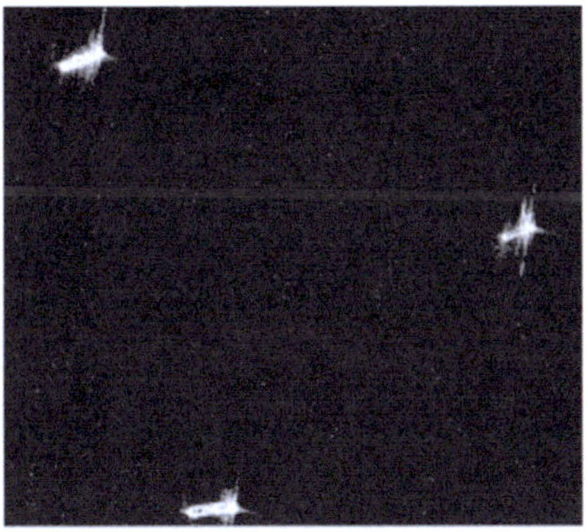

(a) Original image

(b) Partial slice iterative sidelobe correction results

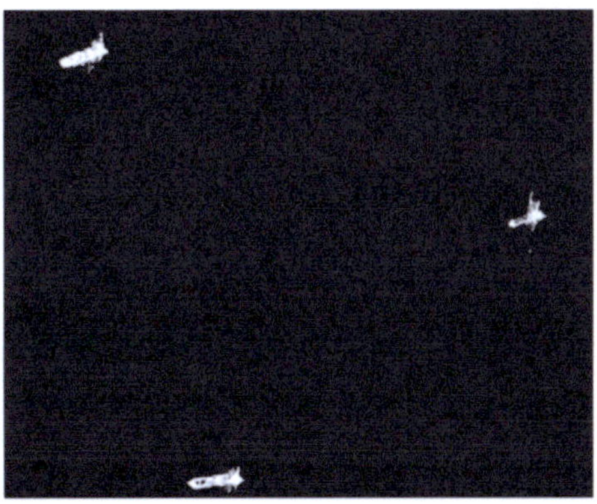

(c) Final test results

Fig. 7.11 Test results

Table 7.2 Comparison of computation time between the traditional method and the proposed algorithm

Sample	Silde CFAR/s	DPG-CFAR/s
1	61.21	4.23
2	70.35	6.25
3	65.37	5.35
4	76.31	7.02
5	68.12	5.59

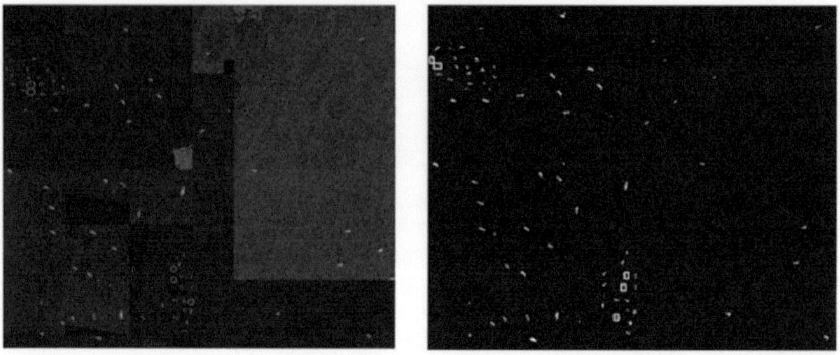

(a) Original and missed experimental images (b) Detection results and false alarms

Fig. 7.12 Misdetection and false alarm results

Experiment 3: Multiple GF-3 images were cropped and stitched together to form a large experimental image with a width of 6250 and a height of 6144. The detection results are shown in Fig. 7.12.

From Fig. 7.11, it can be observed that the proposed iterative sidelobe correction method effectively removes the cross sidelobe effect in SAR images, resulting in more accurate extraction of geometric information.

From Table 7.2, it can be seen that compared to traditional detection methods, the proposed algorithm has lower computational complexity and faster speed. The sliding window CFAR in traditional detection algorithms is the most time-consuming part, accounting for more than 80% of the total computation time, involving a large number of multiplications and accumulations. In contrast, the detection algorithm proposed in this chapter considers real-time application requirements and adopts the global CFAR algorithm, simplifying the computational steps and improving the computational efficiency by approximately 10 to 100 times.

From Fig. 7.12, it can be seen that there are a total of 94 ships in the large experimental image. The missed detections are indicated by red circles in Fig. 7.12a, and the false alarms are indicated by yellow boxes in Fig. 7.12b, resulting in 5 missed detections and 6 false alarms. The missed detection rate is 6.4%, and the false alarm rate is 5.3%. The false alarms are mostly caused by weak and fragmented ship pixels in the image, while the missed detections are mainly due to severe sea conditions, small ships, or weak pixels.

Compared to traditional CFAR detection algorithms, the algorithm introduced in this chapter has the following advantages:

(1) Adoption of global CFAR significantly reduces computational complexity.
(2) Utilization of the iterative sidelobe correction model removes sidelobe effects and enables more accurate extraction of target geometric features compared to traditional feature extraction algorithms.

7.3.2 Building Object Detection Method Based on Interferometric Shadows

Different radar waves have different penetration capabilities, but in general, longer wavelengths have stronger penetration capabilities and shorter wavelengths have higher positioning accuracy. Taking vegetation and foliage penetration as an example, VHF (Very High Frequency) radar with a carrier frequency of 50 MHz has excellent forest penetration capabilities, while the P-band (carrier frequency of 400 MHz) and L-band (carrier frequency of 1250 MHz) can also perform certain ground exploration tasks. SAR is an active detection device that emits modulated electromagnetic waves in the direction of the antenna array. When there is occlusion in the scene, SAR images will exhibit shadows. As shown in Fig. 7.13, L_s represents the length of the shadow on thc ground, and Ls represents the corresponding length of the shadow on the slant range plane.

Based on the measured data shown in Fig. 7.14, it can be observed that building shadows are apparent in the Ka-band, exhibiting clear contour information and low intensity within the shadowed regions.

1. Phase Shadow Analysis

In the context of interferometric phase, the analysis of shadow characteristics begins with examining the statistical properties of the SAR interferogram's phase and modeling its probability density function. The accuracy of the interferometric phase is

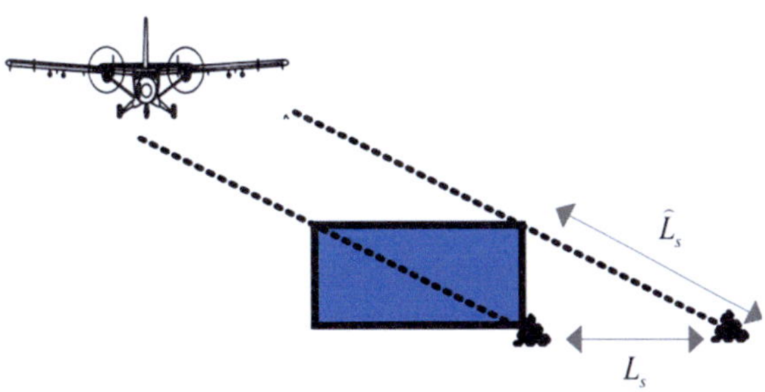

Fig. 7.13 SAR shadow illustration

Fig. 7.14 Architectural shadow in Ka band SAR images

influenced by various independent factors. To achieve high coherence, the following factors affecting decorrelation need to be considered:

(1) Temporal Decorrelation: This includes explicit changes in the scene before and after observation, such as urban changes over years, vegetation growth over months, seismic activities, landslides, and changing water flows, migrating glaciers, etc., over daily intervals. These changes introduce temporal variability in the corresponding regions of the interferogram, resulting in decreased coherence.
(2) Radar Receiver Noise: The noise in the radar receiver chain is primarily attributed to thermal noise. When the echo signal from the target is weak, it can be overwhelmed by internal receiver noise. This noise poses significant limitations on the effective range in traditional pulse radars, but in the SAR context, it directly affects the signal-to-noise ratio of the interferometric phase and consequently the image.
(3) Processing Decorrelation: Both the hardware signal path and the signal processing after digital sampling have a significant impact on the noise characteristics of the interferometric phase. The choice of slant range model and the order of the phase polynomial in the processing flow are crucial. For example, basic frequency-wavenumber algorithms cannot achieve high-quality imaging in slant viewing.
(4) Spatial Decorrelation: Spatial decorrelation arises from differences in viewing geometry and is distinct from other sources of decorrelation. It is inherent to the interferometric principle and can lead to the failure of phase correspondence between two SAR images when the viewing geometry difference is too large.

In Chap. 5, the coherence coefficient defined for along-track interferometry is denoted by γ, and the cross-track coherence coefficient is defined by the equation:

$$\gamma = \gamma_s \cdot \gamma_n \cdot \gamma_p \tag{7.35}$$

In the equation, γ_s represents the coherence coefficient of the scene during the acquisition time, γ_n is the noise coherence coefficient, and γ_p is the processing coherence coefficient. In the analysis scenario of this chapter, dynamic targets are not considered, so the temporal coherence of the scene is assumed to be constant and equal to 1. Considering that the target scene echoes are subject to complex Gaussian modulation with a power spectral density of $2\sigma^2$. The transfer functions of the processing flow are denoted as H_1 and H_2, which produce signal outputs y_1 and y_2 respectively, given the same input x. Taking into account the noise components carried by the processing output signals, denoted as n_1 and n_2, the signal-to-noise ratios are obtained as SNR_1 and SNR_2. The interference pattern is defined as $z_1 \cdot z_2^*$, and the interference phase is given by

$$\Delta\phi = \arg\left(z_1 \cdot z_2^*\right) \tag{7.36}$$

where $z_1 = y_1 + n_1$; $z_2 = y_2 + n_2$. The autocorrelation functions of the original signal x with noise component n_1 are given by:

$$R_{xx}(\tau) = 2\sigma^2\delta(\tau) \tag{7.37}$$

$$R_{n_1n_1}(\tau) = 2\sigma_1^2 \int \left|H_{1p}\right|^2 \exp(-2\pi j f \tau) df \tag{7.38}$$

Here, $2\sigma_1^2$ is the power density of noise component 1, and H_{1p} is the transfer function that introduces noise n_1. The signal-to-noise ratio is then given by:

$$SNR_1 = \frac{\sigma \int_{-\infty}^{+\infty} \left|H_p\right|^2}{\sigma_1 \int_{-\infty}^{+\infty} \left|H_{1p}\right|^2} \tag{7.39}$$

where $\phi_0 = \arg(\gamma)$.The probability density function of the interference phase $p(\Delta\phi)$ is described as:

$$pdf(\Delta\phi) = pdf_0(\Delta\phi - \phi_0) \tag{7.40}$$

$$pdf_0(\phi) = \frac{1-|\gamma|^2}{2\pi} \frac{1}{1-|\gamma|^2\cos^2\phi} \left\{ 1 + \frac{|\gamma|\cos\phi \arccos(-|\gamma|\cos\phi)}{\sqrt{1-|\gamma|^2\cos^2\phi}} \right\} \tag{7.41}$$

Considering that $pdf_0(\phi)$ is defined within the circular phase, its expectation is:

$$E(\phi) = \int_{-\pi}^{+\pi} \phi pdf_0(\phi)d(\phi) = 0 \tag{7.42}$$

Therefore, the phase offset $E(\Delta\phi) = \phi_0 = \arg(\gamma)$; phase variance is given by:

$$E\{(\phi_0 - \Delta\phi)^2\} = \int_{-\pi}^{+\pi} \phi^2 pdf_0(\phi)d(\phi) = \sigma_\phi^2 \tag{7.43}$$

Since the phase expectation ϕ_0 only affects the distribution center and not the phase variance, we only analyze the unbiased quantity $\phi = \Delta\phi - \phi_0$. The coherence coefficient γ_n and processing coherence coefficient γ_p are modulated by signal-to-noise ratios SNR and transfer function H., respectively, given by:

$$\gamma_n = |\gamma_n| = \frac{1}{\sqrt{(1 + SNR_1^{-1})(1 + SNR_2^{-1})}} \tag{7.44}$$

$$\gamma_p = \frac{\int_{-\infty}^{+\infty} H_1(f)H_2^*(f)df}{\sqrt{\int_{-\infty}^{+\infty} |H_1(f)|^2 |H_2(f)|^2 df}} \tag{7.45}$$

The variance of the interference phase depends only on the magnitude of the coherence coefficient. The simulation results are shown in Fig. 7.15.

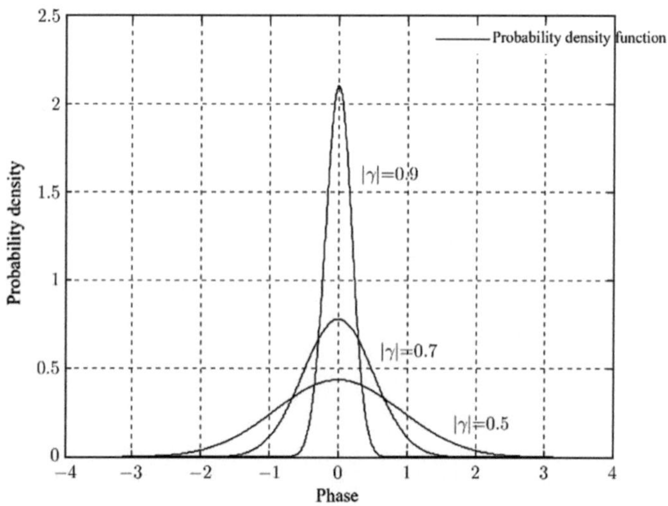

Fig. 7.15 Probability density function

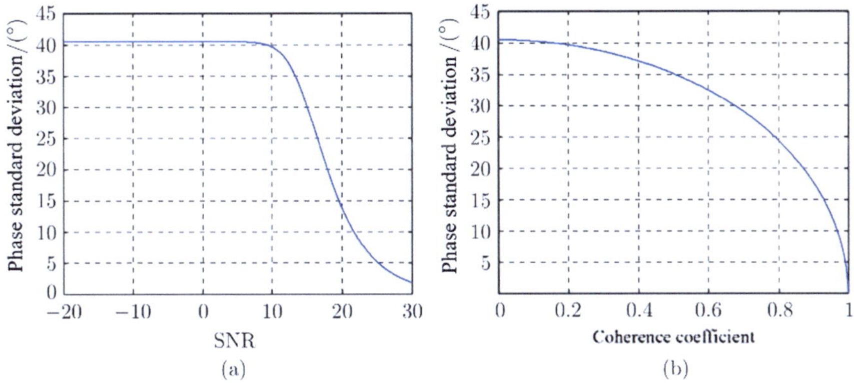

Fig. 7.16 Andard deviation curve with changes in **a** signal-to-noise radio and **b** coherence coeffisient

2. Building Shadow Detection Method Based on Dual Thresholds

When analyzing only thermal noise, the simulation analysis shows the phase standard deviation as a function of signal-to-noise ratio or coherence coefficient, as shown in Fig. 7.16. The analysis reveals that in low coherence scenarios, there is significant phase fluctuation. If the mean phase can be removed, power metrics can be used to distinguish high coherence regions from low coherence regions.

In the measured data images, it can be observed that shadows in the Ka-band exhibit distinct phase characteristics even after removing the first-order phase gradient. The shadow regions, characterized by a zero-mean Gaussian distribution, also possess a relatively strong overall intensity, as shown in Fig. 7.26. These shadow areas exhibit higher energy compared to the weaker coherent echo energy.

This feature is achieved due to the ultra-short centimeter-level baselines of the millimeter-wave multi-channel system. In general, building heights are smaller than the ambiguity height $h < h_{\text{amb}}$. The ambiguity height of a single-pass ranging system can be expressed as:

$$h_{\text{amb}} = \frac{\lambda R_0 \sin\theta}{B\cos(\theta - \alpha_0)} \tag{7.46}$$

where R_0 is the slant range length, θ is the incidence angle, B is the baseline, and α_0 is the antenna's normal viewing angle. The ambiguity height refers to the elevation value of one cycle of phase change caused by terrain variation after removing the flat earth phase. In the case of long baselines, areas with significant height variations can exhibit intensity characteristics with variance close to a Gaussian distribution due to large slope-induced phase wrapping. Through calibration parameter design analysis, it is possible to achieve ambiguity height values exceeding 200 m at an incidence angle of 60°, as observed in the measured data in Fig. 7.17, where kilometer-level

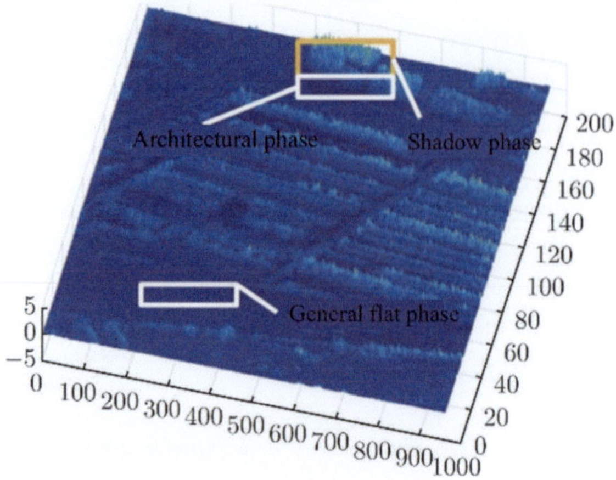

Fig. 7.17 Schematic diagram of interference phase intensity

effective width does not experience phase ambiguity after removing flat earth phase and low-order phase gradient.

By calculating the measured data, the gradient-removed phase is used as a measure. As shown in Figs. 7.18 and 7.19, the shadow areas exhibit significantly higher echoes (45.81 dB) compared to the road scenes. This provides strong evidence for using millimeter-wave shadow data as a basis for building target identification.

However, the echo phase from background scenes also exhibits randomness and contains strong energy regions. Considering the temporal characteristics of shadow regions, where the intensity is only about the background noise level, a dual-threshold shadow detection method for extracting millimeter-wave building information is proposed. The method determines shadow regions based on temporal intensity thresholds and phase energy thresholds. An adaptive mean filtering, based on intensity, is performed to achieve region expansion, enlarging the shadow areas while reducing noise and improving coherence. The output result is obtained by combining threshold decision and temporal intensity determination to create a shadow mask.

The purpose of adaptive filtering is to increase the boundary distinguishability of shadow regions and improve image smoothness. The specific steps include neighborhood summation, squaring of point values, and removal of small values until there are no noisy points within the block mask. Unlike general denoising filtering methods, it does not preserve detailed phase information. The specific description of the enhanced filtering method for "shadow noise" is as follows.

① Firstly, the flat earth phase, first-order phase gradient, and second-order phase gradient compensation are removed within the region of interest. Then, the mean of the absolute phase values is selected as the initial threshold.

② The amplitude mean of the absolute phase values can represent the approximate power trend of the quantity. If filtering is performed with the threshold value at this stage, it will result in a large range of global noise points. Therefore, a

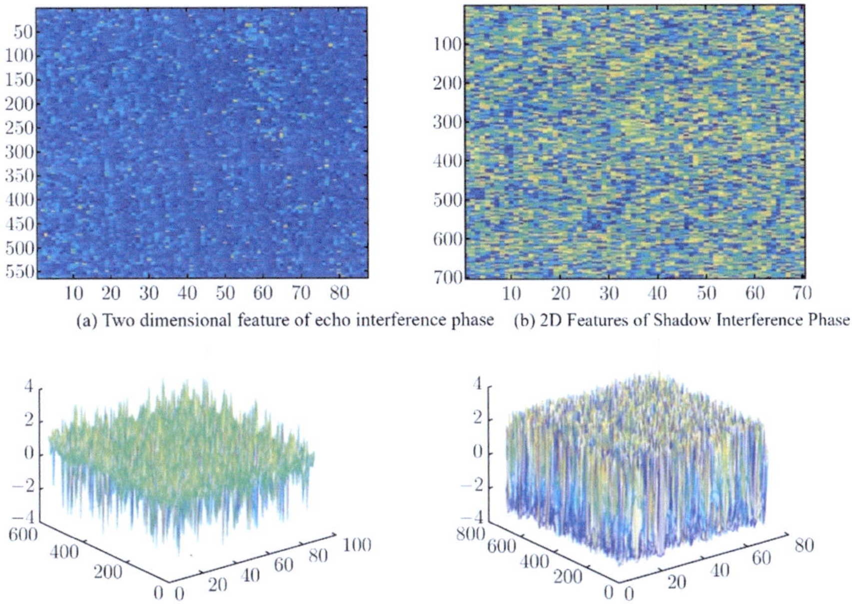

(a) Two dimensional feature of echo interference phase (b) 2D Features of Shadow Interference Phase

(c) 3D features of echo interference phase (d) Shadow Interference Phase 3D Features

Fig. 7.18 Phase characteristics

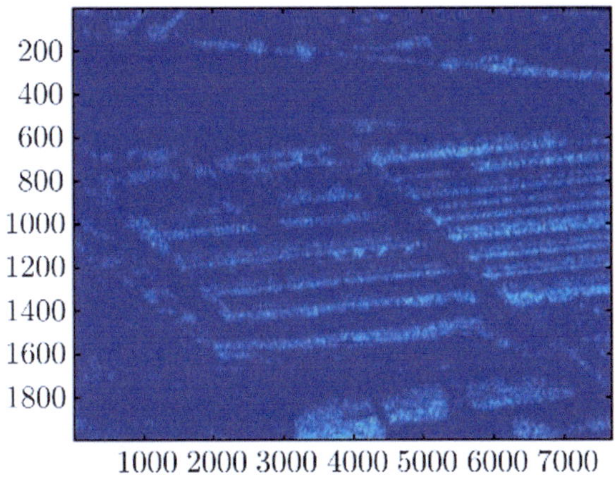

Fig. 7.19 Interference phase intensity

regular mean or median filtering with a small rectangular window is applied to achieve phase smoothing.

③ Finally, the threshold value detection is performed to retain only the "high-energy" shadow regions. The shadow enhancement result is shown in Fig. 7.20.

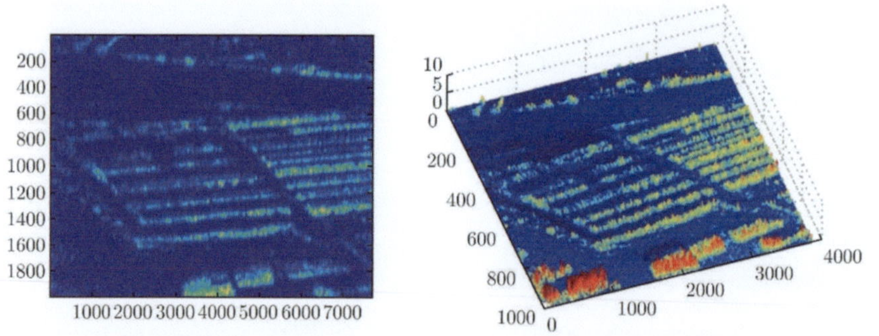

(a) Filtered phase strength (b) Three dimensional schematic of filtered phase strength

Fig. 7.20 Adaptive filtering for shadow enhancement

The workflow of this dual-threshold building shadow detection method is summarized in Fig. 7.21.

Through interference shadow enhancement filtering and threshold extraction using dual thresholds, the potential targets in the mask are identified and located using morphological clustering. Finally, area-based filtering is applied using a threshold of 500 m^2 to extract buildings, achieving satisfactory extraction results. The results are shown in Fig. 7.22

Large shadow regions are presented in Table 7.3, where it can be observed that there are more than 10 regions with shadow areas exceeding 3300 m^2 in the observation area, mainly consisting of standalone buildings and adjacent building clusters. Due to the filtering and clustering process, the actual detection area may be larger than the original shadow region, so the actual shadow area will be smaller than the calculated value. The horizontal and vertical positions are extracted from the region's center point, as shown in the following table. The total number of points in the horizontal axis of the image is 900, and the total number of points in the vertical axis is 1200.

7.4 Inversion of Ocean Information Based on Millimeter-Wave Synthetic Aperture Radar

Millimeter-wave SAR (Synthetic Aperture Radar)/InSAR (Interferometric SAR) in the Ka-band has the characteristics of low penetration, short interferometric baselines, and high phase sensitivity. It enables wide-area marine surveys and detailed inspections, as well as the extraction of marine dynamic topology information and velocity field inversion. Therefore, research on marine information inversion techniques based on millimeter-wave SAR systems/data should be conducted. This includes the design of inversion schemes and quantitative analysis for marine elements such as sea surface currents.

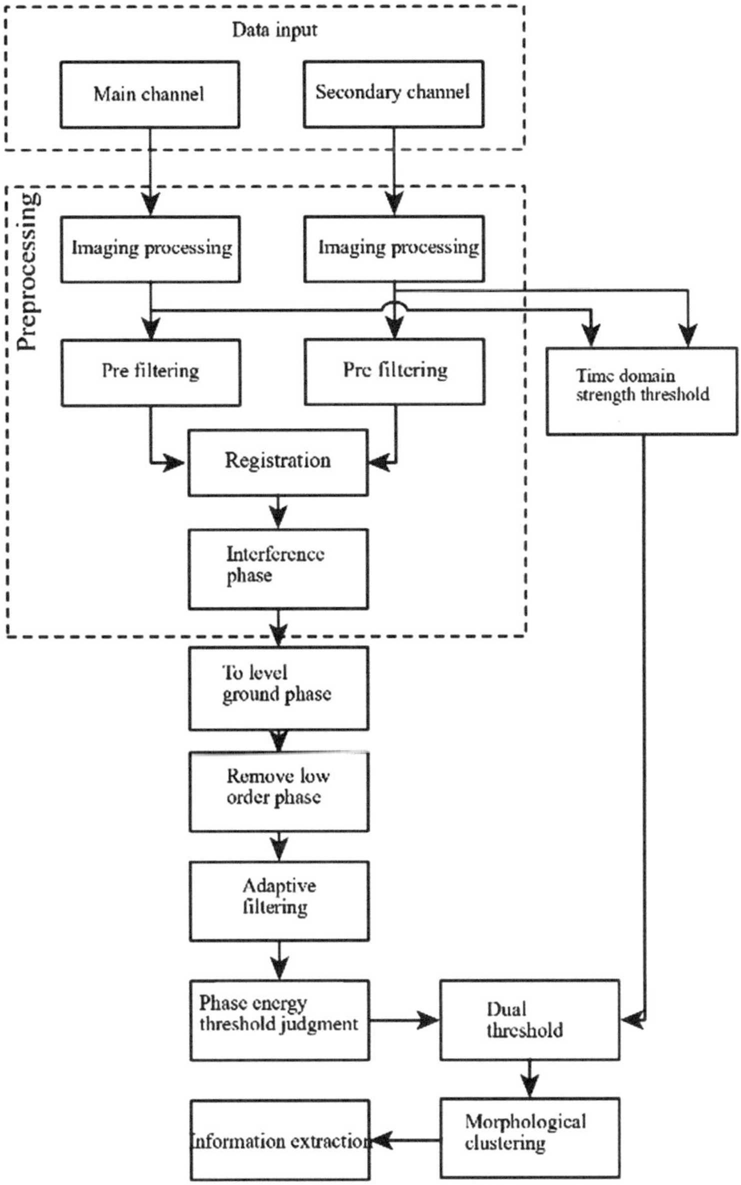

Fig. 7.21 Flow chart of double threshold shadow detection

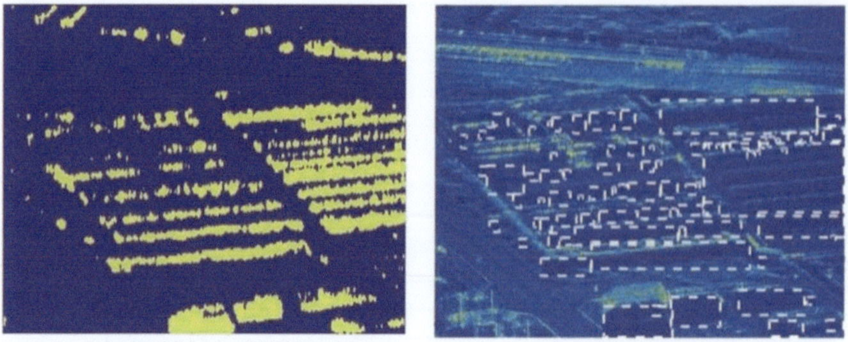

(a) Double threshold mask (b) Building Shadow Recognition

Fig. 7.22 Extracted result of building shadow information

Table 7.3 Target parameter extraction

Shadow numbering	Shaded area area/m^2	Balance/vertical position	
1	26,235	800	573
2	12,336	763	811
3	12,072	337	432
4	9450	475	61
5	7992	518	114
6	6747	818	143
7	5352	399	624
8	4962	548	314
9	4314	489	146
10	3345	765	643

Furthermore, SAR has the advantage of working continuously and under all weather conditions, which cannot be replaced by optical remote sensing. It has always been an important means for marine monitoring. In the development and utilization of various marine activities, the phenomena and activity status of the ocean are associated with multiple aspects of economic interests.

7.4.1 Methods for Sea Surface Current Inversion

The ocean covers more than 70% of the Earth's surface and plays a significant role in the global climate, biological evolution, and prevention of meteorological disasters through the transfer of dynamic and thermodynamic energy. Although the longwave geostrophic component in the oceanic environment can be estimated using sea surface height inversion based on widely deployed buoys and water stations, there is currently no mature spaceborne technique for measuring oceanic currents. The use of traditional SAR-based repeat-pass or distributed interferometry for measuring oceanic currents is challenging due to the decorrelation of moving targets. Therefore, developing flow field measurement methods and systems based on single satellite multi-channel interferometry is of great significance. Millimeter-wave SAR in the Ka-band has attracted extensive attention in the field of Earth observation due to its advantages in short baselines and high sensitivity to surface scattering characteristics under an interferometric configuration. It can achieve near-optimal performance with a single Ka-band satellite, which requires distributed interferometric baselines in other bands such as L-band and C-band. However, according to the radar equation, the millimeter-wave band suffers from high attenuation, and factors such as transmitter efficiency and atmospheric losses have hindered its large-scale application. In this regard, millimeter-wave systems based on the DBF-SCORE (Digital Beamforming—Single Channel, One Reference) technique can avoid the need for peak transmit power in the tens of kilowatts, thereby achieving excellent measurement results. It has the potential to address the wide-area interpretation of submesoscale (1 km–10 km) vortices and turbulence velocity vectors from space.

This subsection focuses on the precise interpretation technology of ocean currents based on the millimeter-wave band. It enables the wide-area vector inversion of geostrophic and non-geostrophic flows through multi-channel dual-beam practice. Additionally, the analysis of beam broadening in the DBF configuration and other error sources is conducted, and the suitable on-board processing architecture and ground processing flow for interferometric configurations are clarified. The DBF-SCORE technique is a wide-swath imaging solution based on multi-channel processing, which has been described in detail in Chap. 3, Sect. 7.2. It first equips the transmission antenna with a small aperture size corresponding to the beam width and multiple small aperture receiving antennas at the receiving end to achieve full coverage of wide scenes. The signal quality is then enhanced through the reception-side scanning, i.e., SCORE, and theoretical gain improvement is achieved through channel-wise phase alignment synthesis. Figure 7.23 shows the scan reception and dual-beam scanning diagram used for omnidirectional velocity measurement in the range profile.

1. Beam Broadening Loss

Multi-channel interferometry can be effective in weak echo situations. However, according to the literature, in the ideal model, the echo beam width will be broadened after the projection of slant range to ground range due to the influence of incident angle and pulse width. The broadening quantity is given by:

$$\chi_\theta(\theta) = \frac{t_p c}{2R_a\left(\frac{R_a\cos(\theta)}{\sqrt{R_e^2-(R_a\sin(\theta))^2}} - 1\right)\sin(\theta)} \tag{7.47}$$

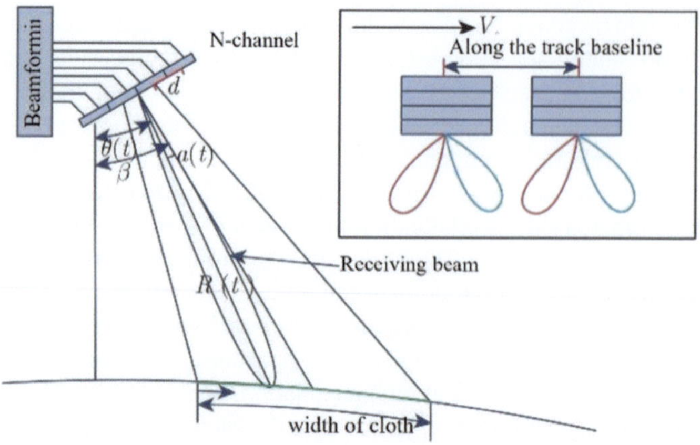

Fig. 7.23 Schematic diagram of scanning and receiving for comnidirectional flow velocity measurement

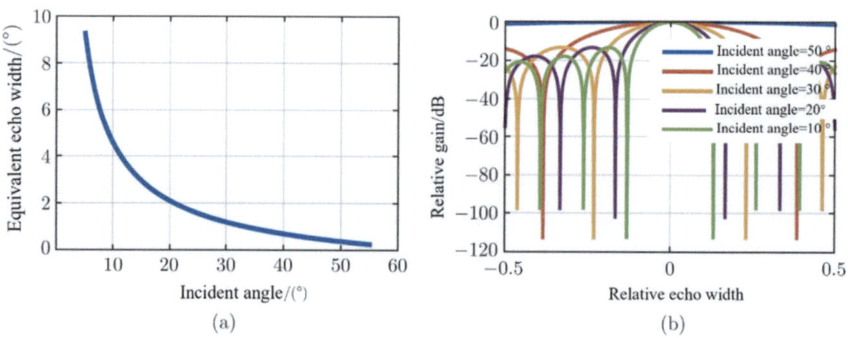

Fig. 7.24 Echo equivalent beam broadening

When the beam width exceeds the main lobe of the synthesized beam, there will be a gain loss. Moreover, when the pulse extension loss (PEL) modulation depth is too large, additional phase modulation will occur due to amplitude modulation. Figure 7.24a shows the echo broadening as a function of incident angle for a fixed pulse width, and Fig. 7.24b shows the corresponding equivalent DBF antenna reception pattern for the broadened beam.

Wave Angle Error The coefficients used in DBF-SCORE technology are directly related to the angles. If there is a large error in the wave angle information, it is equivalent to a misalignment between the narrow beam scanning and the actual echo, resulting in additional phase modulation. The relationship is given by:

$$\omega_n(t) = \exp\left(-j\frac{2\pi d_n \sin(\alpha(t))}{\lambda}\right) \tag{7.48}$$

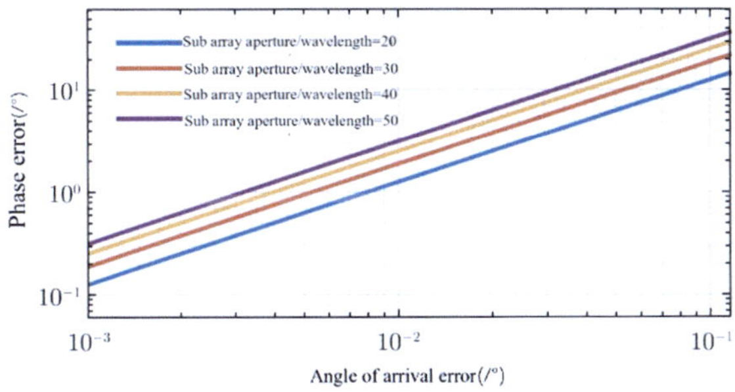

Fig. 7.25 Arrival angle error and corresponding phase error

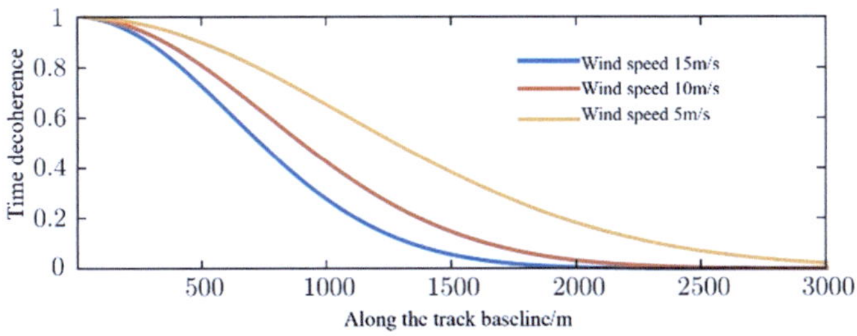

Fig. 7.26 Decoherence factor caused by baseline along the track

Figure 7.25 shows the relationship between wave angle error and corresponding phase error.

When the wave angle estimation error occurs, it will also cause the misalignment of the center angle of the echo's equivalent beam direction diagram. This will result in a decrease in gain and even degradation of pulse compression performance, leading to secondary phase error modulation.

Analysis of Measurement Architecture The fixed amplitude and phase errors in the internal links of the multi-channel system will affect the synthesis performance. The channel intensity in the echo is both noise and useful signal. In single-channel processing, amplifying the entire digital signal does not improve any indicators. However, the intensity differences between channels will affect the contribution of signal-to-noise ratio after signal synthesis. Therefore, amplitude equalization should be performed in the absence of prior information. Similarly, fixed phase calibration is also based on the assumption that the noise characteristics of the single-channel signal are good, and the delay is slowly varying, so that the phase at a certain point or segment can be used as a reference for phase calibration.

In the interferometric configuration shown in Fig. 7.23, the multiple channels of each antenna array on the slant range plane perform one-dimensional high-resolution and gain improvement. On the azimuth plane, high resolution is obtained through synthetic aperture, and interferometric phase extraction is performed on the time baseline. The coherence time of SAR data on the sea surface is directly related to the wind speed, and the calculation of the decorrelation factor caused by the baseline length under different wind speeds is shown in Fig. 7.26, indicating that distributed systems or high-resolution ATI are not suitable for high-precision wind speed measurement under strong wind speeds.

Considering the implementation of one of the two antennas in Fig. 7.27, the phase performance under echo broadening, wave angle, and fixed amplitude and phase errors is analyzed. The fixed amplitude and phase errors can be calibrated by a high-precision internal calibration device. Figure 7.27 shows two implementation architectures of beam synthesis before and after pulse compression. LNA represents a low-noise amplifier, IF is the intermediate frequency reference signal, E_N is the calibration amount for fixed amplitude and phase errors, $\omega_N(t)$. is the SCORE weighting factor, and $H(f)$ is the frequency-domain matched filter.

For the architecture shown in Fig. 7.27a, the traditional SCORE architecture, which performs synthesis first, the shaping of the wideband chirp signal's beam pattern is significantly affected by narrow beam modulation. The optimal solution is to use frequency-independent weighting coefficients for modulation. However,

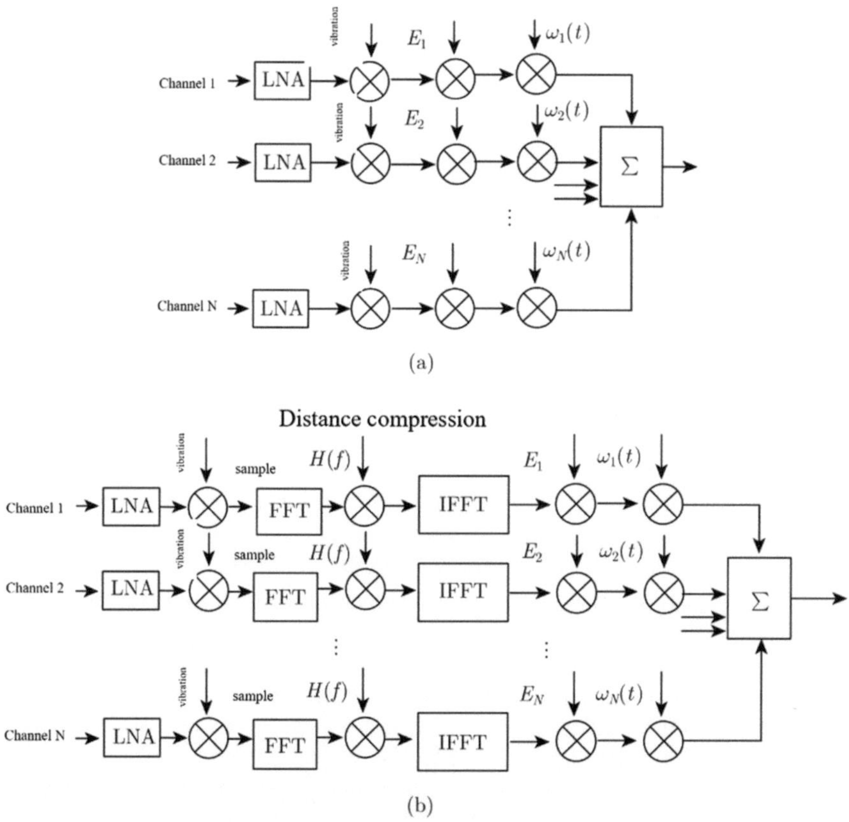

Fig. 7.27 Comparison of beam synthesis implementation architectures

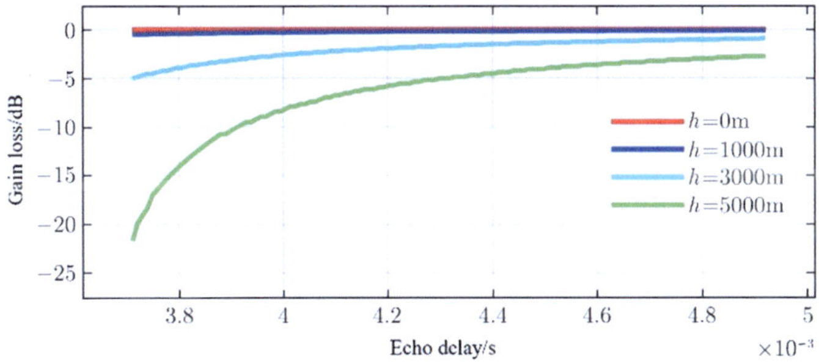

Fig. 7.28 Gain loss caused by terrain mismatch

this operation is computationally expensive. For the architecture in Fig. 7.27b, the signal processed by pulse compression undergoes range gain modulation and resolution acquisition, followed by coherent accumulation. The difference here is that the receiving direction diagram is essentially a synthesized narrow beam in the former case, while it is a wide beam with the original small antenna aperture in the latter case. Therefore, for high-resolution or high-precision interferometric requirements, the latter has an advantage.

Since it is a single-satellite interferometry, the phase modulation term caused by wave angle estimation error is removed when the difference is taken between the two arrays, and it does not constitute residual phase. However, the gain loss directly increases the RMS phase error. Simulation results in Fig. 7.28 show that for a sea surface with fluctuations much smaller than kilometers, wave angle estimation can be omitted. The symbol "h" represents the difference in the sea surface.

The determination of the geostrophic component of the upper layer flow field in the ocean relies on the dynamic terrain data of the sea surface. The zonal and meridional velocities, v and u, are given by:

$$v = -\frac{g}{2\gamma\sin(\varphi)}\frac{\partial D}{\partial x} \tag{7.49}$$

$$u = -\frac{g}{2\gamma\sin(\varphi)}\frac{\partial D}{\partial y} \tag{7.50}$$

where g is the acceleration due to gravity, γ is the Earth's rotation rate, φ is the latitude, and x and y are the coordinate scales. D is the average sea surface height, and N is the geoid height. Therefore, the composition of the measured wide-area flow field data after flow field inversion can be achieved through precise sea surface mapping to obtain average sea surface elevation, followed by differencing with the geoid height to obtain the dynamic terrain of the sea surface, and then further fitting models to obtain geostrophic currents. Currently, there is no direct spaceborne mapping source for non-geostrophic components, which can be extracted from the full-directional flow field of along-track interferometry after the removal of fitted data. It is necessary to clarify to what extent the accuracy of sea surface elevation is meaningful. Based on the accuracy of existing MDT models and the measurement capabilities of gravity satellites and altimetry satellites for geoid and ocean dynamic surfaces, it can be seen that wide-width data at the kilometer scale can be effective.

The application of small-aperture single-satellite high-coherence multi-baseline data has the capability to achieve large-scale wide-area sea surface mapping. Based on the above analysis, a along-track interferometry method based on DBF-SCORE is proposed. It extracts velocity information in the target area through pulse compression, DBF synthesis, imaging, registration, phase extraction, and removal of geostrophic components. The flowchart is shown in Fig. 7.29

In the ideal orthogonal beam case, the interference phase caused by the velocity of the target is given by:

$$(v_{fo}, v_{ba}) = \left(\frac{2\pi}{\lambda}\frac{B_x}{V_p\phi_{fo}}, \frac{2\pi}{\lambda}\frac{B_x}{V_p\phi_{ba}}\right) \tag{7.51}$$

v_{fo} and v_{ba} are the orthogonal velocity components, B_x is the along-track baseline, V_p is the satellite velocity, ϕ_{fo} and ϕ_{ba} are the interferometric phases for the forward and backward beams, respectively.

By performing interferometry on two SAR images obtained from the forward and backward antennas, and obtaining the phase of the interferometric complex image, the velocity can be directly derived through linear transformation using the phase. However, it should be noted that the flow field obtained at this time is only the Doppler velocity projected onto the radar line of sight. To obtain the sea surface flow field, the influence of the long-wave orbital velocity and Bragg phase velocity should be removed. Among them, for situations where high accuracy is not required, the removal of long-wave orbital velocity can be achieved by averaging over a large area. However, the use of spatial averaging to remove the long-wave orbital velocity is based on the periodicity of the large-scale wave orbital velocity. Therefore, it is necessary to accurately know the precise wavelength information of the observed sea surface in order to perform spatial averaging. However, based on the literature, few people estimate the wavelength before performing spatial averaging. In addition, since the sea surface may have inconsistent wave lengths at different spatial positions, directly using determined spatial averaging for processing often leaves a residual term in the estimated velocity, which affects the accuracy of the flow field measurement.

Bragg phase velocity can be calculated using the following equation:

$$c_p = \sqrt{\frac{g}{\left|\vec{k}_b\right|} + \frac{T\left|\vec{k}_b\right|}{\rho}} \tag{7.52}$$

where g is the acceleration due to gravity, T is the surface tension, ρ is the density of seawater, and $\left|\vec{k}_b\right|$ is the wave number of the Bragg wave. In general, there will be two Bragg waves propagating toward the radar and away from the radar. The relative

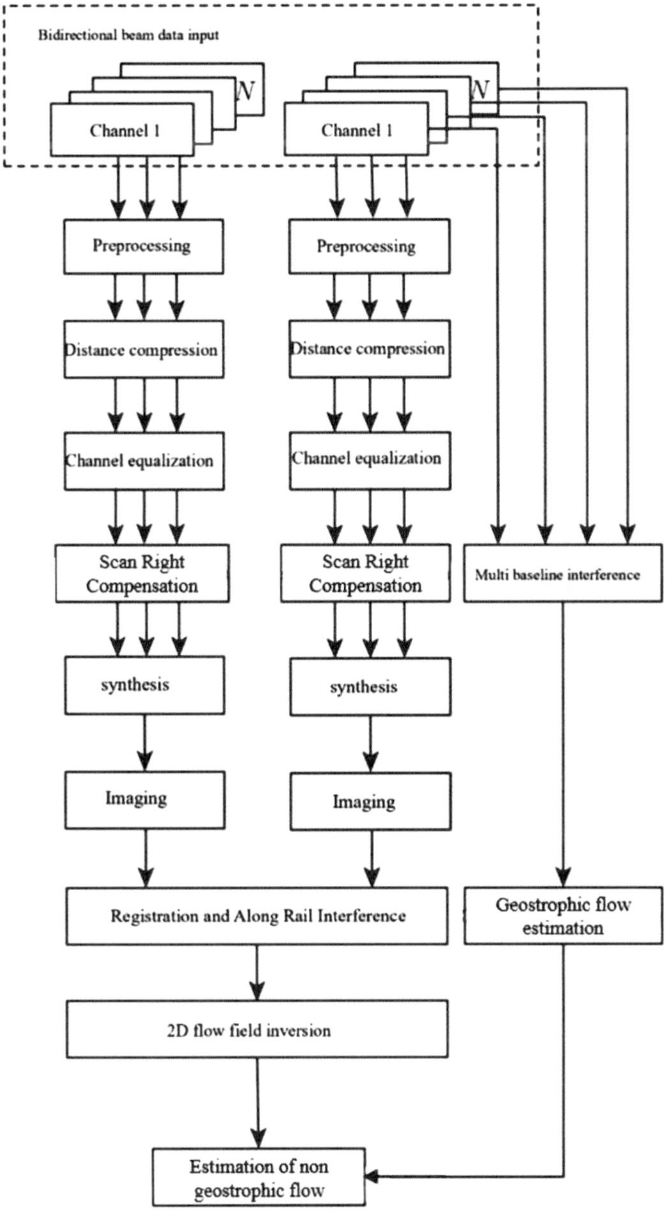

Fig. 7.29 Flow field processing process

contribution of the two waves depends on the wind direction. The removal of Bragg wave phase velocity is based on the principle that waves with similar frequencies have the same propagation speed. This is achieved by coherent removal to eliminate the influence.

Simulations are conducted with a 200 km two-dimensional bandwidth and velocity range between 0 and 3 m/s, including velocity components caused by sea surface height fluctuations. The selected velocity range falls within the ambiguity velocity. As shown in the following figures, first, the all-directional velocity measurements for two implementation methods with a 20° incident angle are performed. It can be observed that the gain reduction caused by PEL directly affects the velocity measurement accuracy. Under consistent sea surface NRCS parameters, the average error of the velocity measurements in Fig. 7.30a is 4.7%, while in Fig. 7.30.

Further reducing the incident angle while maintaining the same pulse width, the beam broadening increases at a smaller incident angle, and it can be seen that the velocity parameter accuracy after pulse compression and DBF processing is significantly better than that of the baseband echo directly undergoing DBF measurements, which can no longer achieve effective velocity measurements (Fig. 7.31).

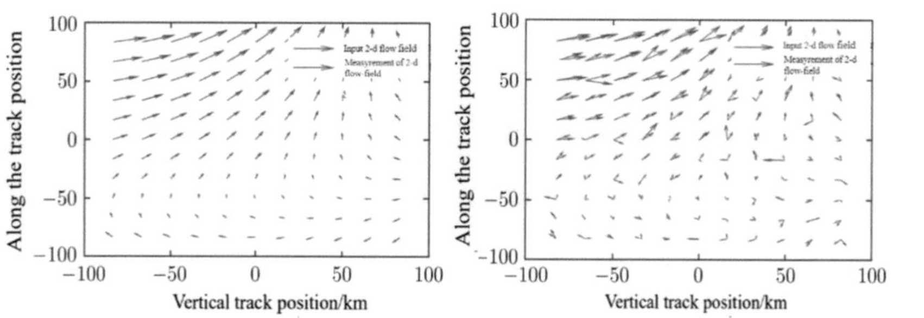

(a) Echo Direct DBF Inversion of Flow Field (b) DBF inversion of flow field after pulse compression

Fig. 7.30 2D flow field inversion results (20° incident angle)

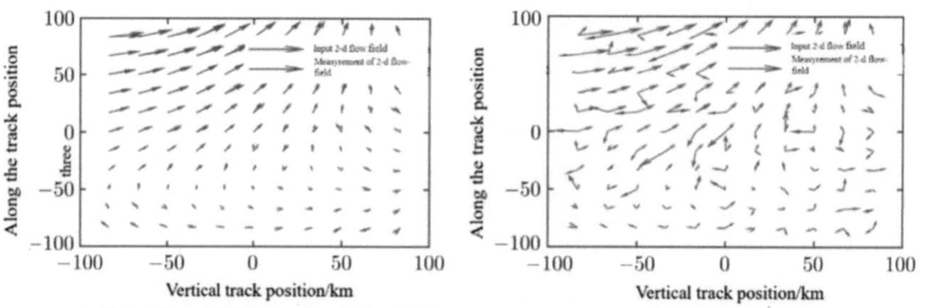

(a) Echo Direct DBF Inversion of Flow Field (b) DBF inversion of flow field after pulse compression

Fig. 7.31 2D flow field inversion results (10° incident angle)

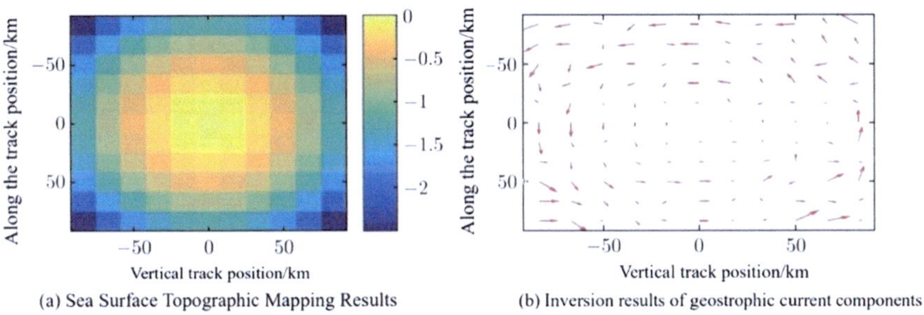

Fig. 7.32 Surveying and mapping results of interesting baseline and geostrophic current components

By performing multi-baseline interferometric measurements, the terrain fluctuation in a 40,000 km^2 area is obtained (assuming the absolute elevation can be calibrated through an external model, here only considering relative elevation accuracy), as shown in Fig. 7.32.

By measuring the average sea surface elevation and differencing it with the geoid height, the geostrophic component of the flow field caused by MDT is obtained. By differencing the total velocity vector, the geostrophic component and non-geostrophic component can be fitted, as shown in Fig. 7.33.

DBF-SAR can enhance the interferometric measurement performance of the system in weak scattering environments. In terms of specific implementation, real-time pulse compression of phase errors in multi-channel radar systems followed by synthesis can improve along-track interferometric performance under small incident angles or long pulse widths. By performing two-dimensional flow field inversion, channel-to-channel interferometric extraction can be used to extract geostrophic components, achieving quantitative and fine inversion of products. Under a 10° incident angle with the same pulse width, the velocity accuracy can be improved from 11.5 to 4.7%. M4S model is used for the analysis of ocean current velocities. The model process is shown in Fig. 7.34. M4S removes the influences of various factors

Fig. 7.33 Inverted non geostrophic current component

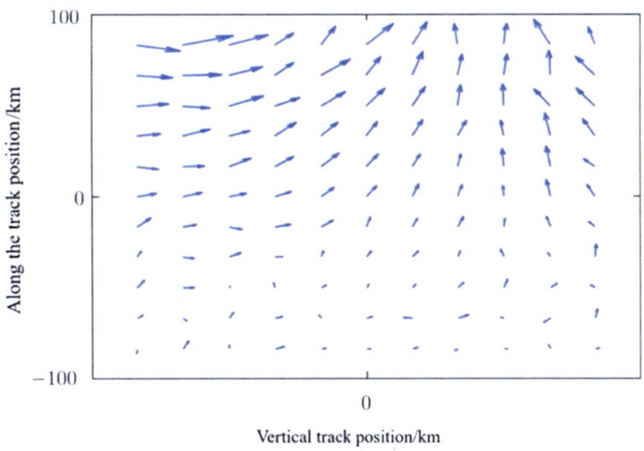

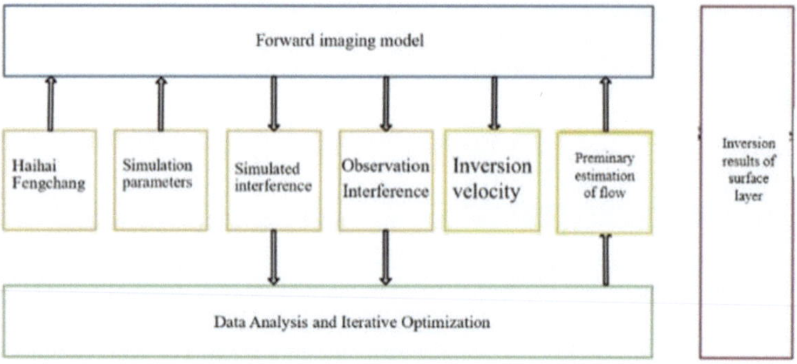

Fig. 7.34 M4S iterative inversion of surface flow chart

through iteration, making the simulated interferometric phase consistent with the actual interferometric phase, and obtaining the "optimal flow field."

The implementation steps of this model are shown in Fig. 7.35. The inversion of the flow field does not require precise prior knowledge of Bragg phase velocity and long-wave orbital velocity, avoiding errors introduced by the wave motion model itself.

7.4.2 Sea Surface Height Inversion Method

The rigorous expression for the cross-covariance of complex reflection coefficients in SAR interferometry for ocean observation is given by:

$$\left\langle r_1(x,t,\tau) r_2^*(x,t,\Delta\tau)\right\rangle = \sigma(x,t)\exp\left[i\frac{2\pi\,\Delta t}{\tau_c}\right]\exp\left[-\frac{\Delta t^2}{2\tau_c^2}\right] \exp[-2ikU_0(x,t)\Delta t] \tag{7.53}$$

Here, $\Delta t = B/2V$ is the temporal baseline, U_0 is the radial velocity component of the rough sea surface, τ_c is the scene coherence time, and σ is the backscattering coefficient. The first phase term represents the phase noise added by the mean sea surface velocity U_0. According to imaging requirements, the phase noise needs to be much smaller than the average phase term, i.e.:

$$|2k \cdot U_0 \Delta t| \gg \left|\frac{2\pi\,\Delta t}{\tau_c}\right| \tag{7.54}$$

This leads to the scene coherence time limitation as shown in Fig. 7.36:

$$\tau_c \gg \frac{\lambda}{2|U_0|} \tag{7.55}$$

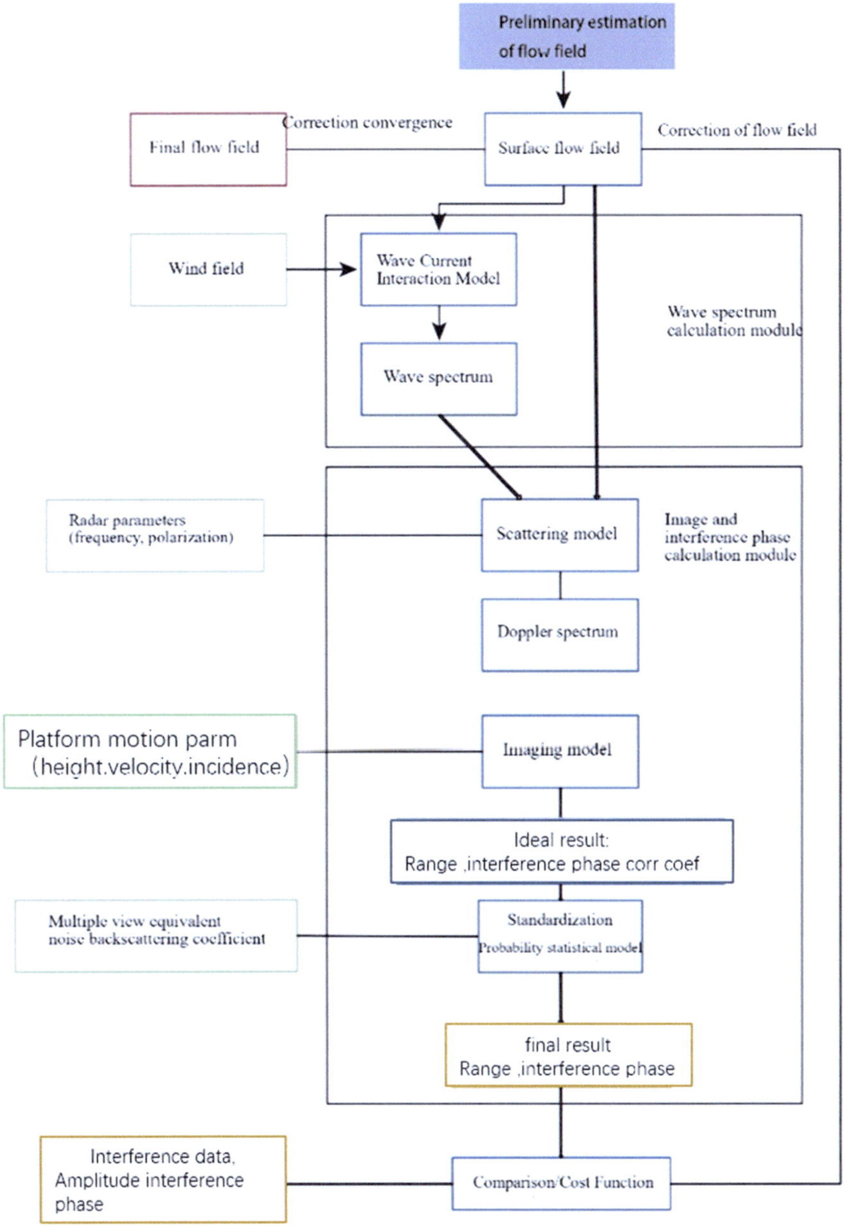

Fig. 7.35 Flow field inversion M4S model

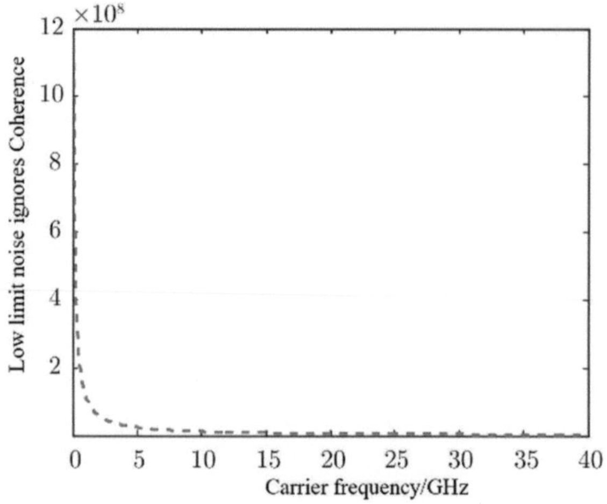

Fig. 7.36 Noise ignores coherence time frequency relationship

$$\sigma'(x, t, \Delta t) = \sigma_0\left(-\Delta t^2/2\tau_c^2\right) \tag{7.56}$$

The loss term in the radar backscattering cross-section caused by the time interval in the interferometric process is described by the phase $\exp\left[-\Delta t^2/2\tau_c^2\right]$, and the backscattering coefficient can be expressed as:

In order to extract interferometric information from the phase, the effective signal $\sigma'(x, t, \Delta t)$ needs to be higher than the noise level. As shown in Fig. 7.37, the scene coherence time needs to satisfy:

$$\tau_c \gg \frac{B}{2V}\frac{1}{(2\ln 2)^{1/2}} \tag{7.57}$$

To address this, a sea surface elevation inversion process based on millimeter-wave single-pass interferometry is proposed in Fig. 7.38. It avoids the phase modulation and echo intensity loss caused by the flow field under long temporal baselines. The inversion process includes data preprocessing, Doppler centroid estimation, two-dimensional focusing for image-based wind speed estimation, cloud-rain-ice region determination, and wave height estimation based on coherence and signal-to-noise ratio. After on-board and external reference data calibration, high-precision wide-swath sea surface elevation data calibration is performed under point calibration. Oceanic elements such as gravity anomalies, wave height, eddies, wave peaks, and geostrophic flow field are inverted based on dynamic topological information.

In this regard, this book carries out verification of millimeter-wave interferometric technology based on a single flight platform with multiple receiving apertures. As shown in Fig. 7.39, the elevation of the lake water surface is obtained, and there are certain fluctuations in the internal wave height due to factors such as wind fields.

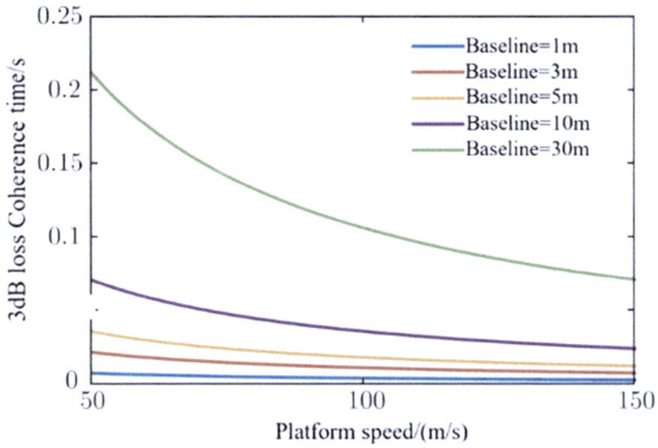

Fig. 7.37 3dB loss coherent time

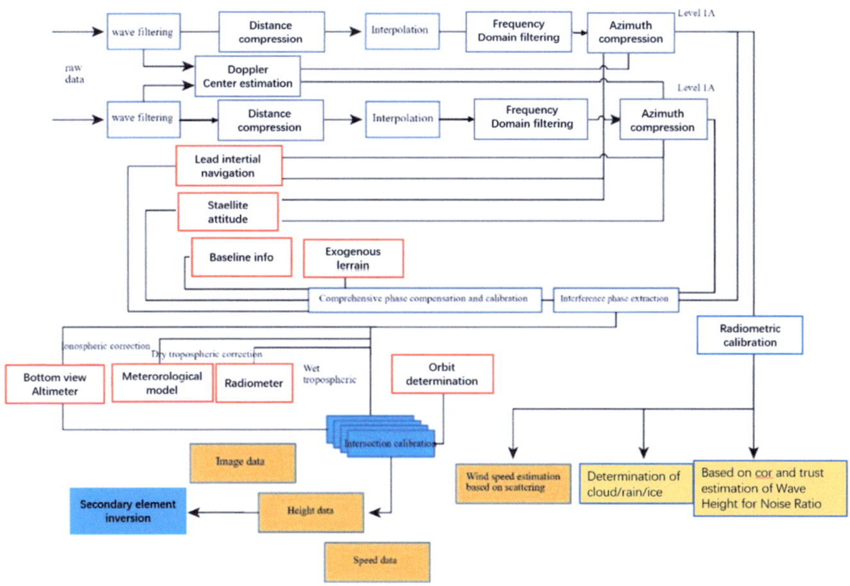

Fig. 7.38 Sea surface elevation inversion on space-based single interference

To achieve high-precision interferometry requirements based on millimeter-wave SAR payload data acquisition and multi-source collaborative error correction, especially for wide-swath altimetry applications, calibration methods are needed to assist in achieving the desired indicators. Specific technical means include: (1) Using radiometers to achieve wet tropospheric correction and developing quantitative correction algorithms for dry troposphere and ionosphere; (2) High-precision precise orbit determination; (3) Improving delay error and reducing interferometric phase error through internal calibration; (4) System error correction through external

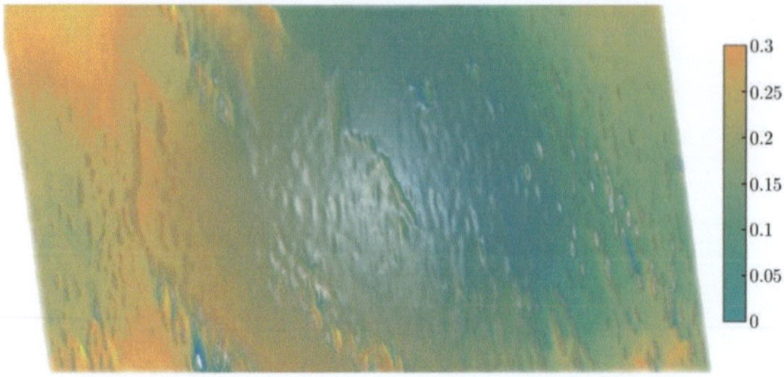

Fig. 7.39 Verification of airborne single flight surface interferometry altimeter technology

calibration. The specific analysis of residual elevation data plane calibration achieved by data ascending and descending tracks is shown in Fig. 7.40.

As shown in Fig. 6.12 in Chap. 6, the basic assumption of the cross-calibration technique is that baseline tilt errors will result in sine modulation of the height measurement. In wide-swath ocean mapping systems, small incidence angles are generally used, resulting in an approximation of sine modulation as

$$\sin\theta \approx \theta \tag{7.58}$$

Therefore, tilt errors can easily be equivalently transformed into a linear bias in the vertical track dimension, referred to as "lift", and the tilt error within the swath is illustrated in Fig. 7.41. Based on this, the wide measurement intersection area formed at ascending and descending tracks can form the simplest and most effective tilt error correction area.

Since the error quantity has continuity in both time and space distribution, it can be separated from the effective signal in spectral analysis. The measured value $H_{obs}(x, t)$ can be decomposed into the true height signal $H_{real}(x, t)$ and the coherent error term:

$$H_{obs}(x, t) = H_{real}(x, t) + xR(t) + x^2/a \cdot \delta B(t)/B + \varepsilon(x, t) \tag{7.59}$$

where t is defined as the azimuth time, x is the distance from the ground position, $R(t)$ is the antenna tilt. Simulated sea surface elevation data with long, medium, and short waves are generated, and centimeter-level accuracy bottom-view altimeter data is obtained along the track. The simulated sea surface and sea surface elevation profiles are shown in Fig. 7.42.

(1) Direct calibration The direct method calculates the large-scale variation of the true topography H_{real} in H_{obs} using H_{ref} (external information). If we use Y as the difference:

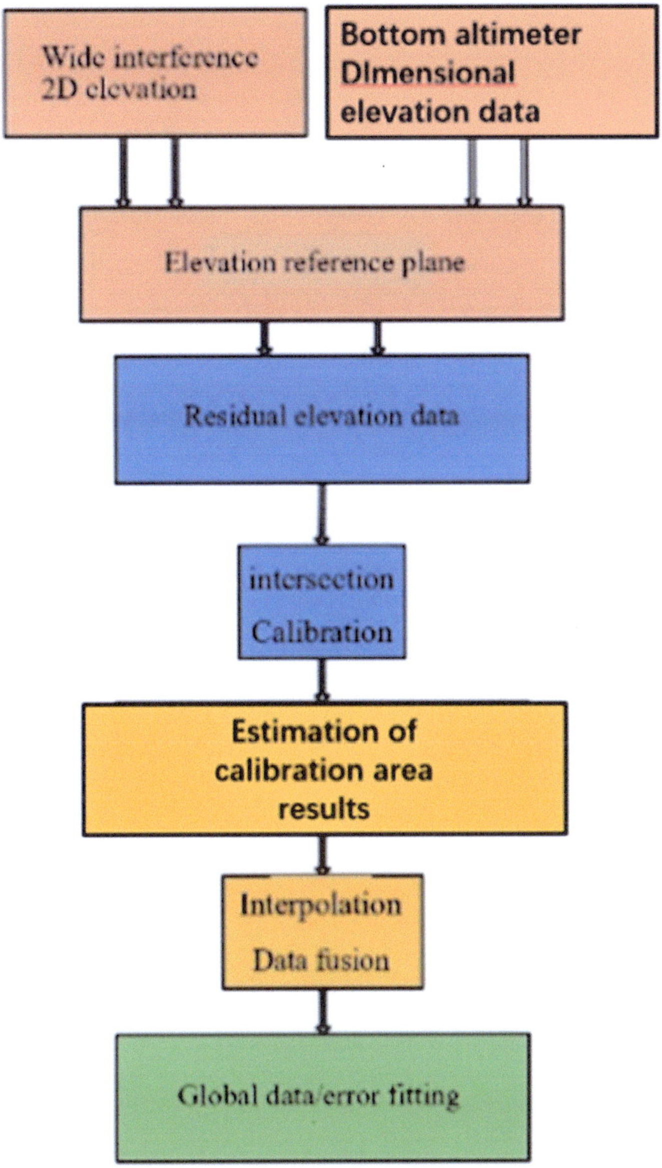

Fig. 7.40 SAR and altimeter collaborative calibration process

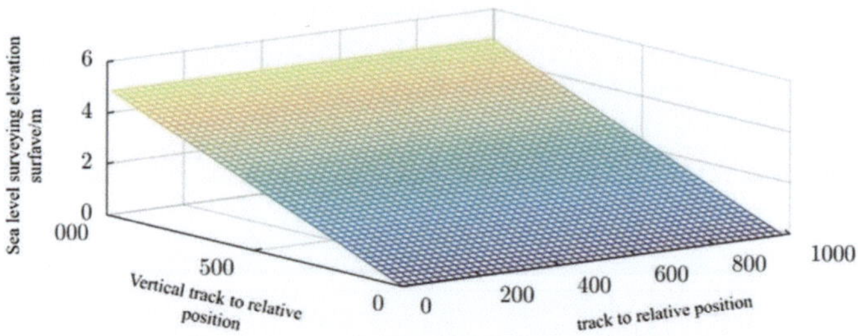

Fig. 7.41 Distribution of inclination error along vertical spacing direction

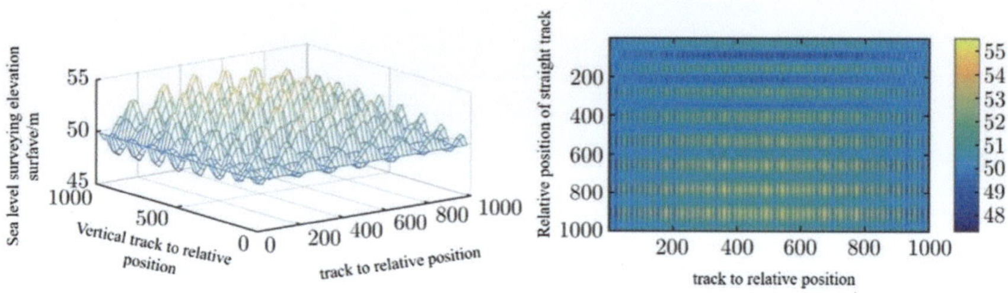

Fig. 7.42 Simulated sea surface data

$$
\begin{aligned}
Y(x,t) &= H_{obs}(x,t) - H_{ref}(x,t) \\
&= x.R(t) + x^2/H.\delta B(t)/B + \delta H(x,t) + \varepsilon(x,t)
\end{aligned}
\tag{7.60}
$$

Using $H_{ref}(x, t)$ allows the data to be shifted, allowing the error characteristics to be manifested only in the residual term. The residuals before and after calibration are shown in Fig. 7.43. Figure 7.43: Residuals before (left) and after (right) direct calibration.

(2) Intersection point calibration As shown in Fig. 7.44, the diamond-shaped intersection area of the ascending and descending tracks consists of four small regions and one bottom-view altimeter intersection point. If the time difference between

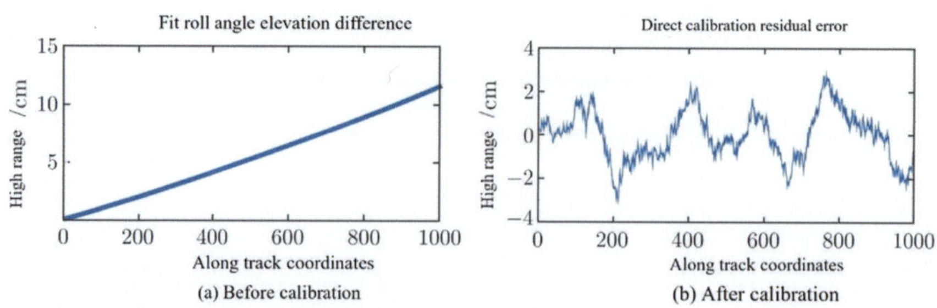

Fig. 7.43 Comparison before and after direct calibration

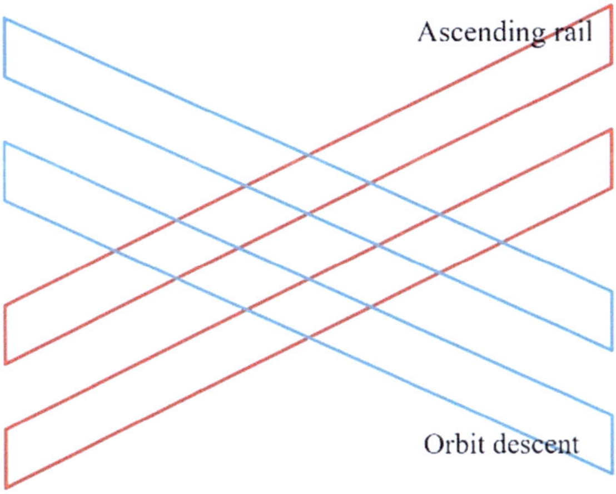

Fig. 7.44 Schematic diagram of track intersection

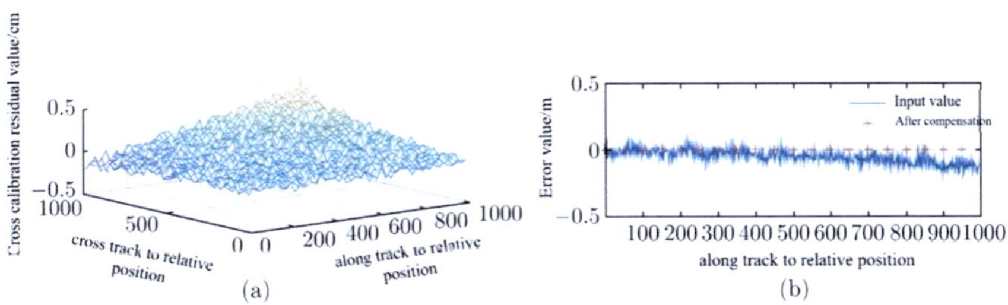

Fig. 7.45 Comparison of cross calibration residual error (**a**) and along track error (**b**) before and after

the two track data is small enough, each measurement can be assumed to share a portion of the true value H_{real} and the error ε. However, $R(t)$ and $\delta B(t)$ change over time between two observations, and the two tracks have different relative crossover distances (even if the target area has the same latitude and longitude).

The residual after intersection calibration and the comparison of along-track errors are shown in Fig. 7.45

Interferometric SAR data can obtain first-order product sea surface height, and by appropriately integrating other sensor data, dynamic sea surface height, significant wave height, and gravity anomalies can be inverted. Finally, assimilation methods can be used to invert fourth-order product oceanographic elements such as thermoclines, convergence zones, and seabed topography.

References

1. Tang, M., Lin, T., & Wen, G. (2011). A review of ship detection methods in remote sensing images. Journal of Computer Applications Research, 28(1), 29–36.
2. Chen, J. (2021). A super-resolution SAR image water area segmentation algorithm and its application. Journal of Electronics & Information Technology, 43(3), 700–707.
3. Huang, W., Zhou, C., & Li, D. (1997). Current situation and demand of SAR ocean applications in China. China Aerospace, (12), 5–9.
4. Li, H. Y., Song, H. J., et al. (2014). A modification to the complex-valued MRF modeling filter of interferometric SAR phase. IEEE Geoscience and Remote Sensing Letters, 12(3), 681–685.
5. Wang, H., Zhang, H., Dai, S., et al. (2018). Azimuth multichannel GMTI based on Ka-band DBF-SCORE SAR system. IEEE Geoscience and Remote Sensing Letters, 1–5.
6. Sun, X., Wang, Z. R., & Sun, Y. R. (2019). AIR-SARShip-1.0: High-resolution SAR ship detection dataset. Journal of Radar, 8, 2019.
7. Li, T. (2019). Research on ship target detection method based on SAR images. Xi'an University of Electronic Science and Technology. (in Chinese)
8. Fan, Q. J. (2016). Research on ship target detection and identification technology in spaceborne SAR images. National University of Defense Technology. (in Chinese)
9. Zhang, L. (2018). Research and implementation of ship target detection in GF-3 SAR images. Xi'an University of Electronic Science and Technology. (in Chinese)

MIX
Papier aus verantwortungsvollen Quellen
Paper from responsible sources
FSC® C105338

If you have any concerns about our products,
you can contact us on
ProductSafety@springernature.com

In case Publisher is established outside the EU,
the EU authorized representative is:
Springer Nature Customer Service Center GmbH
Europaplatz 3, 69115 Heidelberg, Germany

Printed by Libri Plureos GmbH
in Hamburg, Germany